RURAL EUROPE

RURAL EUROPE

Identity and Change

KEITH HOGGART
King's College, London

HENRY BULLER
Université de Paris VII

RICHARD BLACK
University of Sussex

Routledge
Taylor & Francis Group

LONDON AND NEW YORK

First edition published 1995 by Hodder Arnold
Co-published in the USA by Oxford University Press

Published 2014 by Routledge
2 Park Square, Milton Park, Abingdon, Oxon OX14 4RN
711 Third Avenue, New York, NY, 10017, USA

First issued in hardback 2017

Routledge is an imprint of the Taylor & Francis Group, an informa business

British Library Cataloguing in Publication Data
A catalogue record for this book is available from the British Library

Library of Congress Cataloging-in-Publication Data
Hoggart, Keith
    Rural Europe: identity and change/Keith Hoggart, Henry Buller, and Richard Black.
    264 p. Includes bibliographical references and index.
    ISBN 0-470-23571-3
    1. Europe--Rural conditions. 2. Social change--Europe.
I. Buller, Henry. II. Black, Richard, 1964– . III. Title.
HN373.H63 1995
307.72'094--dc20   95-4832
                                          CIP

ISBN 13: 978-0-3405-9699-9 (pbk)
ISBN 13: 978-1-1384-6589-3 (hbk)

Contents

Figures

Tables

Preface

Given the spate of recent books on rural places, societies, economies, identities and development, and indeed on the geography and political economy of contemporary Europe, to launch yet another text on this crowded scene is a daunting task. But within all this endeavour, the lack of a volume that combines the themes of 'rural' and 'Europe' is startling; still more so if one considers the role of rural transformation in underpinning the historic development of European capitalism, and the variety of forms this took over a diverse continent. It seems to us that despite a considerable literature on identity and change in Europe, there is an absence of comparative studies on rural areas. Thus despite the fact that agricultural policy, for example, represents the arena in which the strongest process of European integration has occurred, national considerations remain to the fore in the literature on problems facing rural areas, with material that specifically focuses on rural issues being fragmented.

This fragmentation provides a starting point for the general thrust of our argument in this book; for if academic endeavour is largely centred on particular states, so too is government policy-making, civil society, social identity and even, despite globalization, the organization of economic affairs. This national focus reflects not only on the continued importance of states in policy debates, but also on the distinctive national traditions, experiences and problems that continue to exist in Europe's rural areas. Without wishing to argue that national-specificities have special causal powers, indeed the role of global and Europe-wide socioeconomic forces of change appear ever stronger in our view, comparison between national trajectories does provide a neglected starting point from which to explore socioeconomic diversity and change in rural Europe. It is to analyzing the form and assessing the strength of these nation-specific forces that we will be directing our attention. In doing so we will cover issues of socioeconomic standing and cultural identity, which will be followed by explorations of transnational policy formulation, before focusing on the character of the economic, social and environmental transformations that have occurred and are occurring within the rural areas of the European Community. We identify where relevant cross-national trends, which in their focus on wider transnational regions also speak to the diversity of socioeconomic status and opportunities for 'development' within a Europe that is still characterized by inequality rather than integration.

In the genesis and writing of this book, numerous colleagues have provided indirect and direct insights into processes of transformation in rural Europe. To thank them all would take-up almost as much space as the book itself. However, we would like to thank in particular colleagues in Geography at King's College London and the Université de Paris VII, along with co-workers at the CNRS Groupe de Recherches sur les Mutations des Sociétés Européenes and the Centre for Rural Economy of the University of Newcastle-upon-Tyne. Amongst those who have given particular help and encouragement, we owe a debt to Maryvonne Bodiguel, Josefina Cruz Villalón, Marcel Jollivet, Pierre Lenormand, Philip Lowe, Neil Ward and Rebecca Pritchard, who all, in their own way, contributed to this volume. We would also like to thank Roma Beaumont and Gordon Reynell for their hard work in preparing the maps for the text. As with all such endeavours, we take responsibility for the inaccuracies and interpretations contained within this book.

Glossary

B	Belgique/België (Belgium)
CAP	Common Agricultural Policy (of the European Community)
CEC	Commission of the European Communities
DK	Danmark (Denmark)
D	(Deutschland) Germany
E	(España) Spain
EAGGF	European Agricultural Guidance and Guarantee Fund (also known as FEOGA)
EC	European Community
EC6	The six original members of the Community (Belgium, France, (West) Germany, Italy, Luxembourg and the Netherlands)
EC9	The nine countries that made up the EC between 1973 and 1980 (the original six plus Denmark, Ireland and the United Kingdom)
EC10	The EC9 plus Greece (the EC member states from 1981 to 1985)
EC12	the EC10 plus Portugal and Spain (the EC member states from 1986 to 1994)
ECU	European Currency Unit
EFTA	European Free Trade Association
Ell	Elláda (Greece)
EMS	European Monetary System
ESA	Environmentally Sensitive Area
ERDF	European Regional Development Fund
ESF	European Social Fund
F	France

FAO	Food and Agriculture Organization (of the United Nations)
GATT	General Agreement on Tariffs and Trade
GDP	Gross Domestic Product
GNP	Gross National Product
ha	hectare
HMSO	Her Majesty's Stationery Office (Britain)
I	Italy
IMP	Integrated Mediterranean Programme
INSEE	Institut National de la Statistique et des Études Economiques (France)
IRL	Ireland
L	Luxembourg
LFA	Less Favoured Area
MAFF	Ministry of Agriculture, Fisheries and Food (UK)
MCA	Monetary Compensatory Amount
MOPU	Ministerio de Obras Publicas y Urbanismo (Spain)
N	Nederland (Netherlands)
NIC	newly industrializing country
O	Österreich (Austria)
OECD	Organization for Economic Cooperation and Development
P	Portugal
R&D	Research and Development
SEM	Single European Market
SME	small and medium sized enterprises
Su	Suomi (Finland)
Sv	Sverge (Sweden)
UAA	Utilized Agricultural Area
UK	United Kingdom
WTO	World Tourism Organization

1

Globalization, nations and rural transformation

Eight years ago, two of the present authors published a volume that sought to examine rural development trajectories as generalized processes (Hoggart and Buller, 1987). In *Rural development* we provided an interpretation of change in rural areas that was essentially grounded in Britain and the United States, yet which drew no serious contrasts between these nations, nor sought to locate the distinctive experiences of their country areas in terms of the dissimilar structures and processes that are found in other places. This is not the only occasion on which we have provided commentaries on 'the rural' which offer too generalized an interpretation of the processes and patterns we have described (e.g. Hoggart, 1990). In more recent years, however, we have become increasingly aware or, perhaps more accurately, have forced ourselves to be more forthright in recognizing cross-national differences in causal processes (e.g. Buller, 1992). This recognition is not particularly original, but it is our contention that acceptance of national differences receives insufficient attention within the rural studies literature. Analysts do recognize that such differences exist, but in conceptualizing and writing on 'the rural' they too often soft-peddle in introducing this understanding into their contributions. As such, within writings in the English language, there is a danger that undue weight is given to an Anglo-American view of rurality and rural development. The likely effect of this is not simply to weaken, or at least isolate, the potency of processes, structures and conceptualizations that are presented in rural studies in other nations, but also to make Anglo-American images more shallow, for the limited geographical base of their supposed generalities is insufficiently recognized. In writing this volume, one of our primary objectives is to draw attention to the existence of cross-national similarities and differences in rural development, in order to emphasize the variety of rural trends and identities that exist in the European Community today.

The fact that we have focused on Europe highlights a further underpinning of our efforts. We contend that, even as the fortunes of rural areas are

increasingly determined by transnational processes and structures, to understand these more 'global' causal forces we need to appreciate how they are
grounded in nation-specific practices. In terms of the articulation of these
forces, this is perhaps most evident in international political relations. As one
illustration, when examining EC legislation a consideration that should be
foremost in interpretations is the understanding that '... [most] European
policy is not made in the Community institutions, but rather in the national
parliaments ...', so 'Europe-wide' legislation reflects a compromise between
national objectives (Wenturis, 1994, 235).[1] Of course, when articulated
through governments, a nation-specific impetus is (relatively) easy to identify
and interpret, if only because there is an institutional framework which ties
many actions to single nations.[2] Yet distinct national contributions to forces
of stability and change are also manifest in the way transnational processes
affecting capital and civil society are penetrated by national mores and
practical adjustments. Providing insight on these phenomena for the EC
appears to us to be particularly appropriate, for the present balance of work
on EC affairs favours generalized accounts of European integration (e.g. Cole
and Cole, 1993; Dawson, 1993) or examinations of events in specific
countries that are largely taken in isolation (Brouwer *et al* ., 1991; Rodwin
and Sazanami, 1991). The intention here is not to provide a country-by-
country descriptive account, as Parker (1979) has offered. Rather, our aim is
to provide commentary that will help elucidate distinctions in national
traditions that contribute toward an uneven pattern of rural development
within the European Community. These traditions are seen as interacting
with EC initiatives and transnational socioeconomic impulses to produce a
complex European rural geography.

Stated in this way, our intentions might be read as a postmodern emphasis
on distinction and divergence that stands in opposition to works on European
integration and globalization processes. This is not our aim. We are persuaded that a deeper understanding of the words and concepts that are used
by people in their daily conversations can shed light on the meaning of
rurality (Halfacree, 1993). It can also contribute to theorization of behavioural differences. For example, valuable insights can be gained from analyses

1 The geographical remit for this text is the European Community rather than Europe as a
   whole, hence the quotation marks around 'Europe' in this sentence. Most commonly,
   however, there are no quotation marks around Europe in the text, although our meaning
   for this word is the EC.
2 In no sense should this be taken to mean that a clear separation exists between the private
   and public sectors in absolute terms. Neither should it be read as meaning that single
   public sector institutions have clearly defined boundaries, which distinguish them from
   other (private or public) organizations. It also does not mean that there is unanimity in
   the goals and operational styles of public sector units. All this was recognized long before
   the privatization of state institutions and roles became a major political issue, and applies
   irrespective of conceptualizations like corporatism or instrumental-marxism which see
   strong private-public sector links (e.g. Self, 1977). The key issue is that relative to the
   processes and structures of capital and even civil society, the state is more strongly
   grounded in a national territory and derives much of the stimulus for its behaviour from
   addressing political issues that arise in connection with that territory.

of the role rural householders believe women 'should' take in the household or on farm enterprises, even though most of this work does not trace its roots to postmodernism (Hillebrand and Blom, 1993; Haugen, 1994; Little, 1994). Indeed, there is a fear that the post modern emphasis on 'difference' has the potential to generate largely uncoordinated descriptive accounts, which do not take us far in understanding the genesis of the conditions they describe. Certainly, as yet, that work which has focused on individuals has generally provided uncontextualized, descriptive commentaries that offer little theoretical advancement. In good measure this arises because general trends in human behaviour are not autonomously generated from personalized actions, but are conditioned by the past and present structured circumstances of social existence.

Irrespective of whether individuals or small groups possess unique interpretations of their lived environment, we are poorly served by seeking to aggregate individual behaviour into local or national trends. The whole is much more than the sum of its parts. A simple illustration of this is seen in links between social class and electoral support for British political parties. At the level of single voters, links between class and political allegiance have weakened over time. At the level of electoral constituencies, though, they have strengthened. Manual workers might be less likely to vote Labour now than in the past, but people living in constituencies with a large manual labour population are even more inclined to support Labour (Warde, 1986). In similar vein, while analysis of personal actions provide informative insights on social differentiation, as with Schmitt's (1994) work on differences in farming practices on female- and male-headed production units, such distinctions are heavily conditioned by long-established social structures that generate not only regional disparities in gender divisions of labour (Sackmann and Häussermann, 1994) but even cross-national divergences in gender roles (Pfau-Effinger, 1994). Put simply, personal behaviour is grounded in enduring social structures that are not static, but whose changing nature is informed by previous social practices (Sahlins, 1981).[3]

As such, lifestyles, worldviews and conceptions of rurality, along with the behaviour patterns that we can observe in rural areas, can be quite different or even contested in single places. However, as the totality constitutes more than its individual elements, understanding dissimilarities in paths of rural change still requires that we theorize dominant societal structures and processes. In this volume we explore such structured circumstances, by asking whether socioeconomic trends in rural Europe follow trajectories that are infused with nation-specific causal tendencies. Of necessity, we have to start such an analysis by recognizing that a multiplicity of causal processes and structures have been instrumental in bringing rural Europe to its present

3 Major changes in social order can upset this pattern, as with a major influx of in-migrants that 'swamp' the local population (as in city commuter hinterlands). But even here it is the complex interaction of trends in a multitude of different decision arenas that gives areas their peculiar character.

variegated form. These forces of continuity and change operate at local, national and transnational scales. We offer no scoring of causal priorities. Our task is more limited: it simply inquires into the extent to which understanding national-level differences helps us appreciate the socioeconomic characteristics, trajectories and identities of countryside areas in the European Community.

Theorizing nation-specific ruralities

When we first considered writing this volume, one of our intentions was to provide a strong theoretical underpinning to our assessment of dissimilar conceptions of rurality in Europe and, more particularly, of dissimilar processes of rural development. As this project progressed, this aim became ever more distant, as we realized that the problems we confronted were far too complex and under-researched for clear lines of causal accounting to be presented. It follows that any notion of theorizing in this text is more about moving toward a position where theorization is possible, rather than offering rounded explanation. As such, we cover items that need clarification, as much as we highlight issues of agreement.

The complexities of developing theoretical insight on rural development processes should not be under-stated. Socioeconomic change in rural areas inevitably covers the whole gamut of societal transformations that nations experience, yet requires a theoretician to find common cause across places whose likeness arises simply from either a shared low population density or consensus amongst its residents that their home place is 'rural'. By itself low population density carries little theoretical weight, for we know that places of relatively sparse population incorporate a variety of socioeconomic trends, as well as experiencing or encountering dissimilar development processes and pressures (e.g. Hoggart, 1988; Hodge and Monk, 1991). Moreover, personalized views on rurality inevitably centre on a sub-plot of rural change processes. As with any sub-plot, the story line can be both fascinating and of critical importance, but of itself it is not enough; assuming the intention is not simply to understand social change in rural areas but also to theorize the causal mechanisms that lie behind them. Once again, what we are suggesting is not innovative nor new. Yet it needs emphasizing, for, in common with much research in the social sciences, rural studies are passing through something of a transitional phase. Dissatisfaction has grown over the inadequacies of prior research practices, leading to calls for new interpretive paradigms. What we fear is a repeat of the all too common occurrence of too readily rejecting the strengths of past practices owing to the appeal of a new research emphasis. This is a fear that we are not alone in recognizing (e.g. Murdoch and Pratt, 1993).

Cynics might argue that one of the characteristics of academic enterprise is for researchers to promote new areas of research endeavour, accompanying their introduction with over-exuberant attacks on previous practices to justify a change to a new mode of analysis. Certainly, at least two tendencies increase the impetus to over-emphasize the need for change in analytical approach. The first arises from the vast literature that now exists on most aspects of academic research. In combination with the growing recognition of a need for more inter-disciplinary understanding, plus sharply increasing academic workloads, the chances of a single piece of work drawing attention to itself are lessened, unless the message that is projected is somewhat 'different'. Arguably, in presenting such differences, more attention accrues if the scale or depth of the critique of past practices is intense. The chances of such critiques drawing succour are also enhanced by the prevailing character of research communities in the social sciences. While it would be an exaggeration to divide academic investigators into either a theoretical or an empirical camp, even a cursory glance at publication records and journal entries indicates that most researchers either favour more abstract theoretical interpretations of human behaviour or engage in detailed examinations of that behaviour. For us, this division sets up a tension, in which more empirically-oriented investigators, through the very nature of in-depth analyses, are prone to realize weakness in *existing* theoretical perspectives. This has the consequence of leaving these researchers open to the influence of an innovative theoretical perspective that seems to offer 'solutions' to flaws in existing paradigms. What makes this problem worse for rural studies is the fact that the vast majority of theoretical insights are formulated with a view to addressing national or city-dominated processes and structures. This generates added difficulties for rural analysts, since the formulations are most commonly visualized at a broader geographical scale than their study sites (as for most theorizing on uneven development processes, especially when the international dimension is included). Alternatively, interpretations are produced that are of uncertain relevance for rural areas, either due to their city-focus or because they are formulated for activities that rarely occur in the countryside. It comes as little surprise, then, that rural research has often been described as theoretically retarded (Cloke, 1989a), although there appears to be chagrin amongst rural researchers over the labelling of their research field as a laggardly enterprise.

Nonetheless, the changing content of contemporary social science research would seem to place rural studies in an advantageous position. In part this is due to altered public consciousness about (so-called) rural issues, as well as to new trends in behaviour that have brought new roles for the countryside. There is no doubt, for instance, that the public has become increasingly concerned over issues of environmental degradation, with agriculture a major focus for such concerns (Comolet, 1989; Pitman, 1992; Rude and Frederiksen, 1994). At the same time, much of the European countryside is being energized by contemporary trends that affect activity spheres ranging from manufacturing to tourism, as well as from day-trips to commuting and migration (e.g. Fielding, 1982; Cross, 1990; Harrison, 1991; Chevalier, 1993). Such changes have been instrumental in increasing research attention on rural issues. Public interest in the countryside has grown not simply from its changing role in production processes and its growing importance as a site for consumption, but also from its increasing significance in policy terms. In this respect, it is particularly relevant to note trends such as greater interest in environmental issues and a growing recognition of the dramatic socioeconomic changes that the Single European Market could herald (e.g. Boussard, 1990; Grahl and Teague, 1990).

None of these trends specifically requires a change in theoretical or conceptual orientations. What underscores such analytical transformations are general adjustments in the social sciences as a whole. Here two particular tendencies favour rural investigations. The first arises from a growing interest in analyses of consumption; particularly as these relate to self-identity and place-generation (Sack, 1992; Bocock, 1993). In this mould there is an emerging interest in what rurality means for different people and how conceptions of 'the rural' are used for political purposes (Lowe, 1989; Mormont, 1990; Cloke and Milbourne, 1992; Halfacree, 1993). Accompanying this there is now much disenchantment with structuralist accounts of societal change, with analysts concerned over their potentially functionalist and reductionist tendencies, as well as with the difficulties of relating general abstract principles to concrete events (Wickham, 1988; Lovering, 1989). Here, as with 'identity studies', we detect a shift in the emphasis of rural investigations toward locally-grounded interpretations of rurality.

> A seductive avenue for researchers has been an analysis of the processes of internationalization (both political and economic) and their potential spatial consequences. But top-down empirical demonstrations of the consequences of international tendencies for national, regional and local structures, which begin with descriptions of global tendencies and attempt to predict local responses, remain problematic. More compelling ... is the reverse approach: seeking evidence of local action and local systems of relationships in the formation of rural localities in a more internationalized world. This perspective challenges what we regard as a former, unreflexive approach of structuralist concepts to rural change in which the distinctive role of locality and rurality in the economic restructuring and urban experience of society in the late twentieth century was to easily dismissed (Marsden *et al* ., 1993, 172).

In their emphasis on intermediate-level conceptualizations, the ideas of Marsden and colleagues are compelling. However, we are concerned about the phrasing of their argument. As the above passage indicates, the tone of criticism of structuralist approaches often implies little more than an inversion in perspective. Thus recognition that structures are not transposed directly into individual behaviour leads us to criticize structuralist modes of explanation. But there is a risk that the babble of voices fermented by attempts to theorize by building up from the individual may lead to an impasse. Beside, it can be seriously questioned as to whether such an enterprise is at all feasible without insights on linkages between local and global being informed by an accepted theorization of structured circumstances that influence locally-grounded behaviour. In effect, while there is much merit in rejecting attempts to impose a structural explanation on the actual experience of individual behaviour, this in no sense means that we can aggregate the actions of human agents and produce an understandable theoretical whole. To be intelligible, interpretations of local actions must be informed by theorizations at the global and the national levels (Taylor, 1993).

This is not to doubt that people who live in close proximity can generate theoretically significant social action. In so far as such territorial units attain meaning for their residents and provoke communal action, they merit serious theoretical interest (Urry, 1987). However, despite an abundance of research attention in the search for such locality effects, significant question marks still surround the issue of what these actually are (Jonas, 1988; Duncan, 1989). Local sentiments can be strong, but it is unclear whether they simply provide a particular flavour to action, or represent wider ideals around which social action is organized and valued, that set one locality off against another (Cohen, 1982; Bouquet, 1986). What is lacking is clear evidence that locally-grounded socioeconomic 'units' carry sufficient weight to be a *primary* determinant of rural change.

In contrast, a multitude of studies can be cited which show that localities have long been penetrated by transnational forces. Moreover, while these can interact with local societies to create geographical differences in behaviour (e.g. MacDonald, 1963), it is at the national level that institutional and ideological capacities are more capable of imposing a distinctive character on global trends. It follows that it is at the transnational level that we should start our search for understanding socioeconomic transitions in rural areas, as it is at this level that the reality of most fundamental societal adjustment is commonly dictated (Taylor, 1993). Examples of this are provided by the impact of New World grain production on the demographic structure and agrarian economy of late-nineteenth century Europe (e.g. Beckett, 1986; Borchardt, 1991), or by the effects of the First World War on the rural landscape (Clout, 1993a). More recent international impulses include trade policies (Cloke, 1989b; Philip, 1990; Giannitsis, 1994), the new international division of labour in manufacturing (Kiljunen, 1992; CEC 1994c), European Community membership (Mykolenko *et al* ., 1987; Maraveyas, 1994) and international migration (King, 1993; Buller and Hoggart, 1994a). While local arenas are important venues for exploring the precise impact of societal changes, the power that local communities have to direct their own futures is limited, despite their perhaps unique and certainly adaptable responses to extralocal forces. Indeed, even when local people are critical agents in local socioeconomic change, their primary role often results directly from their inability to control outside forces (Black, 1992). Only in restricted circumstances, such as under highly centralized local power structures, do we find local inputs dominating development trajectories (e.g. Kendall, 1963; Wilson, 1992). This dominance is generally short-lived, for its guiding principles almost inevitably operate within a context of benign acceptance on the part of stronger extralocal forces. Despite much publicized opposition to 'outside' intrusions into local affairs (e.g. Lowe and Goyder, 1983), most extralocal effects, such as national laws and regulations, impose themselves in a quiet but forceful manner that brings major changes even to those places that seem neglected by or immune from outside power interests (e.g. Bird, 1982).

Drawing on Taylor's (1993) insights from world-systems theory, this points towards the need for conceptualizations of societal processes to be distinct at different geographical levels, even though there is need for integration across these levels. For Taylor, the global scale is the level of *reality*. In principle, we can empathize with this view in terms of standards of living; the continent in which you live has the most profound association with average incomes, work experiences, service provision and lifestyles. Viewed in process terms, as European expansion overseas bound distant territories more tightly into the nineteenth century world capitalist market, and as transport improvements made long distant trade in perishable farm commodities feasible, the demise of European agriculture was in danger of being decided by global production relationships (Peet, 1969). In much the same way, the creation of new competitive advantages in newly industrializing countries is currently undermining manufacturing in Europe (CEC, 1994c). Yet if these forces for change had been left to their own devices, the consequences for any single part of rural Europe could have been quite different from what actually occurred or is occurring.

The missing factor is that states exist, and impose themselves between global processes and ultimate local impacts. Certainly, theorization on 'the state' is contested, with significant disputes over the extent to which state processes and agencies have autonomy from the dictates of capital accumulation (Evans *et al* ., 1985; Block, 1987). Nevertheless, there is widespread recognition, despite disagreements over scope and degree, that the state is not simply an arm of capital (Dunleavy and O'Leary, 1987; Mann, 1988). Indeed, although market competition characterizes capitalist economies, the state is often required to coordinate, regulate and promote in the interests of particular fractions (or all) of its producers. This point is starkly transparent in the agricultural sector but is also important for large manufacturing corporations (O'Connor, 1973; Block, 1987). The reason state agencies are able to do this is because they (generally) have the support of their national populations. Where this is not the case, these institutions are unable or at least less able to be effective. Most commonly such agencies have legitimacy to act on behalf of a broader population through recourse to the appeal of nationalism. Smith (1991, viii) describes this as '... the most compelling identity myth in the modern world'. It is no surprise, then, that Taylor (1993) identifies the nation as a second explanatory scale, and seeks to capture its tone by labelling it as the level of *ideology*. Although state actions are penetrated by power relationships, it is the ideological acceptance that the state operates in the best interests of its citizens that imbues state actions with a legitimacy to insert itself between the individual and worldwide forces for change. Effectively, the national level introduces a reality of its own. And this does not simply involve state institutions, but also incorporates cultural and business organizations which gain legitimacy in their own spheres through their national associations. Nonetheless, this authority is highly circumscribed by global economic, political and social power relations, as well as by the necessity to maintain the sanctity of a nationalist ideological appeal (Mann, 1988).

In the modern world, we struggle to find geographical units below the level of the state that consistently display, even over limited territories, shared meaning that merits viewing them as causal 'agents' of theoretical relevance (Duncan, 1989). Regional or local governments clearly have institutional structures that could develop a substantial identity, and this does occur when these carry a regional separatist or nationalist appeal, as in the Basque lands (Pais Vasco), Corse or Scotland. However, generally they lack either a strong identity or effective power, being subordinate to national governments. Ethnic and religious groups likewise carry some appeal, but their strength is partly conditioned by their frequent coincidence with nationalist sentiments or state support, and their limited ability to coordinate either economic or political initiatives weakens their organizational role. For the vast bulk of the population, therefore, the third scale at which their lives are organized is that of their locality. This is the world in which they experience the reality of global forces and the translating effects of national inputs. Where such territorial units share a broadly-based identity amongst their residents (not a one-off coming together to resist an 'outside' force), they can be significant foci for group action and can impose a (somewhat) distinctive imprint on broader societal forces. Yet, even if we allow that such local territorial groups do not have to be homogeneous to be meaningful, that they need only possess coherence for a limited array of issues, and that they can mean different things to people in the same area (Cohen, 1982; Jonas, 1988), there is little that leads us to suppose that such units have major causal effects in a European context (Savage *et al*., 1987). Indeed, because of the limited autonomy of such units, at least in terms of their determining effects on major trends in local living conditions, along with theoretical uncertainties about what constitutes the essence of a 'locality', we would agree with a host of other researchers and warn against too ready an acceptance of the claim that localities are significant causal factors that impose themselves between the individual and global processes of socioeconomic change (Warde, 1985; Newby, 1986; Urry, 1987; Duncan, 1989). In no sense does this mean that local events and practices are not critical to understanding socioeconomic change. But the fact that local events are significant should not be extended into the expectation that socioeconomic processes can be theorized in terms of the 'unique' contribution of 'autonomous' local social practices. In a limited number of cases this might be so, but the primary dimensions of behaviour are structured at a higher level than the locality.

This does not weaken the case for undertaking locality-based studies, nor for increasing our understanding of how production and reproduction practices coalesce in peculiar ways in specific places (Newby, 1986). What we are asserting is not that local social processes are unimportant, but that we need to position them with regard to national and transnational causal forces. We also need to be careful over labelling processes as 'local' when they might be general in occurrence. For the purposes of this book, the role of the national level is a primary issue for analysis. By placing the spotlight on national issues we are not proposing that they be afforded ascendancy in causal priorities; indeed, we make no claims about the manner in which they are situated within a causal framework. What we do argue is that in theoretical formulations of development processes, particularly but by no means exclusively as these relate to rural areas, there has been a tendency to miss out the causal imprint of the national scale. One result of passing over this causal level is that we assume too readily that global processes are generalizable across space, save for locally inspired deviations. Quite feasibly the search for global generalizations is a chimera, for global forces might be heavily conditioned by national processes.

Global processes and state positions

In reality the interactions of global and nation-specific processes are two-way. It is not the case that countries simply 'receive' international impulses and adjust to them in their own specific manner. However, the extent to which countries are able to make their own imprint on broader cultural, economic, political or social forces is grounded, in some measure, in the standing of a country within the global political economy. In this sense, there is no point in conceptualizing a superpower in the same way as a small, poor country. The superpower does not simply stand in a stronger position in terms of resisting unwarranted 'outside' influences but also occupies a substantially different role in instigating and perpetuating forces for change that are identified as 'global'. Nor are national imprints on broader forces simply decided by governments. When we refer to a nation or a superpower, we include a wide range of human agents. In terms of structuring global development processes since 1945, for instance, we can readily point to the role of the US government in creating international organizations like the

UN, the IMF and the World Bank, in a manner that favoured US interests (Cox, 1987; Blake and Walters, 1992). We should also note that even the largest corporations generally have their operations directed from and grounded in one country, with compromises on overseas corporate operations being struck with governments that host multinational investment (Scott, 1985; Gill and Law, 1988). But economic and political institutions exert power across much broader arenas than their own 'sector'. For one thing, the consumption of products carries meaning that creates both place and personal identities (Sack, 1992). Hence, the globalization of Coca Cola and MacDonalds is not simply important for the economic transactions this embodies, but also because their consumption is an element in the globalization of consumption tastes and thereby of personal identities (Bocock, 1993). The consuming of physical products is only one aspect of this process, for global flows of information and ideas are likewise dominated by a few nations, so their cultural values have more chance of being interpreted as 'the norm'. The dissemination of US values by Hollywood and CNN provide obvious examples here, but we should rephrase a point made earlier to identify its significance in this context, for the association of the English language with globally dominant economic and political power relationships cannot be dissociated from an over-reliance on Anglo-American ideas and conceptualizations in the academic world. One consequence of this is that a particular view on the character and processes of change in rural areas tends to dominate. In less powerful countries such subservience leads to fears about a loss of national identity and autonomy (e.g. Grant, 1965). The corollary is that more dominant countries should feel more capable of retaining national traditions.

This points to our need to position Europe in its global context. What is meant by Europe obviously has some bearing on the conclusion that is reached in this regard. However, we do not wish to become distracted by disagreement over what Europe means to different people. For the purposes of this text, Europe refers to the countries of the European Community. This definition in itself has been subject to change, with the original six member states (Belgium, France, Italy, Luxembourg, Netherlands and West Germany) being joined by three new members in 1973 (Britain, Denmark and Ireland), by Greece in 1981, Portugal and Spain in 1986, the former East Germany in 1990 and by Austria, Finland and Sweden in 1995. It might be argued that, on the basis of similarities in socioeconomic standing, countries such as Norway and Switzerland should be included under a European rubric (the former almost becoming a member in 1973 and 1995). But this is not a critical issue for our purposes. In adding or subtracting a few nations to the European Community we do not substantially alter the place of 'Europe' in the global political economy. And given a European Community of 15 nations, such minor adjustments do not create a significant enhancement (or subtraction) to the degree of internal variation within 'Europe'.

The role of the EC as a central player on the global scene is not difficult to demonstrate. If we examine production figures for the EC12, for instance, we find that the Community accounted for 39.2% of global exports in 1992 (42.5% if the three new members from 1995 are added) and 39.6% of global imports (42.9% with the new members; GATT, 1993). In agriculture it is not simply the production capacity of the EC that symbolizes its importance but also the role of the EC in farm commodity trade. Much of the controversy surrounding the recent Uruguay Round of GATT negotiations stems from protectionist policies in the EC that have changed the Community from a net importer to a net exporter of farm products (Philip, 1990; Ingersent *et al*., 1994a), with self-sufficiency now having been attained in a wide range of commodities (see Table 1.1). We could give further examples of the eco-nomic power and the cultural and political importance of the EC, but as good an example as any is provided by the agricultural sector. This is not simply because this economic arena is most obviously allied to rural areas, but also arises from the central role that agriculture has played in progress towards European integration (Arter, 1993). One result is that, despite complaints over added national policy interventions (Tangermann, 1992), agriculture is a genuinely European policy field. In this sphere, states have allowed their

Table 1.1 Percentage self-sufficiency in agricultural commodities in the EC, 1991

	All grain	Rice	Potatoes	Sugar	Wine	Cheese	Butter	All meat
Belgium/ Luxembourg	53	0	150	224	0	39	99	150
Denmark	151	0	95	223	0	363	165	315
France	226	27	100	213	93	116	98	105
Germany	127	0	90	129	58	91	101	95
Greece	140	105	95	88	139	83	27	71
Ireland	99	0	77	173	0	352	1225	299
Italy	84	256	84	94	114	86	75	73
Netherlands	28	0	164	167	0	272	248	230
Portugal	51	65	72	1	123	99	107	90
Spain	99	139	92	79	118	85	170	96
UK	124	0	92	53	0	67	60	87
EC	129	84	101	123	96	105	111	105
Austria	108	0	96	99	95	118	102	107
Finland	*	0	*	70	0	*	123	110
Sweden	146	0	86	99	0	84	176	106

Source: CEC (1994d, 264-265). Note: * no data.

national autonomy to be diminished in order to further European cooperation. This certainly cannot be said for many other policy arenas, where the EC
has been slow to harmonize policies across its member states (Williams,
1991).

A key reason for this arises from the innate disagreements that emerge
when national governments must compromise on their most obvious and
immediate self-interest in order to advance a longer term or communal goal.
This dimension of EC policy-making was very evident in the formation of the
Community. As Milward (1992, 283) indicates, while it has become somewhat fashionable to explain the Common Agricultural Policy (CAP) as a
Franco-German deal, in which the Germans ceded to French farm exports in
order to gain access to a larger market for manufactured goods, in reality
French demands for a capacity to export farm produce had more to do with
the French government's inability to address internal problems of farm over-
production than providing a counterbalance to German manufacturing gains.
This interaction between internal and Community politics has remained with
us since, with internal divisions within the British Conservative Party providing a prime example. Another example concerns reported attempts by then
French prime minister Edouard Balladur to obstruct EC legislation in order
to win right-wing support for his candidature for the French presidency. In
some measure, different interpretations of what is in Europe's best interest
are bound to be conditioned by the dissimilar circumstances of national
policy-making. It is no surprise that Italy welcomed the southern enlargement
of the European Community, given the expectation that this would shift
agricultural policy directions away from favouring north European products
(Featherstone, 1989). But short-term political expediency, dissimilar views
on ideal long-term solutions and uneven productive capacities do not account
for the complexity of cross-state value differences. In addition, there are
genuine divergences in popular views that have built-up over the centuries.
These emerge from the history of socioeconomic change within a territory,
combined with the experience of state-building and disintegration, plus the
influence of political power relationships. That such value differences are still
apparent across nations is evident from the dissimilar attitudes of EC citizens
toward European integration (see Table 1.2, page 14). Yet such values are
subject to change, with Greece providing a clear example of a nation in which
many citizens (and political leaders) had severe reservations about EC
membership (Featherstone, 1989), but where support for the Community is
now strong, despite some notably adverse economic effects after accession
(Tsoukalis, 1993; Giannitsis, 1994). A somewhat similar process appears to
have occurred in Portugal, although Bacalhau (1993) suggests that the
transition here was from indifference to support. Indeed, evidence points
toward a convergence of value dispositions within Europe, with growing
trust for one another amongst citizens of different EC countries (Reif, 1993).

This provides one indication of the basis on which Europe has come to
occupy a central role in the world political economy. Organizations repre-

Table 1.2 Indicators of differences in citizen beliefs across EC nations (% agreeing with statement)

	Belgium	Denmark	France	Germany	Greece	Ireland	Italy	Luxembourg	Netherlands	Portugal	Spain	United Kingdom
Membership of the EC is a good thing (Nov/Dec 1993)[1]	59	58	55	54	73	73	68	72	80	59	54	43
Membership of the EC is a bad thing (Nov/Dec 1993)[1]	9	23	14	13	4	8	7	6	5	12	14	22
Single European Market is a good thing (Sept/Oct 1992)[1]	39	43	31	39	48	55	52	35	44	51	38	32
Single European Market is a bad thing (Sept/Oct 1992)[1]	11	16	12	12	11	6	8	13	9	4	11	18
An EC charter on fundamental social rights would be a good thing (Sept/Oct 1992)[1]	54	52	60	62	72	70	75	58	75	70	61	57
An EC charter on fundamental social rights would be a bad thing (Sept/Oct 1992)[1]	6	25	14	9	3	6	3	10	8	3	4	21
The CAP is a good thing[2]	27	15	28	18	55	48	25	25	26	42	27	16
The CAP is a bad thing[2]	15	43	34	44	7	9	15	25	32	5	15	49
Believe in equal rights for men and women[3]	35	*	42	22	*	58	44	*	38	49	38	60
Are very proud of their nationality[3]	26	*	32	17	*	76	41	*	22	*	45	52

Source: [1] CEC (1994e); [2] CEC (1989a); [3] Ashford and Timms (1990) Note: * no data

senting the EC might not be able to dictate their wishes, but they are important actors that need to be given due regard. It is perhaps not external restraints that most limit the scope for action of EC institutions but pressures from within the EC itself that relate to dissimilar values and interests. To understand the grounds for these divergences, it is not sufficient to focus simply on the national positions of the EC member states, for it is rare that there is overwhelming agreement within a country over its own interests and goals. In particular, given our assertion that the national scale has received insufficient attention in research on rural development, we must ask what we mean by a national interest or worldview, and whether such views have the required coherence to provide a base for distinctive rural trajectories?

Unity in national values?

The notion of national values that underscores our examination of rural Europe is broader than that found in much geographical writing on cross-national differences in 'the countryside'. Most commonly, work of this kind has had its primary focus on notions of landscape; on the visual appeal and character of rural habitats, and how images of rurality are carried into city and suburban design. A good recent example of this is Michael Bunce's (1994) comparison of English and North American images of the countryside. The distinguishing features of our approach, compared with Bunce's, are found in evaluations of both product and production process. For the former, Bunce is primarily concerned with landscape features, whereas our interest focuses more on impacts on people; on their living standards and lifestyles. Linked to this, our approach focuses more on processes than on end-products. Thus, following a well established tradition of geographical interpretation that has produced some admirable insights (e.g. Lowenthal, 1991), Bunce takes 'the landscape' as his object of inquiry and provides an interpretation of national orientations toward the countryside that is couched in terms of cultural origins and present-day symbolic meanings. Having its roots associated with cultural geography, this approach is in the tradition of an important strand in rural geography research. However, if we wish to understand outcomes for people, rather than for landscape as such, we need to adopt a broader vision of national differences; one that inquiries into the underlying causal processes that produce particular (national) sociocultural practices.

Penrose (1993) reminds us of what is involved here in her insistence that nations are not natural entities but social constructions. As such, national identities and cultural values do not emerge as an inevitable outcome but are contested. Thus Penrose argues that: 'the *general* category of nation is reified by the attempts of *specific* groups of people occupying *specific* places to demonstrate that they constitute an example of this *general* category' (Penrose, 1993, 31). People are not equally able to contribute to the formulation of national identities and values. There is contestation over the nature of national characters and over what 'the nation' actually is (e.g. Williams, 1980). These contestations are a product of, and help reinforce through legitimation, specific power relationships, as well as emerging from conflicts over the distribution of societal 'goods', whether tangible or intangible. It follows that if the central idea of 'the nation' changes, then this is likely to reflect an altered distribution of power and material gain. Hence, specific conceptions of a nation embody 'unique' benefit structures. One practice that is embodied in the reality of 'the nation as ideology' is the use of national symbols and slogans to further vested power interests (Taylor, 1993). Functionaries might well wish to project the image that state actions are in the best interests of all citizens, but the winners and losers from state processes are not randomly distributed (O'Connor, 1973; Dunleavy and O'Leary, 1987). As a consequence, we can expect particular images of rurality, and specific 'uses' for the countryside, to depend upon national power distributions. This implies that the strength and acceptance of images of a desirable rurality depends in some measure upon the degree of unity over and within a nation's leadership.

Significant power relationships, meanwhile, are not restricted to state institutions. In the sphere of production, the structure and history of capitalist development is grounded in real social formations and conflicts that take their essence from national settings (Katznelson, 1986; Pfau-Effinger, 1994). Viewed from this perspective, the state is best conceptualized not as an 'actor', whether in the international or in the national sphere, but as an instrument (Taylor, 1989), even if this 'instrument' is a process rather than an object. As an instrument it reflects the conflicts, compromises and alliances that dominate power relationships. When compromises emerge, they are partly decided by the external situation of the nation, with the world-system perspective emphasizing the practical limits within which national leaders operate. However, this should not distract us from recognition that national sentiments are fundamentally compromises between national elites that require support from the general population to maintain their integrity. Nationalism provides a key mechanism in this process, as it binds people through a sense of 'imagined' community, with this deep sense of horizontal comradeship reducing tensions that emerge from inequities in society (Anderson, 1983).

Of course, nationalism creates tensions in its own right, not simply across states but also within them. This arises because the collective identity that surrounds nationalism is not automatically transposed onto the state. Most evidently, this is seen in cases where members of a population are prone to define themselves as a 'nation' that has been subsumed under an 'alien' state. Europe has a significant number of these groups (Williams, 1980), and their existence can remain hidden for some time, as conflict in the former Yugoslavia indicates. On the surface, such regional or national separatist sentiments differ in political activism, in the relative socioeconomic standing of their territorial base (compared with the nation's core) and in the primary characteristic that distinguishes them from other populations in a state. However, even though they might be a long way short of being capable of forcing a separation from their present state authority, the distinctive values of such groups raises the prospect of dissimilar sentiments and identities existing across rural areas. This is readily seen in France, where inland, remote rural areas of Corse provide the symbolic core of a Corsican identity, in contrast to the Parisian focus that lies at the centre of national identity on the mainland (Winchester, 1993). Similar distinctions can be identified in Britain, where the imagined country idyll is a particularly English phenomenon, and especially a southern English conception, which should not be conflated with the distinctive self-imagery that is associated with rural Scotland (Penrose, 1993).

Our view is that we must understand the position that images of 'the countryside' or 'rurality' carry in national identities if we are to appreciate cross-national differences in rural trajectories. In furthering our understanding we need to recognize that such identities are not decided simply by state institutions, not even in association with distinctive patterns of production. Although production differences do distinguish peoples, they both inform and are informed by consumption (Bourdieu, 1984; Glennie and Thrift, 1992). And while consumption differs significantly across Europe (see Table 1.3), it does not intersect with state and production processes in an unambiguous way (Featherstone, 1990). Put simply, the 'rural' is not only a sphere of production but also a site of consumption. The pattern of consumption, as with any process of consumption, helps define the consumer (Bocock, 1993). Most immediately these consumers are rural residents but more generally the nation as a whole gains identity and meaning from the manner in which rural space is used (or abused). Yet the cultural symbolism that is embodied in notions of rurality is open to manipulation by what Bourdieu (1984) refers to as cultural intermediaries; that is, by opinion formers who, for whatever reasons, perpetuate a particular image of the countryside. In nineteenth century Britain, the cosy image of the countryside

Table 1.3 Food consumption in EC countries, 1991

Product consumed (kg/capita)	Belgium Luxembourg	Denmark	France	Germany	Greece	Ireland	Italy	Netherlands	Portugal	Spain	United Kingdom
Cereals (without rice)	72[a]	71	74	73[a]	106[a]	95[a]	110	55	85	71	74
Potatoes	97[a]	57	72	75[a]	89	142	39	87	107	106[a]	94
Vegetables	103	80[b]	124[c]	81[a]	229	102[a]	182	99	125	192	65[c]
Fruit	57	49[b]	58[c]	61[a]	42	*	77	38	67	66	38[c]
Wine	39	22	65	26[a]	25	4	62	14	57	43	10
Milk products	8	14	9	93[b]	6	18	6	13	9	10	12
Eggs	14	14	15	15[b]	11	11	11	11	8	15	11
Meat (without offal)	94	98	101	94[b]	71	87	84	87	69	95	71
(of which beef)	22	19	24	21[b]	19	17	22	20	15	13	19
(of which pigmeat)	48	65	37	58[b]	21	38	32	44	27	49	24
Oils and fats	26	41[b]	22[c]	21[b]	33[c]	23[c]	31[b]	38[c]	23[c]	32	30[b]

Source: CEC (1994f, T168–T169). Note: * no data ; [a] 1990/91 ; [b] 1989/90 ; [c] 1988/89.

that emerged owed much to the promotion of a wholesome (national) identity icon which helped underscore imperialist expansion (Daniels, 1993). At the same time this proved a foil for concern over environmental and social despoilation emanating from factory towns (Lowe, 1989). This idealization, which came to be so wholeheartedly accepted by the middle and then working classes, clearly benefited the urban wealthy and the landowning aristocracy; both because of the distraction caused by this contrived imagery over the 'heart' of England (Lowenthal, 1991; Daniels, 1993) and from the presumed idyllic existence of poorly housed, terribly paid and significantly overworked farm labourers (Newby, 1977). In similar vein, although offering a quite distinctive symbolism, in France '... the political mystique of the traditional rural way of life [acts] as a symbol of the nation's strength and solidarity' (Thompson, 1970, 127). This also is no 'natural' phenomenon, for members of the old nobility and the conservative bourgeoisie who owned rural estates were instrumental in forging the myth of peasant dominance and unity in the countryside as part of their campaign to strengthen state support for farm production. They were also successful in deflecting attention away from the large share of national landholdings that fell in their own hands (Tarrow, 1977; Tracy, 1989). As these two examples illustrate, national images of rurality emerge from power relationships (see also Mormont, 1990). The forces and social groups that promoted such relationships often come to be replaced by others, which change or capture dominant representations. Thus, in Britain, the middle classes have become critical to the continuance of an idealized countryside imagery (Jenkins, 1992; Savage *et al* ., 1992). It follows that if we are to understand the character of rural Europe, we must investigate the historical paths through which such areas have passed, as well as the pressures, problems and opportunities that have confronted them in recent decades.

Structure of this book

In order to explore distinctions and similarities in the treatment of rural areas within the European Community, we need first to understand distinctions in the 'objective' circumstances of rural regions within nations. This we turn to in Chapter 2, where divergent views on rurality and its characteristics are examined. Here the well-known existence of disparity in rural conditions within countries is confirmed, but so too is the presence of areas in different countries with shared development problems and potentialities. Complexity in the socioeconomic circumstances of 'rural Europe' is intensified by dissimilarities in the 'ideologies of rurality' that are found within the Community. The basis for these different conceptions, which lead to distinctive preferences for rural transformation and adaptation, are explored in Chapter 3. Ideological structures have long historical roots, as well as wide public support. Both are grounded in the enduring character of power relations within states. Ideologies of rurality set frameworks in which national governments, along with private sector institutions, pressure groups and citizens at large, accept specific trends in the countryside. Chapter 3 draws attention to rural traditions within the European states, seeking both to highlight peculiarities in national positions and identify shared sentiments across states.

In Chapter 4 we draw on this base to explore processes of change in policy towards the countryside. Here we focus on the transnational level and identify how national orientations commonly conflict with international pressures for rural transformation and policy change. The institutionalized structure of the EC itself enforces pressures favouring uniform responses toward agriculture, the environment and rural areas. These do not result simply from an internal dynamic, but also from the manner in which European nations coordinate their actions in response to outside 'threats', as seen in the Uruguay Round of GATT negotiations. Intra-European issues are obviously of significance as well, with some sharing of objectives and policy criteria for the CAP, environmental regulations, and allocations from regional and social funds. Yet countries often find these attempts to impose uniformity, at best, irksome or, at worst, economically and politically damaging to their governing parties. Hence, in considering the transnational element in rural development trajectories we need to recognize the dissimilar objectives of nations in their interactions with European Community institutions, as well as in cross-national conflicts such as the British-French 'lamb war' and the British-German beef dispute. In effect, in this chapter the central issue is how individual states interact, collectively and individually, in ways that influence the development potential of rural areas.

The seeming 'political' focus of this chapter is followed by a more overtly economic tone in Chapter 5, in which our attention is directed at production practices that change rural economies. For this we start by considering globalization processes, as seen in the industrialization of agriculture and the emergence of new economic landscapes under the impress of so-called post-Fordist economic restructuring, and through the actions of transnational institutions like multinational corporations. Set in this framework, new economic roles for rural areas are identified as the economic significance of agriculture has declined, new rural manufacturing spaces have developed and service sectors have been transformed. Once again the form these changes have taken is not uniform across nations. But does the character of these changes carry particular national shades, as well as obvious local colourings?

The nature of change in civil society is then explored in Chapter 6. Here we examine forms of social organization in rural Europe that are of longstanding, inquiring into forces of modernization, class division and population change that have produced new social arrangements and practices that vary both across localities and nations. Again we note how global processes of (social) change have impinged differentially on rural areas across

Europe, as well as assessing the manner in which national specificities are allied to social form and change. Focusing more explicitly on the physical context of rural life, Chapter 7 then draws attention to landscape and environmental issues in rural development. Here the dictates of socioeconomic restructuring and technological advancement that drive and support adjustments in the country environment are considered. The manner in which change in the rural environment is threatened by agricultural restructuring, the commodification of rurality and social and institutional processes allied to environmentalism are all explored in this chapter.

2

The rural mosaic: diversity in Europe

On a European scale there is little chance of reaching consensus on what is meant by 'rural'. This arises for two main reasons. First, the very notion of what is 'rural' is not agreed. Also, the rather different cultural, demographic, environmental, political and socioeconomic circumstances that face European countries lead their residents, or more formally their governments, to emphasize dissimilar attributes as key characteristics of 'their rurality'. What is most apparent from a review of different national approaches to administrative definitions of rural is that rural areas are more readily identifiable as non-urban space than as rural *per se* . It is both more straightforward and more convenient to establish definitions of urban areas, based on population size or building density, than to attempt to identify the defining parameters of rural space. This in itself says much about the chaotic nature of rural both as an 'operational category' and as a 'symbolic construction' (Urry, 1984; Bodson, 1993).

Official designations of the rural

States such as Denmark, France, Ireland, Italy, Spain and Sweden all classify as rural those administrative units that fall below defined population thresholds for urban zones. The thresholds adopted, however, are markedly different between states. While the conventions for Italy and Spain are for urban areas to have a population of 10 000 inhabitants, with rural areas *de facto* where the population falls below this threshold, in Denmark and Sweden the requirement for urban areas is for at least 200 people where homes are not more than 50–200 metres apart (Pumain and Sant-Julien, 1992; Persson and Westholm, 1994). In France, the national statistics service defines urban communes as those with 2000 inhabitants, while the basis for the distinction between aggregate urban areas and aggregate rural areas in Ireland is set at 100 inhabitants. Such major differences point to the multiplicity of rural contexts within Europe. These reflect both the historical development of the countryside, and the physical environmental and socio-political factors that play a key role in the 'construction' of rural space. In very simple terms, Mediterranean definitions of rural space tend to employ high population thresholds because of their distinctive pattern of rural land-use evolution. Owing to the mountainous and upland nature of much of the land surface, alongside the large landholdings that characterize some parts (like southern Spain), rural populations are relatively concentrated around urban hinterlands. What is recognized as rural, in a productive as well as a demographic sense, tends therefore to have a more concentrated population distribution than in other regions. In Greece, for example, figures calculated at a national level, which yield an average rural population density of 30 inhabitants per km^2, are misleading in that they fail to account for the fact that only 30% of rural Greece is farmed. A more appropriate rural population density can be calculated at around 100 inhabitants per km^2 (Kolyvas and Gotsinas, 1990). By way of contrast, nations such as France or Denmark, which have historically sustained dispersed, low density but demographically important rural populations, have adopted far lower population thresholds for defining urban centres. What is important in the Franco-Greek comparison is that whereas the former's concept of rural is founded upon the assumption that human activities, and particularly agriculture, have enabled the colonization and occupation of virtually the entire national territory (which consequently is seen as being 'rural'), the Greek notion, which is equally rooted in the historical coupling of rural with agriculture, makes an implicit distinction between productive, occupied 'rural' space and a more extensive, non-productive mountainous space (Sivignon, 1989).

This functional approach to defining rural space has parallels in a number of other European states. In the heavily urbanized states of Belgium, Britain, Germany and the Netherlands, the differentiation between urban and rural is essentially one based upon and reinforced by land-use planning policy for which rural areas are close to being taken as synonymous with either agriculture or landscape protection (CEPFAR, 1990). In a circular and self-reinforcing process, rural areas become the targets of rural plans which are subsequently used as the basis for enumerating rural areas.

Significantly, the majority of European states recognize the need to go beyond a simple urban-rural dichotomy in their spatial classifications. Thus, it is not uncommon to find a multiplicity of rural typologies, with some being created and implemented by national planning or statistics agencies (like the *Institut National de Statistique* in Belgium, the *Bundesforschungsanstalt für Landeskunde und Raumordnung* in Germany, the *Istituto Centrale di Statistica* in Italy and the Office of Population Censuses and Surveys in Britain), with others arising from the fruits of academic labour (e.g. Hodge and Monk, 1991; Clout, 1993b; Bontron 1994). What is of interest here is that these subdivisions display pronounced cross-national similarities both in the choice of criteria used to define rural types (most commonly including the relative size of the agricultural workforce, average commuting distances, service provision and population densities) and in the 'types' that ultimately result from the classification procedures.[1] And while these classifications reveal complexity (e.g. Hodge and Monk, 1991; Clout, 1993b), it is clear that the notion that rural and urban can be positioned along a unidimensional scale retains potency in administrative classifications.

At the level of the European Community, rural policy is similarly founded upon a discreet and implicit differentiation of rural zones according to their position with respect to urban centres (CEC, 1988). Nevertheless, the Commission has continually shied away from explicitly defining the rural space(s) of Europe. Thus, in a written response to a question from the European Parliament, the Commission maintained that:

> At the present time, there is neither a geographical delimitation of rural space nor a harmonized definition of rural population within the Community. Under these conditions, all estimations of the rural population, either by Member States or by regions, can have only an indicative value and cannot thereby serve as the basis for comparisons of any significance (CEC, 1990b, authors' translation).

Ultimately, rural administrative definitions provide us with an excellent example of the tension between national rural traditions and contemporary rural processes. Hence, while Britain largely did away with the administrative distinction between urban and rural districts in 1974, and so recognized the irrelevance, in administrative terms, of a nineteenth century construction based essentially upon the different service needs of densely populated towns and an emptier countryside, France has clung with tenaciously to its communal structure (Klatzmann, 1994). This is just as much of a nineteenth century construction, although it is one that was founded, inversely, upon the need to administer a predominantly rural and agricultural population. Although commentators have long maintained that the urban-rural division inherent in the French classification has lost its pertinence (e.g. Berger and Rouzier, 1977), government attempts to merge communes have not been a success. Furthermore, attempts to create a more viable and contemporary spatial classification, the *zones de peuplement industriel ou urbain*, though useful as a basis for assessing the degree of urban penetration within the French countryside, are currently being abandoned as they now cover over 90% of the French territory (Lévy, 1994).

1 As one illustration, Clout (1993b) provides a sixfold categorization that draws on the work of other European investigators, such as Meus and associates (1990). This recognizes the following types of rural areas: dynamic rural places, which display a high degree of economic and social wellbeing, in which the economy is based on (1) truly commercial agriculture, (2) peri-urban activities or (3) catering for recreation and tourism; economically lagging rural areas, which are distinguished by (4) their poor services and facilities or (5) their marginal geographical location, with its associated high costs of transportation; and, (6) rural areas that are in the process of being abandoned by their populations.

A rurality of the people?

It would be convenient for this book if these divergent articulations of rurality at the state level merely reflected different popular conceptualizations of rurality. However, the links between national portrayals of 'rural' and any underlying understanding of what this constitutes in the popular imagination are not direct and simple. Most evidently for official definitions, such links are subjected to the distorting effect of administrative convenience. Thus pre-1974 statistics on rural England are based on an aggregation of Rural District Council figures; and with the demise of that administrative category in 1974 a vacuum has been left in officially agreed delimitations. Theoretically, the links are no less troublesome, given the near binary divide of different approaches to conceptualizing rurality. At one end of the scale there are those who view 'rural' as an abstract concept; where rural areas are distinctive 'entities' from other (urban) places. At the other end of the scale we find those who see rurality as being defined personally; where rural is a mental construct which, for the individual, encompasses reality as he or she sees it (Halfacree, 1993). For some, this latter approach might seem to hold the key to understanding what is 'really rural', for it focuses on how 'rural' is experienced by those who integrate visions of rurality into their everyday lives. Certainly, it appears to provide rural studies with a stronger theoretical footing; following more than a decade in which the 'chaotic' standing of rural as a theoretical conceptualization had reduced its utility to little more than a convenience (Urry, 1984).

> There is now, surely, a general awareness that what constitutes 'rural' is wholly a matter of convenience and that arid and abstract definitional exercises are of little utility (Newby, 1986, 209).

Even so, the expectation that a theoretical future can be found for the concept rural in personalized definitions needs to be handled with care. This vision might have substance if the intention is to understand the activities of residents in low density areas, or even countryside attitudes, housing strategies and recreational pursuits of city residents. But this will not get us far in understanding the structural constraints and opportunities arising from processes operating in society at large. This is because the 'directing' nature of such forces often does not arise from any explicit rural characteristic, but emerges from the attributes of places that interact with decisions of a sectoral nature. An example is provided by cuts in public expenditure, which often affect country locations owing to the higher costs of service provision in low density locales (Caldock and Wenger, 1992). Where national policy-makers do infuse their decisions with an explicit rural vision, there is no guarantee that this coincides with that held by countryside dwellers. Thus national politicians and bureaucrats may demarcate lower income rural areas as an economic and/or social 'problem' (e.g. Schweizer, 1988), even though their characterization of the 'problem' is not shared by rural residents. Even if there is a coincidence of imagery, which should not be expected to be commonplace, given that agents at international, national and local levels are involved, as well as those from the public, private, voluntary and household sectors, the manner in which images are translated into action is not straightforward. Rather, it is an outcome of power relationships, in which considerations of rurality carry different weights, both across agents and in different decision contexts. Policy-makers also possess dissimilar conceptions of what is meant by, or is desirable in, rural areas. It hardly needs saying that the balance of each of these is likely to vary from one nation to another.

There is also the question, well recognized by some who see merit in researching individual images of rurality (Murdoch and Pratt, 1993), that country dwellers behave in different ways, not simply because they do not share the same personal vision of rurality but because they aspire to dissimilar goals, and are unequally able to attain the objectives they desire. Hence, the peace and quiet that some find an enchanting feature of 'their' countryside is a key factor that leads others to leave.

> The desertion of what they may regard as 'traditional' culture in peripheral rural areas is often assisted by their inhabitants' views of the economic disadvantages of geographical peripherality and their tangible expression in the decision to migrate. But a community's positive orientation to its apparent collective distinctiveness may well lead its members to determined efforts to make it flourish by regulating their own behaviour in strict accordance with its principles ... (Cohen, 1982, 5).

As such, while an individual's concept of rurality can provide prompts for personal or collective action (Lowe, 1989; Mormont, 1990), for collective action to occur there must be shared features in these images. In so far as action is stimulated by a shared rurality, or more accurately the need to protect or support it or its residents and institutions, we might say that 'rurality' has causal powers. However, such specific instances of causation are not easily extended to general processes of societal change, even within rural areas.

The need for care arises at two different levels. Within locations it is visible in social class, ethnic or gender divisions. In Agnone in southern Italy, for instance, Douglass (1984, 13) found that between social classes '...the chasm is so great in the minds of so many that it practically assumes caste-like overtones'. One consequence is the occupation of distinct spaces by social groups, with peasants avoiding village-centred events, given that they '...are ignored, ridiculed, or patronized, depending on the situation'. This village – open countryside split is evident in other settings (e.g. Pitt-Rivers, 1960; Davis, 1973; Schweizer, 1988), although country dwellers do not always occupy the inferior social position (as in eastern Ireland; Bax, 1976). Nevertheless, the existence of distinctive social groups can prompt forms of social organization, and through them shared orientations towards 'the rural', that distinguish farmers from nonfarmers (Silverman, 1965), religious groups (Golde, 1975) and recent arrivals from longer-term residents (Forsythe, 1980; Wild, 1983; Chevalier, 1993). Perhaps most obviously, we find people occupying different social worlds due to class differences, that render opportunities within and commitments to particular places uneven (Pahl, 1965; Cutileiro, 1971; Douglass, 1984; Cloke and Thrift, 1987). Differentiation also occurs through the socialization processes and behavioural outcomes that commonly orient or restrict the activity spaces of women (Davis, 1973; Schweizer, 1988; Little, 1994).

In effect, the social representations of rural that are associated with different groups of people are micro-cultures. The critical link this has with their behaviour comes from the fact that 'culture' is learnt (Jenkins, 1992). In part, given that the routine behaviour of people is partially determined by the 'objective' features of their social world (Bourdieu, 1984), we would expect to find disparity in concepts of rurality between those who have always lived in cities and life-long countryside dwellers. However, as 'people *learn* to consume culture and this education is differentiated by social class' (Jenkins, 1992, 138), along with, amongst other things, gender and race, we should not expect common views by place of residence. An exception is where unity is an outcome of shared cultural experiences, as was likely to be the case in so-called 'traditional' rural communities (Arensberg, 1937; Pitt-Rivers, 1971) or where it results from power relationships that draw weaker groups into acceptance of more dominant ideologies, as, for example, with deferential social structures (Newby, 1975, 1977). This does not mean that personal behaviour is determined simply by objective socio-demographic features (Featherstone, 1990). Personal identities are certainly expressed through consumption, but enjoyment of a 'commodity' (rurality) only partly results from its physical consumption (living, visiting or working in rural areas). It also arises from what this consumption says about the consumer (Bocock, 1993). Consumers can be expected to project (and personally take) different images from the process of consumption, not simply because their consumption is driven by dissimilar aims (home, work, play) but also because they use consumption differentially to forge an identity for themselves.

From this it should be clear that, even though personal conceptions of rural can be powerful causal agents in certain behavioural forms, the complexity of aggregating, or finding common threads through, diverse images of rurality means that the chances of finding a strong unity in personalized visions of 'the rural' *within* localities is open to question. Indeed, although this condition might hold in some places, there is enough evidence of intra-local divisions to contradict any expectation that this will generally hold across places (Mormont, 1990). Certainly, some consistent threads should be found, at least at a regional level, if not nationally. But studies of longstanding, conducted throughout Europe, have established that the once-heralded social isolation of village communities declined long ago (e.g. Keur and Keur, 1955; Brody, 1973; Wylie, 1974; Jenkins, 1979). Today, international and national forces make significant inroads into local identities. It would therefore be overly sentimental and simplistic to conceptualize rural identity within nations as an outcome of the aggregation of local actions and attitudes. More likely, this identity emerges from power relationships, especially at the national level, in which specific groups utilize imagery to project messages that favour themselves. As such, notions of what encapsulates 'the rural' cannot be disentangled from issues such as agricultural lobbying to strengthen farm incomes (von Cramon-Taubadel, 1993), commercial advertisers' desires to identify their products with wholesome countryside endeavours (e.g. Burgess, 1982), or home owners' attempts to thwart change to 'their rurality' by appeal to the merits of preserving national heritage and the environment (Lowe and Goyder, 1983). In turn, such images only capture one part of rural reality. Representational icons might help win support (or sales) for particular projects, but they are commonly contradicted by pressures favouring other economic, social or political interests, particularly over issues of capital accumulation and electoral support. This dissonance helps explain why French farmers, well-versed in the reality of their socioeconomic situation, find that their conception of little public support is confounded by opinion polls which portray them as a dynamic population group, that is a major national asset (Boussard, 1990). Likewise, being less aware of the changing reality of rural Britain, city dwellers appear to be swayed by a wholesome imagery of the countryside, which 'blinds' them to its altered state and the damaging effects of present-day socioeconomic trends to their cherished land (e.g. Young, 1988).

There is, then, dissonance between image and reality. But even more important is cross-national unevenness in the intensity with which 'rural' appears as a cultural marker, that attracts public support for its maintenance and enhancement. Fundamental to these differences is the extent to which 'rural' is a symbol, whose character projects a message about consumers of that 'entity' and, more broadly, about national identity. In the same way that the state introduces decision criteria that contradict central tenets of capital accumulation (Block, 1987), so acceptance of rurality as a cultural icon imposes restraints on the logic of capital accumulation. As such, acquiring cultural status for 'rurality' is inevitably contested. It takes time to achieve, as well as requiring broad support for its sustenance. Of particular importance here is the role played by the middle classes. Given their combined purchasing (and potential political) power, which can outweigh upper class imprints and places them in a more assertive position that the working classes, the middle classes hold a key position in determining tenets of popular culture, and hence prevailing visions of rurality (Jenkins, 1992). Yet the middle classes are not comprised of autonomous visionaries. Their tastes are both conditioned by and manipulated by others (Bourdieu, 1984). Arguably, therefore, they commonly bestow legitimacy on popular culture rather than determining its precepts. Nevertheless, they are not passive recipients, as an interaction effect operates between 'manipulators' and 'manipulated' which creates limits and behavioural opportunities for both. In a capitalist democracy, a strong combination of economic and political power means that the cues the middle classes take-up then broaden the appeal of consumption items. As such, the diffusion of certain cultural norms from the upper classes to the middle classes has important implications for the future of the countryside (Glennie and Thrift, 1992). The manicured gardens and country mansions of the British aristocracy provide one illustration of a well-grounded dimension of English landscape tastes (Lowenthal and Prince, 1965). Meanwhile, the predisposition to accept such cultural icons has provided the so-called cultural intermediaries of fashion, design, the media and the advertising industry with an important symbol for promoting capitalist interests and/or attaching themselves to marketable commodities. In northern European nations at least, the countryside is undoubtedly one such commodity (Cloke and Goodwin, 1992; Bunce, 1994).

What we should recognize is that, as with tourist destinations (Cohen, 1988), the fact that middle class consumers are more numerous than their upper class forebears commonly dilutes the experience they seek; '... in the drive to accumulate the objects of the new ambient culture, there is only a gradual realization that these objects are fundamentally inauthentic' (Gane, 1991, 77). Effectively, then, we return to the tension that exists between the countryside as a site of production and as a venue for consumption. The productivist core of the countryside is often little appreciated, and its impact little understood, by those who see rural areas primarily as a site of consumption. Hence, while urban-based middle class groups might have a stranglehold on popular conceptions of rurality in some countries, this does not mean that this image fits the reality of country lives. Even in Britain, where 'the countryside' is deeply infiltrated with a consumption-driven symbolism, change in rural environments continues to be dominated by production criteria (Lowe *et al.* , 1986).

Yet it is in Britain, with the status afforded to its landed aristocracy and the burgeoning nineteenth century appeal for an alternative to the landscape desecration and the ruthless, impersonal treatment of workers that accompanied urban-industrial growth, that the strongest appeal for a consumption-related rural imagery holds sway (Lowe, 1989; Edwardes, 1991; Lowenthal, 1991). Although some countries in northern Europe experienced similar urges from the Industrial Revolution, pressures resulting from population growth and land shortage (Denmark, Ireland, the Netherlands), the lesser social standing of the aristocracy (France), internal divisions between regions (Belgium, Germany) and struggles against both physical and political restraints (Finland, the Netherlands), all conspired to produce a less intense impetus for such attachments elsewhere (Rying, 1988; Cannadine, 1990; Abrahams, 1991). And while this consumption-derived imagery has uneven roots in the temperate lands of northern Europe, its presence is even less substantial in southern Europe. Here the rank poverty and continuing economic problems of many rural (and urban) places was hardly conducive to a consumerist vision. In Italian regions like Sardegna (Sardinia) and the Mezzogiorno (Davis, 1979; Schweizer, 1988), in Andalucía and Extremadura (Brenan, 1950; Salmon, 1992), and in much of Greece (Manganara, 1977; Economou, 1993) and Portugal (Cutileiro, 1971; Black, 1992), the majority of past rural residents had lives which centred more on ensuring an income above base requirements than worrying about consuming surpluses. The same point holds for some peripheral regions in northern Europe, such as western Ireland and northern Scandinavia (Walsh, 1986; Abrahams, 1991).

Impasse?

What all this points to is the difficulty of reaching any single definition of rurality. Personalized identities are likely to be specific to a limited range of places and perhaps to particular social groups, as well as varying with the issue that is motivating an area's residents (or an outside group) to think of rurality. But if personalized conceptions of rural are not conducive to a single interpretation of what rural is, then the problems this raises pale beside those that exist if we seek an abstract, universal definition that is intended to help explain differences in behaviour (Hoggart, 1990). While rose-covered cottages in places of one or at most two short streets might be central to many English people's conception of a rural settlement, in southern Italy and southern Spain we find spartan towns of perhaps as much as 30 000 people occupying critical roles in the agrarian economy, with the open countryside often being largely bereft of inhabitants (e.g. Jansen, 1991). To us, these locales are equally rural, since these southern European agro-towns have hearts and souls that are intricately woven into the fabric of a rural-centred economy. Any stifling of inputs from the open countryside inevitably produces economic decline and most likely social disintegration. Adopting this as a working premiss, where does this take us in our quest for a Europe-wide definition of rural to guide this book? Quite simply, in so far as we are intent on providing some measure of an 'objective' description of rural areas across Europe, it leads us to adopt the only definition that is sustainable, no matter what the country under investigation; namely, population density. In our view, the one core idea that infiltrates the way people from a variety of nations and social backgrounds see rural areas is that of open country areas punctuated by periodic settlements. Provided that their primary *raison d'être* arises from integration with that open countryside, we do not feel uncomfortable with the idea that towns can be treated as part of a rural area, although 'rural region' might be a more appropriate term in such instances. It also does not concern us that many rural areas do not have an economic base in agriculture or forestry. On the contrary we see the association of rurality with agriculture as a contingent relationship, with rural features being equally

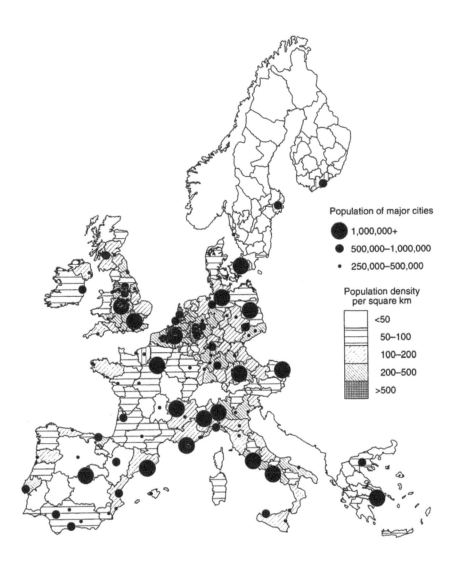

Figure 2.1 Population density in the EC in 1992. Source: compiled from:
a) CEC (1994c, 17); b) CEC DG-XVI (1991, 29); c) Statistiska Centralbyra
(1994, 28); d) Tilastkeskus Statistiskcenralen (1994, 51); e) Österreichischen
Statistischen Zentralamtes (1993, 17); f) Central Statistical
Office (1994, 180–190) and g) Statistiches Bundesamt (1994, 59–60)

evident when economies are based on small-scale fishing activity (Carter, 1974) or 'rural' tourism (Grafton, 1984). Yet because of the variety that exists in economic bases, in sociocultural behaviour norms and in settlement structures, there is also no magical figure for population density which demarcates the transition from a rural to an urban area. In describing the diverse circumstances of country areas in Europe, therefore, we rely predominantly on population density (and lack of access to major urban centres) to delineate the extensive tracts of European territory which are rural in their settlement configurations (see Figure 2.1, page 30).

The diverse character of rural Europe

Using a definition based primarily on population density, we find that within the European Community some nations, like Greece, Ireland and Portugal, possess a rurality which extends to almost the whole of a nation, outside a few major cities. Indeed, the proportion of the population living in settlements of less than 2000 inhabitants ranges to as high as 57% in Portugal and 32% in Greece (Eurostat, 1990). In other areas, such as northern Denmark, the centre and south west of France, Corse and Sardegna, inland Spain, northern Scotland and central Wales, significant proportions of the national domain are far from a large urban centre, as well as effectively being isolated from the main economic centres of the continent (see Figure 2.2, page 32). With the accession of Austria, Finland and Sweden, the magnitude of large-scale rural tracts in the EC has increased through the addition of a semi-arctic belt, with its own living problems (e.g. Weissglas, 1975; Massa, 1988). The particular nature of this belt has been recognized in the creation of an Objective 6 category for Structural Funds, acknowledging the special difficulties of sparsely populated areas with extreme climatic conditions (CEC, 1994c, 15). In addition, the 1995 extension has brought a substantial increment in the mountainous territory of a Community that already faces considerable rural development problems due to its abundant upland areas (see Figure 2.3, page 34).

In complete contrast with these areas are the urban-centred ruralities of south east England (even much of Britain, where 52% of the population lives in urban centres of at least 100 000 people; Eurostat, 1990), the Paris Basin and the Randstadt (Figure 2.2). Here pressures for urban growth generate quite different forces for change, with demands for rural residences amongst city workers often emphasizing social inequities, in access to housing, changing priorities for service provision and the evolving character of new job opportunities (Kayser, 1991; Cloke and Goodwin, 1992). Such 'accessible' rural areas not only possess different characteristics of change but also offer different images of 'rurality' and rural problems. This difference is

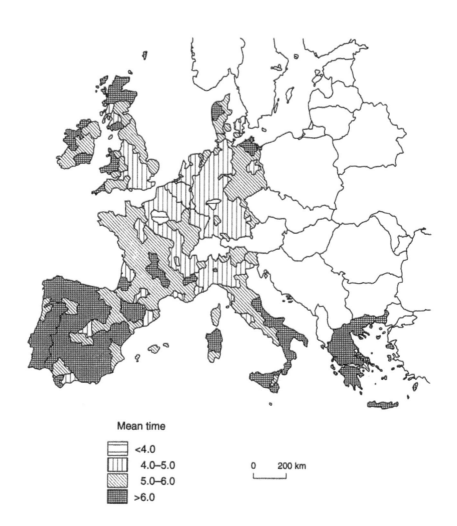

Mean time

☐ <4.0
▥ 4.0–5.0
▨ 5.0–6.0
▦ >6.0

0   200 km

Figure 2.2 Average travel time to 194 economic centres, 1991.
Source: CEC (1994c, 111)

reflected in the recent EC report on *The future of rural society* (CEC, 1988), which separates 'accessible' from more peripheral and very remote areas in its discussion of the challenges facing rural regions.

Of course, it is too simplistic to view basic differences in the countryside simply in terms of urban-centred and remoter areas. In particular, even if we mistakenly take rural as simply the opposite of urban, we have to recognize that the linkages of rural areas to their urban counterparts provide more variety than the presence or absence of strong commuter ties, and that such linkages extend even the most remote rural regions. At one level we see this in the incidence of migration into rural locations, which has not only altered in pace and population composition over the last 20 to 30 years, but also embodies streams which have different implications for recipient communities and future rural development (Forsythe, 1983). Three such streams are return migrants, retirement migrants and new, working age in-migrants. On a probability basis, it is to be expected that return migrants are spread in proportion to population losses in earlier decades; in other words preferentially to areas with weaker rural economic performances. Even if some migrants do not return to their precise area of origin, but prefer places with greater job opportunities (e.g. Unger, 1986), the magnitude of outflow from peripheral rural regions, particularly in southern Europe, has produced substantial return movements, with family callings, failure in migrant destinations and a desire for improved social status being key reasons for returns to home regions (Manganara, 1977; Cavaco, 1993). Not surprisingly, given the significance of these decision criteria, the economic consequences of return migration are generally not considered to be especially positive (e.g. King, 1984); although this might not be a fair description for areas that are already experiencing growth, such as through the expansion of tourism (Mendonsa, 1982; Cavaco, 1993). Thus a wealth of evidence shows that return migrants rarely introduce new development options, for returnees can be those who have not found success in their original destination or wish to utilize the wealth they accumulated there for displays of visible consumption, so enhancing social status in their home communities but making little impression on economic growth (e.g. Lijfering, 1974; Manganara, 1977; King *et al* ., 1985; Gmelch, 1986; Cazorla Pérez, 1989; Black, 1992).

A second major migration linkage between urban and rural areas is constituted by retirement migration, which arguably has a more positive profile in terms of economic consequences. Here in-migrants tend to bring fresh income to their recipient areas, although in places this carries accompanying cost increments for health and social service provision (Caldock and Wenger, 1992; Age Concern England, 1993). In many parts of Europe the spatial arrangement of retirement flows displays different qualities from those of return migrants. This is particularly evident in societies with a predisposition toward rural residence (like Britain), where the bias in retirement migration is toward rural and coastal locations (e.g. Rees, 1992). In France retirement flows used to carry the overtone of a much delayed return

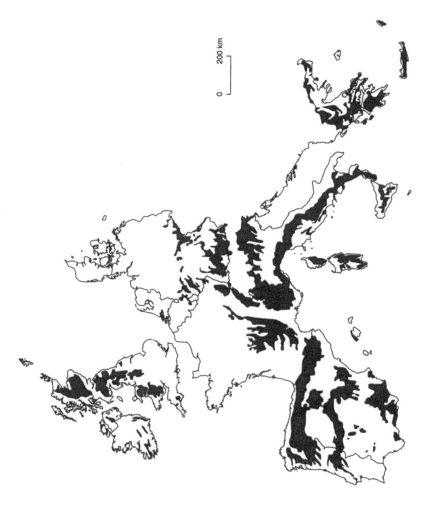

Figure 2.3 European upland areas, as defined by the EC. Source: CEC DG-XVI (1991, 162)

migration, but the tendency today is for more rural-bound elderly migrants to move to places with which they have no historic family connections, with destinations selected for diverse reasons, including aesthetic appeal (Cribier, 1993). This trend is not universal, especially in southern Europe, where the drift towards a rural retirement home has yet to gain hold. Here, elderly home changes are more associated with rural-to-urban flows than with movement in the opposite direction (e.g. Table 2.1). As for inflows of pre-retirement migrants, no uniformity can be assumed. Both in southern Europe and in the more mountainous and peripheral parts of the continent, the reversal of previous depopulation trends was slow to occur and appears to have been short-lived in some countries (Fielding, 1990; Walsh, 1991); although this effect might be age-specific (Table 2.1). So we are more likely to find population inflows that are independent of major shifts in employment close to core economic zones in the continent (e.g. Court, 1989; Bolton and Chalkley, 1990; Buller and Hoggart, 1994a), with this effect being evident within nations as well as across countries (Dematteis and Petsimeris, 1989).

Of course, when we aggregate different causal processes, we find that their combined effect produces spatial patterns that defy easy explanation. This is certainly the case for rural populations of working age, whether their increased presence is migration-induced or not (such in-migrant flows are referred to as counterurbanization by many writers, although this term might more appropriately be restricted to the supplanting of rural depopulation by rural in-migration). There can be no doubt that much in-migration into rural areas is not inspired by consumption or lifestyle desires, although this can be important, but is consequent upon shifts in the geography of economic activity (Fielding, 1982; Cross, 1990). Yet population and employment improvements can occur with no transparent causal priority (Court, 1989) or employment might even follow population deconcentration (Cheshire and Hay, 1989). Certainly, so-called urban-to-rural employment shifts, which

Table 2.1 Net retirement migration rate per 1000 residents by size of settlement, Andalucía 1989–1990

	Age 16–39	Age 40–64	Age 65 or more
Less than 1001 residents	−48.1	4.7	−7.8
1001 to 2500 residents	−30.0	3.6	−2.6
2501 to 5000 residents	−17.3	4.5	0.1
5001 to 10 000 residents	−12.5	5.3	1.9
10 001 to 25 000 residents	2.9	7.3	3.7
25 001 to 100 000 residents	10.9	7.8	6.7
More than 100 000 residents	−5.6	−2.4	−0.3

Source: own compilation from Instituto de Estadística de Andalucía (annual).

have been recognized for some time now (e.g. Keeble *et al* ., 1983), do not have to depend upon attracting in-migrant workers. A particular characteristic of rural locations, in some instances, is the potential to draw on an existing workforce that is underemployed, yet has traditionally been viewed as hard working, accustomed to relatively low wages and little unionized (e.g. Keeble, 1989). Again the absence of uniformity should be noted. The image of a conservative, deferential rural workforce might capture the imagination of certain urban corporate executives but it fails to describe the reality of many rural areas, with radical rural regions being pronounced features of the political map of Italy, Portugal and Spain (MacDonald, 1963; Snowden, 1979; Bermeo, 1986; Collier, 1987). Variety in response is to be expected across rural areas within nations, let alone between nations. Thus countries such as Greece still see a high proportion of their manufacturing expansion squeezed into a small number of congested cities, rather than leading to a manufacturing-led transformation of its rural space (Nikolinakos, 1985; Economou, 1993). Nonetheless, the key issue is one of linkage between urban and rural areas, with increasing challenges to the distinctiveness of rural demographic and employment profiles, which extends in certain cases even to more remote rural areas.

What Greece also warns us against is focusing too much attention on one economic sector. Even though manufacturing has become more concentrated in a few cities, incomes in rural areas have moved towards the average for urban centres. One reason why this has arisen is because tourism has had a positive benefit on rural incomes, with the potential for tourist development being inversely related to that of manufacturing (Economou, 1993). Nonetheless, with tourism increasingly promoted as if it is a magic wand that will speed-up rural economic progress (Council of Europe, 1988; Hannigan, 1994), it is necessary to note sharp distinctions in the style of tourist enterprise (Fúster, 1985; Shaw and Williams, 1994). The coastal and island developments that so characterize Greek and other Mediterranean tourist centres have generated rapid growth in small coastal settlements (e.g. White, 1986), as major corporate enterprises become primary agents of economic change (Vera Rebello *et al* ., 1990). The sun and sand orientation of the mass tourist enterprises that so typify Mediterranean coastal locations has quite different implications for rural areas than individualized tourist endeavours which focus on culture, environment and heritage, as in much of inland France (Shaw and Williams, 1990). In the former case, the consequences are likely to be a complete economic and social transformation, which often leads to urbanization (Pearce, 1987; Burton, 1994). The more gentle impact of the latter, while retaining and often building upon the essentially rural character of an area, has more uncertain impacts, which can be relatively modest (Getz, 1981; PA Cambridge Economic Consultants Ltd., 1987; Hannigan, 1994) but may lead to major economic gain and landscape transformation in what otherwise seems an inhospitable economic environment (e.g. Moore, 1976; Barker, 1982; Grafton, 1984).

The significance of tourist impacts are arguably no less in marginal and remote economic regions, as small absolute income gains produce quite substantial rises in relative wealth in these places (e.g. Moore, 1976). Positive tourist evaluations also give a boost to local self-esteem when population losses threaten social stability (e.g. Messenger, 1969). However, while the subtle influence of light tourist activity allows traditional occupations to be pursued, and adds an extra economic layer to ease potential occupational worries, even small inflows can have a disruptive impact on local cultural and social relations (Council of Europe, 1988; Pigram, 1993). Thus some communities have responded to fears about negative social impacts by resisting the economic charms of even small-scale tourist enterprises (Wallman, 1977; Vincent, 1980). Illustrative of such adverse effects is the manner in which farmers in Fuenterrabia (Spain) gained substantial income hikes on the arrival of mass tourism and yet continued to leave the sector as the social values that tourists encapsulated heightened feelings of social inferiority (Greenwood, 1976). To capture the way in which tourism complicates the geographical picture of development potentialities in rural Europe, we perhaps need simply to note that climate favours southerly locations, while upland and mountainous environments have their own particular attractions, with slight populations in such areas enabling small visitor influxes to bring welcome benefits to small places (Grafton, 1984; PA Cambridge Economic Consultants Ltd, 1987). However, places that rely on single season activities (like skiing) can be highly marginal if they cannot find attractions to extend visitor numbers to other seasons (Getz, 1981; Hannigan, 1994).

Adding agriculture to this picture further emphasizes the complexity of the situation, even though the trend everywhere is toward a declining role in rural economies for this sector (see Table 2.2, page 39). At the present time, there is much variation in the contribution of agriculture to regional economies (see Figure 2.4, page 38). Admittedly, farm income differences are decided partly by dissimilar product mixes, with returns for dairy, pig and poultry operations greater than for traditional Mediterranean products (CEC, 1992a, 1993a). Common Agricultural Policy payments emphasize this fact by providing more support for north European products than for Mediterranean ones (Brown, 1988; Williams, 1991).[2] Yet we should not over-interpret this, for the Dutch have the highest farm incomes in the EC, with a mean average figure per farm about five times that of Greece (OECD, 1994a), and yet only 45% of farm produce in the Netherlands is covered by CAP guarantee prices, compared with an EC average of 60% (Hommes, 1990). More significant are divergences in the characteristics of farm operations (see Table 2.3, page 40), along with larger units of production, a younger and more highly skilled workforce, more intensive production practices and faster increases in farm productivity (see Table 2.4, page 41).

2 According to Franzmeyer and associates (1991), for instance, 70% of 1986-1989 guaranteed commodity payments under the CAP went on beef, veal, cereals, sugar, milk and oilseeds.

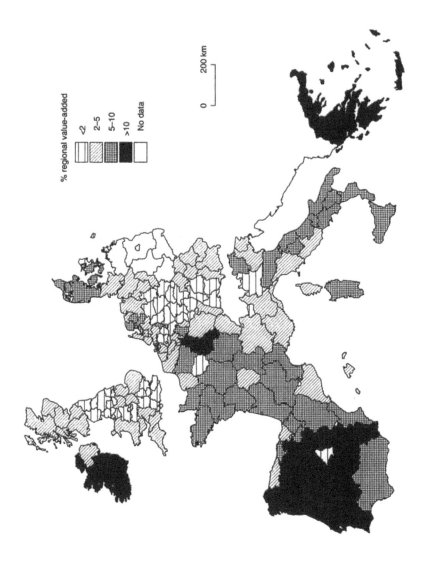

Figure 2.4 Share of agriculture in gross domestic product, 1987–1988. Source: CEC DG-XVI (1991, 153)

Table 2.2 Gross value-added in agriculture as a percentage of GDP, 1977–1990

	1977	1978	1979	1980	1981	1982	1983	1984	1985	1986	1987	1988	1989	1990
Belgium	2.3	2.4	2.1	2.1	2.2	2.3	2.4	2.3	2.1	2.1	1.9	1.8	2.1	1.7
Denmark	4.6	4.7	4.0	4.1	4.4	4.8	4.1	4.7	4.2	4.1	3.4	3.3	3.7	3.5
France	4.2	4.2	4.2	3.7	3.6	4.1	3.7	3.6	3.4	3.2	3.0	2.8	2.9	2.8
Germany	2.2	2.0	1.8	1.6	1.7	1.9	1.6	1.6	1.4	1.4	1.2	1.3	1.3	1.1
Greece	14.0	14.7	13.4	14.7	14.8	14.8	13.5	14.2	14.4	13.7	12.4	13.0	13.2	11.6
Ireland	16.1	15.4	12.7	10.2	9.4	9.5	9.6	9.9	8.3	7.6	8.4	8.8	8.3	7.2
Italy	6.3	6.2	6.0	5.5	5.1	4.9	5.0	4.4	4.2	4.1	3.8	3.4	3.3	3.0
Luxembourg	2.6	2.5	2.4	2.1	2.3	2.9	2.4	2.2	2.2	2.1	2.0	1.8	1.9	1.7
Netherlands	4.0	3.8	3.5	3.4	4.0	4.2	4.0	4.2	4.0	4.1	4.0	3.9	4.3	4.1
Portugal	*	*	*	7.3	6.4	6.6	6.0	6.3	5.9	5.4	5.3	4.1	4.2	4.1
Spain	7.3	7.1	6.2	6.0	4.9	5.4	5.3	5.5	5.3	4.8	4.6	4.7	4.2	4.0
UK	2.1	2.0	1.9	1.7	1.8	1.9	1.6	1.8	1.4	1.4	1.3	1.1	1.2	1.2
Austria	4.5	4.7	4.4	4.5	4.1	3.9	3.7	3.8	3.3	3.3	3.3	3.1	3.1	3.2
Finland	4.5	4.3	4.0	4.1	3.6	3.9	4.2	4.0	3.7	3.6	2.8	2.5	2.7	2.8
Sweden	1.8	1.7	1.4	1.4	1.5	1.5	1.4	1.4	1.2	1.1	1.0	1.0	1.0	0.9

Source: OECD (1994b, 72) Note: * no data.

Table 2.3 Attributes of farms and farmers in European nations

	Utilized Agricultural Area in units of 50 ha or more, 1990, percent (EC = 100)[1]	Farms classified as 'family' farms in 1989, percent[2]	Farms classified as 'non-family' in 1989, percent[2]	Non-family regular employees as percent of all full-time farm workers, 1990[1]	Farmers aged 55 years or more in 1992, percent[3]	Farmers self-employed in 1992, percent[3]
Belgium	57.6	80.5	3.0	*	25.0	91.0
Denmark	97.1	69.6	7.7	48.5	26.8	61.3
France	105.3	78.9	4.1	14.1	27.3	82.5
Germany	58.9	67.7	4.4	13.6	30.8	75.0
Greece	14.4	73.0	2.1	*	36.4	95.8
Ireland	72.8	83.3	2.5	4.6	32.4	86.5
Italy	66.0	79.0	2.9	*	30.5	60.9
Luxembourg	119.6	87.9	1.6	15.4	21.7	91.9
Netherlands	43.2	62.0	10.6	*	17.4	64.0
Portugal	101.0	49.5	9.1	15.7	37.0	82.8
Spain	117.7	63.3	16.5	16.1	32.5	69.7
UK	161.1	41.4	23.6	38.5	23.6	51.6
Austria	61.1	*	*	5.0	*	84.8[c]
Finland	28.6	*	*	1.6	37.0[b]	99.3[b]
Sweden	103.1	*	*	*	37.0[b]	75.0[b]

Source: [1] CEC (1994d); [2] CEC (1993a); [3] OECD (1994b, 25-26) Note: * no data; [a] data for 1989; [b] data for 1990; [c] data for 1991

Table 2.4 Indicators of farm production in European nations

	Yields of wheat per hectare, (000s kg) 1992	Increase in wheat yield per hectare 1972–1992, percent	Utilised agricultural area as arable land, 1992 (EC = 100), percent	Utilised agricultural area as permanent crops 1992 (EC = 100), percent	Ratio of meat exports to imports in 1991, by carcass weight 1992 (100+ = more exported)	Total value of national imports in the food, beverage and tobacco sectors, 1992, percent	Value of national exports from the food, beverage and tobacco sectors, 1992, percent	Ratio of food, beverage and tobacco export to imports by value, 1992 (100+ = more exported)
Belgium	6.62	48.8	109.3	9.7	312.1[4]	10.0[4]	10.4[4]	93.6[4]
Denmark	6.17	40.5	168.3	4.4	1702.8	12.2	25.2	116.1
France	6.40	40.0	107.3	44.3	108.7	9.4	13.9	96.5
Germany	5.98	47.5	123.7	15.7	72.5	9.6	5.1	104.8
Greece	2.47	23.5	84.9	223.9	2.8	13.7	27.8	40.9
Ireland	7.87	97.7	30.8	0.0	1112.5	11.5	24.3	129.9
Italy	3.55	43.1	97.9	219.2	22.6	11.8	6.6	94.4
Luxembourg	5.67	82.9	79.8	17.7	[4]	[4]	[4]	[4]
Netherlands	8.01	85.8	78.1	29.0	601.9	12.4	20.4	95.3
Portugal	0.98	*	117.3	213.0	16.1	11.1	7.1	60.4
Spain	1.94	52.8	105.3	199.7	42.7	11.1	14.3	70.8
UK	6.82	60.8	67.5	3.1	36.7	10.5	8.1	84.3
Austria	4.22	*	74.7	24.0	267.7[3]	4.9	3.3	82.0
Finland	3.62	47.2	174.5	0.0	[5]	5.7	2.4	113.1
Sweden	6.01	76.8	151.9	0.0	83.7	7.1	1.8	112.5

Source: CEC (1994d) Note: * no data; [1] data for 1991; [2] data for 1992; [3] data for 1993; [4] Belgium and Luxembourg are combined; [5] meat imports to Finland are too small for an effective calculation.

Table 2.5 Agricultural incomes in the EC

	Index of farm product prices minus index for farm input prices, 1992 (1985 = 100)[1]	Index of agricultural prices in 1992, deflated by the consumer price index (1985 = 100)[1]	Farm net value added per annual work unit, 1990 (1985 = 100)[2]	Farm net value added per annual work unit, for farms of European size unit 40–100 in 1989 (in ECU)[2]	All farms in the EC with a gross income below 2500 ECU in 1989, percent[2]	All farms in the EC with a gross income over 30 000 ECU in 1989, percent[2]	Average hourly wage for full-time manual workers on farms of 1 to 2 workers (ECU)[1]
Belgium	0.1	80.3	107.1	36 913	0.1	5.3	6.07
Denmark	-7.8	72.6	81.9	39 254	2.5	4.6	9.27
France	-6.1	80.0	129.1	24 827	5.3	19.8	5.37
Germany	-7.8	79.9	108.7	25 067	8.5	13.0	7.68
Greece	-4.1	80.6	87.1	13 203	6.9	0.6	*
Ireland	1.6	84.9	105.7	27 067	2.8	2.4	4.53
Italy	-6.5	80.0	90.8	21 022	28.7	16.8	5.88
Luxembourg	-8.7	85.5	117.6	26 160	0.0	0.2	4.75
Netherlands	-0.8	83.1	113.8	39 067	0.6	13.8	7.05
Portugal	-2.4	65.7	66.7	7917	21.2	2.2	1.39
Spain	-4.1	71.3	78.4	12 927	20.5	12.8	4.26
UK	-7.4	77.9	107.3	19 845	3.2	8.5	4.07

Source: [1] CEC (1994d); [2] CEC (1993a) * no data

In general, countries cannot be placed on a simple scale with regard to production and employment criteria. There is one nation that is characterized by large farms, high productivity and high farm incomes (Britain), with Greece occupying the opposite corner with its smaller production units producing low productivity performances. But between these two there is nothing approaching a unidimensional scale. Small farm units are allied to high farm incomes in some nations (Belgium, Netherlands), while relatively large landholdings are associated with low productive capacity and income in others (Portugal, Spain). This all results in a pattern of cross-national farm income disparity that is far from simple (see Table 2.5, page 42), and not directly related to population density or remoteness. Added to this there are significant differences in farm fortunes within nations. The contrast of the giant *latifundia* in southern regions and meagre *minifundia* in the northern areas of Portugal and Spain provide one example of this, but equal comparisons can be seen in the large farm – small farm divides of northern and southern France and Germany (Mellor, 1978; Brunet, 1992), with disparities between upland and lowland farms in Austria and Britain, and between northern and southern regions in Italy and Finland providing lesser but still notable contrasts (King, 1987; Niessler, 1987; Siiskonen, 1988). Such patterns once again warn against a simple characterization of rural areas as 'accessible' or 'remote', 'urban-centred' or truly 'rural', or indeed as heavily populated or underpopulated, without further analysis.

Rural disparities in socioeconomic problems

When these contrasting and at times contradictory trends of economic and social development are put together, the pattern and character of rural wellbeing in Europe is obviously complex. Of course, simple measures can be taken which can then be compared with population density to derive observations on the rural tone of the patterns observed (Figure 2.1). In terms of general economic performance, some significant associations are readily apparent between low income/production capacity and rural peripherality. Across the continent, low population densities (Figure 2.1), and poor access to major (urban) economic centres (Figure 2.2), are closely aligned with low per capita gross domestic products (see Figure 2.5, page 44). Effectively, given the slight income positions for most residents in some nations (like Finland, Greece, Ireland and Portugal), and for large tracts of others (like eastern Germany, southern Italy and most of Spain), this produces a doughnut impression; with the jam and cream occupying the (high income) centre and rural-centred regions at the periphery of the EC occupying what seems like a stodgy outer-rim (Dunford, 1993). However, the apparent simplicity of this pattern is rendered more intricate when changes in economic fortunes,

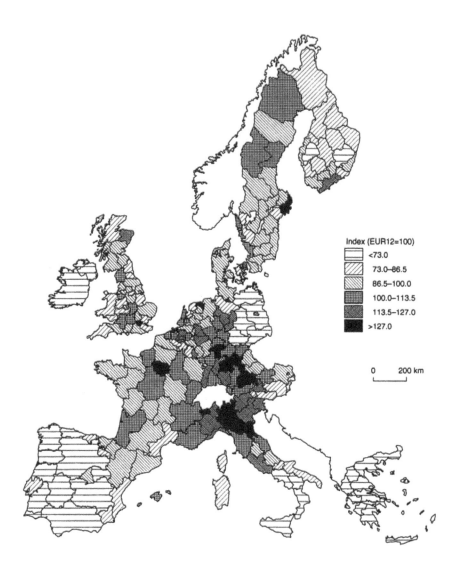

Figure 2.5 Gross domestic product per capita, 1991. Source: CEC (1994c, 36, 159)

demography and social and environmental equity criteria are considered. An examination of the salience of these additional factors forms the focus of this section.

Seen in terms of recent growth in economic performance, a clear divide appears amongst peripheral rural locations. As Figure 2.6, page 46 shows, for Ireland and much of Portugal and Spain, recent years have seen notable improvements in productive capacity, whereas Greece and northern Denmark still lag behind the general pace of change in the EC. These differences are not simply geographical, but express themselves additionally through the processes and character of economic change. The one indicator that by itself is sufficient to highlight this is unemployment. Thus, in contrast with their similar increments in gross domestic product (Figure 2.6), northern Denmark has seen an adverse pattern of unemployment change compared with Greece (see Figure 2.7, page 47), while moderate rates of production change in southern Italy were accompanied by sharp increases in the number out of work. Offering an even sharper contrast, the strong economic performance of many Spanish regions has been accompanied by significant rises in unemployment (Figure 2.7), which distinguishes this nation from Ireland, its high growth counterpart.

What such disparities in production and unemployment change indicate is the way in which, in some measure, economic change is nationally-grounded. Most evidently we can see this in the divergent experiences of the two Iberian economies. Here, indices of general economic standing reveal similarities between these two nations (Figure 2.5). But Spain has recently recorded higher rates of economic growth than Portugal (Figure 2.6), while significantly higher unemployment distance the Spanish experience from that of its neighbour (Figure 2.7). This factor also distinguishes peripheral rural areas with generally high unemployment rates, as in Finland, Ireland and Spain, from those with lower levels, like Greece, Portugal and Sweden (see Figure 2.8, page 48). For Portugal and Spain, this difference cannot be disentangled from national government policies, with the Spanish accepting shorter term unemployment as a by-product of a longer term search for economic improvement through the promotion of capital intensive industries, whereas Portugal has pursued policies more favourable to social equity, but which has led to an emphasis on low technology, low wage enterprises (Brassloff, 1993). As with all national figures, of course, we should be aware that the contrast between these two countries may partly reflect differences in national welfare regulations that define the unemployed, and consequently indicate wider differences than actually exist.

But even if we take such figures at face value, how far the trends they demonstrate can be extended into the future is a matter for conjecture. The expectation of the European Commission is that an arc from south east Spain that passes through southern France and finishes in northern Italy is likely to be a growth region for the future (see Figure 2.9, page 50; CEC DG-XVI, 1991, 14); with other writers viewing these growth impulses, even in

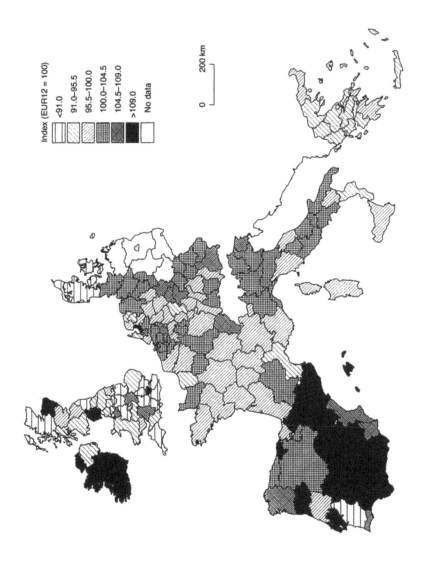

Figure 2.6 Change in gross domestic product per capita, 1986–1991. Source: CEC (1994c, 38)

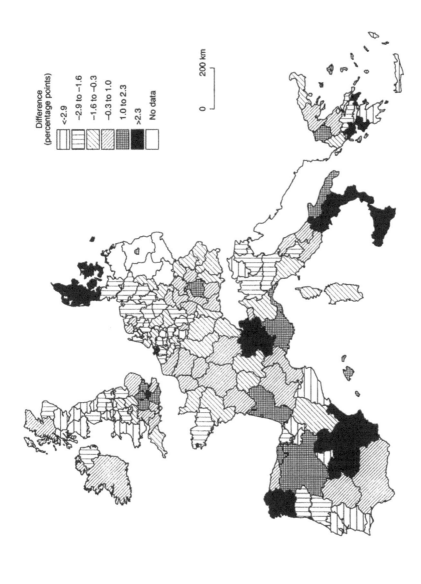

Figure 2.7 Change in unemployment rate, 1986–1993. Source: CEC (1994c, 49)

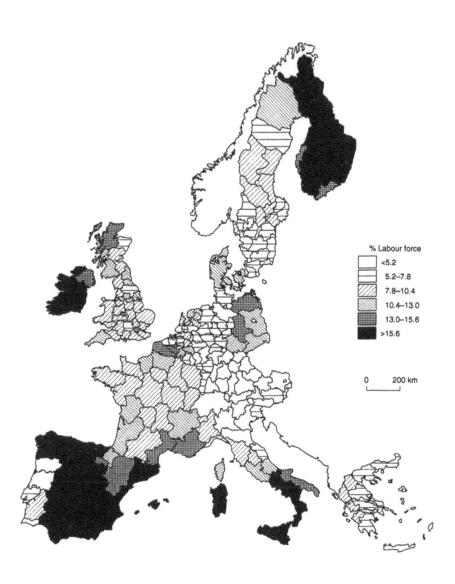

Figure 2.8 Regional unemployment rate, 1993. Source: CEC (1994c, 47, 159)

traditionally retarded economic regions at the margins of this arc, in terms more commonly associated with places like California (e.g. Salmon, 1992). Yet large proportions of these growth areas still fall within the extensive swath of territory that is eligible for EC Objective 1 Structural Fund assistance (see Figure 2.10, page 51); indicating not simply that their economies are still lagging behind (Figure 2.5), but that their growth potential is not expected to challenge the position of Europe's so-called 'vital axis' that runs from south east England through the Benelux countries, northern France, Germany and Switzerland to northern Italy (CEC DG-XVI, 1991, 14; Dunford and Perrons, 1994). And while it is true that southern France and northern Italy do not benefit from Objective 1 assistance, the fact that they draw Objective 5(b) funds points to official recognition that their economies need diversifying. That much of the new high growth belt in southern Europe has yet to reach the mean average GDP per capita of the EC offers one rationale for such funding.

The underlying urgency for economic improvement does not rest simply on this relatively retarded position. In addition, it gains sustenance from the dynamism of regional labour markets, a second important socioeconomic indicator which shows considerable variation across Europe. A critical issue for the EC is that zones of high unemployment often also experience rapid increases in their population and hence labour force (see Figure 2.11, page 53). So, although predictions about future economic conditions are fraught with uncertainties, there seems little doubt that population growth will mean that, out of the EC12, much of Greece, Ireland, Portugal and Spain, along with southern France and southern Italy, will face more intense pressures for job creation by the turn of the century than Britain, Denmark and Germany (CEC, 1994c, 26), notwithstanding a recent decline in birth rates in the former more peripheral countries.

The tensions these pressures create will not be felt uniformly, even for regions with shared expectations about increases in their working age population or roughly equivalent unemployment rates. Primarily, this is because demands for job generation are not simply the result of a desire to address existing mismatches and future growth in the working age population. They also reflect two further factors that should be taken into account when analyzing disparities in Europe's socioeconomic circumstances, namely the demands for new jobs prompted by those requiring new occupations, plus increased economic demands from those who are not in paid employment.

Most evidently, the last of these considerations arises from the ageing of the population. Again this ageing process is not even across Europe. In particular, when we project changes in the population that will be retired into the twenty-first century, we find that some of the poorest rural areas in the European Community will experience the largest gains, so that they end-up with very high proportions of their population in retirement (see Figure 2.12, page 54). Greece is especially notable in this regard, for its age structure by the year 2015 is expected to have much in common with

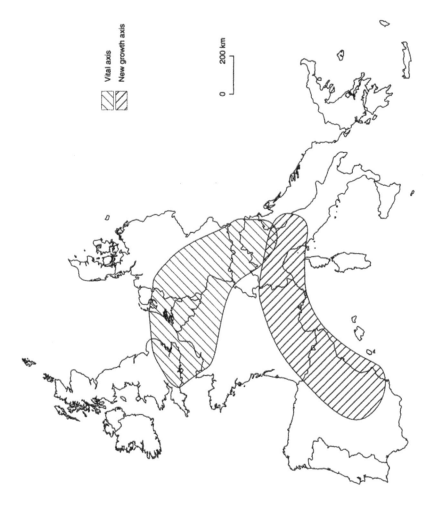

Figure 2.9 The 'vital axis' and 'new growth axis' of the EC. Source: CEC DG-XVI (1991, 14)

Figure 2.10 Areas eligible for EC Objective 1, 2 and 5(b) structural funds. Source: CEC (1994b, 10–11)

wealthier regions in Germany and northern Italy (or indeed the poorer parts of northern Denmark). However, if present trends continue, Greece will reach this position in a much poorer economic state than any of these other nations, or even perhaps of any nation in the EC (see Table 2.6, page 55). Certainly, this effect is not shared by all of the peripheral regions that already suffer from poor income prospects. Most of southern Italy, southern Spain, and especially Ireland, do not have high expectations of a rapid increase in their aged, dependant population; albeit the share of their population that falls into the retired category is expected to rise (CEC DG-XVI, 1991, 42). Of course, one of the problems with taking such projections as predictions is that new population trends and behaviour patterns are inclined to emerge over time; and these often take unexpected pathways. Such adjustments must be taken seriously by the EC, for economic, political and social integration weakens restraints on population migration, especially when changes in social policies and economic conditions across nations increase harmonization. We already know that rural regions are seeing sharp increases in retired household inflows (e.g. Rees, 1992; Cribier, 1993), with favoured destinations expected to see large inward movements over future decades (e.g. compare Figure 2.13, page 56, with the areas of significant retired in-migrant flows in Britain, like East Anglia, the South West and the south coast). These internal movements are increasingly being extended by international home relocations, with recent studies on France and Spain revealing a clear preference for coastal and rural locations in retired migrant destinations (Valero Escandell, 1992; Buller and Hoggart, 1994a). Effectively, then, we could be seeing the early development of a new population geography that could generate its own brand of economic and social problems for the European rural space (Hoggart and Buller, 1995a). This could mean that areas like southern France, southern Spain, upland zones in northern Italy and the Greek islands might see their dependant population increase significantly, not simply as an outcome of intra-national migration processes but also from longer distance cross-border flows. These trends are already evident, and they are no different from the commuter and counterurbanization movements of past decades in raising new tensions and new pressures in rural environments.

If these recipient regions must rely on local funding, then the legacy of years of economic deprivation, which has already resulted in low rates of infrastructural investment (Cutanda and Paricio, 1994), will heighten social pressures. This is so if only because population 'dependency' increases demands for services, with calls for improved services being particularly marked amongst in-migrants who arrive in (remoter) rural areas from more urbanized regions (e.g. Wenger, 1980; Perry *et al* ., 1986). Moreover, this is not a problem simply of adding (extremely young or aged) population increments and thereby generating more service demands. In addition, social tensions are likely to emerge as a consequence of basic socioeconomic trends. In particular, with the EC increasingly drawing back from its earlier almost

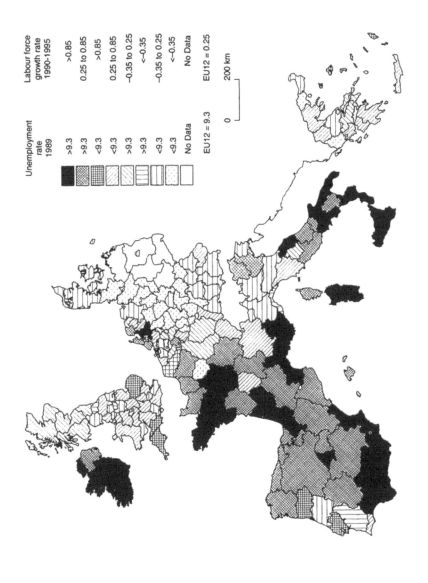

Figure 2.11 Regional job requirements in the EC, 1990–1995. Source: Haverkate and van Haselen (1992, 92)

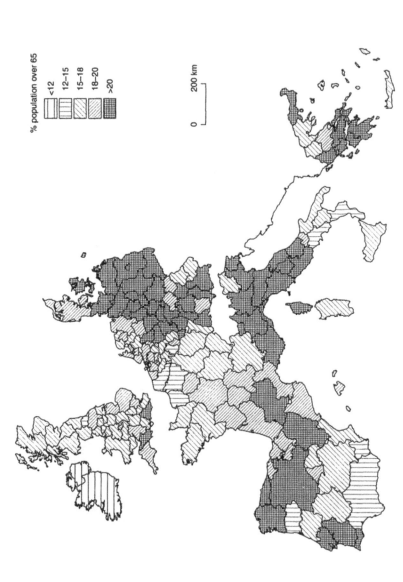

Figure 2.12 Projection for the percentage of the EC population over 65 years in 2015. Source: CEC DG-XVI (1991, 42), with interpolation for the eastern länder of Germany based on CEC (1994c, 30)

Table 2.6 Indicators of national economic performance in the EC

	Change in mean annual currency conversion rate, 1989–1993 (100+ currency weakens against ECU)	Annual mean average balance of trade per capita (000s ECU), 1988–1992	Annual growth of GDP at constant prices 1987–1992, percent	R & D spending as a percentage of GDP, 1991	Employers, self-employed and family workers as percent of civilian workforce, 1992	Index of real wages for manual workers in manufacturing (1985 = 100)	Unemployment rate, 1993, percent (EC12 = 100)
Belgium	93	34[c]	2.8	0.62	17.9	109.4	88.7
Denmark	94	658	1.1	0.75	10.7	112.2	98.1
France	94	−120	2.4	1.38	15.0	*	103.8
Germany	94	596	4.7[b]	1.05	10.0	120.1	67.9
Greece	150	−717	1.8[b]	0.27	47.4	93.5[b]	87.5[b]
Ireland	103	839	5.3	0.44	23.4	117.4	173.6
Italy	122	−5	2.4	0.76	29.1	*	104.7
Luxembourg	93	[c]	3.9	*	10.3	107.6[a]	24.5
Netherlands	93	522	2.7[b]	0.86	11.4	108.0	83.0
Portugal	109	−570	3.5[b]	0.40	25.6	115.0	47.2
Spain	114	−542	3.7[b]	0.54	25.7	113.4[b]	202.8
UK	116	−446	1.6[b]	0.89	13.2	112.1	99.1
Austria	94	−351	3.5	*	13.8	114.7	41.5
Finland	142	247	2.3	*	15.0	118.0	163.2
Sweden	128	474	1.0	*	9.6	*	72.6

Source: CEC (1994d). Note: [a] data for 1990; [b] data for 1991; [c] data for Belgium and Luxembourg are combined, with figures to 1991.

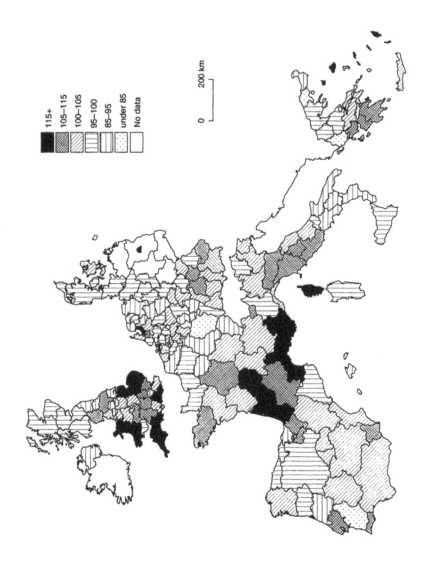

Figure 2.13 Index of inter-regional migration effects to 2015. Haverkate and van Haselen (1992, 80)

open commitment to improving farm incomes and supporting farm populations (Bandarra, 1993), we can expect the economics of farm production to favour larger production units, or, subject to a greater consumer preference for organic products, more efficient ones. If, as expected, this results in yet more labour shedding in the farm sector, we can expect further fundamental realignments in rural economies. These will be far from random in their geographical incidence, for those rural areas that already possess a rapidly ageing population also tend to have production performances per worker in both agriculture and manufacturing that are below the EC mean (see Figure 2.14, page 58), as well as being venues which still have major proportions of their labour force in farming (e.g. Greece, Portugal and northern Spain; see Table 2.7, page 59). This emphasizes the need for successful job creation, if only to stay stationary; let alone to provide the infrastructure for a growing population dependency.

Moreover, if EC social policies are more widely accepted, and especially if liberally-minded governments press for enhanced social equity, then demands for job creation will become even more intense. Although this could better the work prospects of various groups in society, providing more even treatment across race, language and sexual orientation, for instance, for rural areas specifically the most evident implication of such changes will be gender-related. The map of Europe is startling in the uneven manner in which women of working age are differentially present in the paid workforce (see Figure 2.15, page 60). In turn, bald numbers alone do not tell half the story, although they do reveal the very uneven incidence of employment opportunities for women across Europe (see Table 2.8, page 63). Whatever the precise figures on female participation in the paid labour market, there is no doubt that in a great many cases women do occupy the lowest paid and the least secure employment slots (Schweizer, 1988; Little, 1994). In many parts of Europe, a common response to this has been emigration, as younger women have rejected the constraints their mothers accepted to seek work and greater personal independence outside their home area (e.g. Douglass, 1976; Ogden, 1980; Gourdomichalis, 1991). But in so far as societal change is trending in favour of more gender equality, this has significant implications for rural areas in much of Europe, especially, for example, in Ireland, southern Italy and Spain, but also in parts of Germany and Greece. For this will enhance pressures for job creation in places that are already experiencing a surge in demand for new job generation yet have an economic base that is ill-equipped to adapt either quickly enough or with sufficient force to meet the challenge.

It is possible that this evaluation is too harsh. Certainly, societal change favours greater gender equality, if only through the rulings of the European Court and stipulations in the European Social Charter that repeatedly point to 'unacceptable' practices, as seen in a recent glut of rulings against Britain, Denmark and Greece (Council of Europe, 1993). But this is not the only issue that has been projected forcefully into the political arena in recent years.

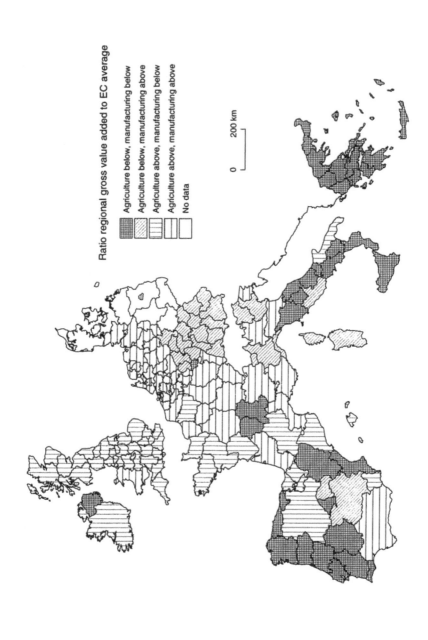

Figure 2.14 Regions classified by gross value added per worker in agriculture and manufacturing, 1988. Source: CEC (1994a, 50)

Table 2.7 Employment in agriculture as a percentage of the civilian workforce, 1977–1990

	1977	1978	1979	1980	1981	1982	1983	1984	1985	1986	1987	1988	1989	1990
Belgium	3.5	3.4	3.3	3.2	3.2	3.1	3.2	3.1	3.1	3.0	3.0	2.8	2.8	2.7
Denmark	7.8	7.9	7.2	7.1	7.3	7.5	7.4	6.7	6.7	5.9	5.7	5.8	5.7	5.6
France	9.5	9.2	9.0	8.7	8.4	8.2	7.9	7.8	7.6	7.3	7.0	6.7	6.4	6.1
Germany	6.0	5.8	5.4	5.3	5.2	5.0	5.0	4.8	4.6	4.4	4.2	4.0	3.7	3.4
Greece	33.2	32.0	30.8	30.3	30.7	28.9	29.9	29.4	28.9	28.5	27.0	26.6	25.3	24.5
Ireland	21.3	20.6	19.6	18.3	17.3	17.0	17.0	16.7	15.9	15.7	15.3	15.4	15.1	15.0
Italy	15.8	15.5	14.9	14.3	13.4	12.4	12.4	11.9	11.2	10.9	10.5	9.9	9.3	9.0
Luxembourg	6.4	6.4	5.8	5.1	5.1	5.1	4.5	4.4	4.4	4.3	4.1	3.4	3.3	3.2
Netherlands	5.3	5.4	5.3	4.9	4.9	5.0	5.0	5.0	4.9	4.8	4.9	4.8	4.7	4.6
Portugal	32.9	31.3	30.5	27.3	26.0	25.2	23.2	23.8	23.9	21.9	22.2	20.7	19.0	17.8
Spain	21.1	20.8	20.0	19.3	18.8	18.6	18.7	18.5	18.3	16.2	15.1	14.4	13.0	11.8
UK	2.8	2.8	2.7	2.6	2.7	2.7	2.7	2.6	2.5	2.5	2.4	2.3	2.1	2.1
Austria	11.8	10.9	10.7	10.5	10.3	10.0	9.9	9.4	9.0	8.7	8.6	8.1	8.0	7.9
Finland	15.1	14.4	13.8	13.5	13.0	13.2	12.7	12.2	11.5	11.0	10.4	9.8	8.9	8.4
Sweden	6.1	6.1	5.8	5.6	5.6	5.6	5.4	5.1	4.8	4.2	3.9	3.8	3.6	3.3

Source: OECD (1994b, 71).

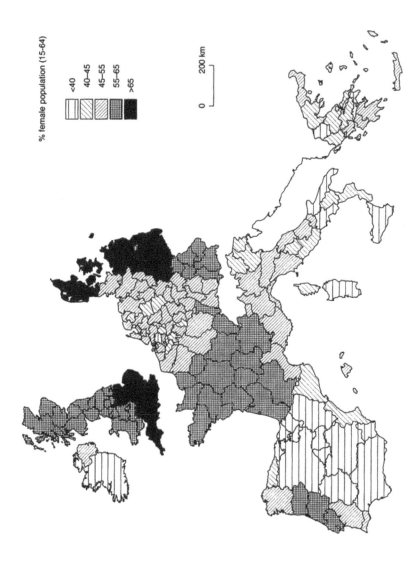

Figure 2.15 Female workforce activity rates, 1990. Source: CEC (1994c, 47, 159)

Whether we look at changing national agendas for political debate, as over environmental issues (Grande, 1989), or repeated local conflicts over appropriate directions for landscape change (Lowe *et al* ., 1986; Chevalier, 1993), the reality of political discourse over the last few decades has been one of growing concern for the quality of the physical environment. The manner in which these debates have become manifest differs across countries (e.g. Buller, 1992). Thus the trend in both public awareness and legislative enactment has, with certain 'cyclical' fluctuations, been progressively directed towards tightening the regulations that govern activities causing environmental damage, favouring the introduction of technologies with less detrimental environmental consequences and encouraging social practices that are more in tune with environmental conservation and enhancement (e.g. Rude and Frederiksen, 1994). But if environmental concerns are to impose tighter restrictions on future economic activity, this should favour new production spaces. This arises partly from the problems created by earlier generations of manufacturing, which even today produces a map of pollution indices that spotlights Europe's nineteenth century industrial zones (CEC DG-XVI, 1991, 117). These environmental problems are now matched by equally problematical environmental damage caused by farming practices; particularly those of highly industrialized farm operations, which the EC favoured for many years, but which are now deemed to have deleterious consequences of unknown durability but certain long-term impact (Pitman, 1992).

Already, the industrialized farming practices of northern Europe, in particular, are responsible for a significant deterioration in environmental standards (see Figure 2.16, page 62). For rural Europe this raises doubts over the viability of a continuance of prevailing farm practices. Even if EC governments are being driven primarily by a desire to reduce tax costs and appease trading partners (Phillips, 1990; Ingersent *et al* ., 1994a), there is no doubt that growing public concern over environmental deterioration provides a handy lever with which to encourage farmers to lower farm output. Perhaps less encouragement is needed for many manufacturers, given changes in the production technologies of many sectors that provide highly efficient, relatively clean, high-tech output. But however positive this general trend might be for Europe as a whole, for individual places the consequences of such transformations could be traumatic. Recent decades have already shown that substantial shifts in economic activity can occur quite rapidly, with the accession of Portugal and Spain to the EC providing an agricultural illustration (Mykolenko *et al* ., 1987; Avillez, 1993; San Juan Mesonada, 1993). A further illustration of this effect is seen in the changing demands and circumstances of manufacturing production that have created new industrial spaces. For much of Europe, it has been intermediate locations between older industrial cores and the periphery that have gained most in this regard (Hadjimichalis, 1994).

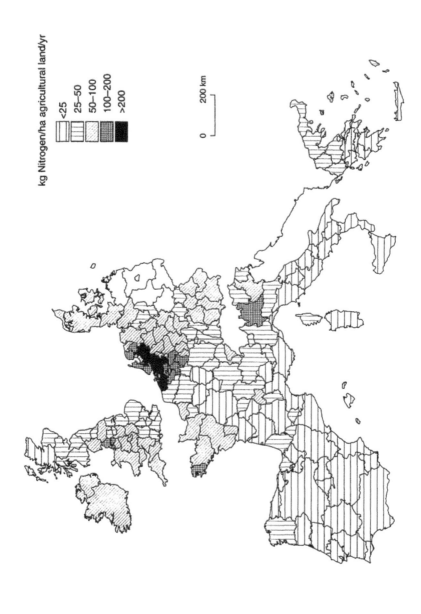

Figure 2.16 Nitrogen produced by animal waste per year, 1985. Source: CEC DG-XVI (1991, 120)

Table 2.8 Indicators of gender differences in employment in the EC

	Unemployment rate for single women, aged 25–49 in 1990[1], percent	Unemployment rate for single men, aged 25–49 in 1990[1], percent	Total female employment in agriculture in 1987, percent[2]	Women's share of employment in agriculture in 1990[2], percent	Ratio labour force participation for single men and women n 1986[3]	Ratio labour force participation rate for married men women in 1986[3]	Ratio female to male average gross hourly earnings of manual industrial workers in 1987[3]
Belgium	10	8	2.5	26.0	76.8	58.4	75.3
Denmark	12	11	3.2	23.1	87.5	88.7	85.6
France	13	11	6.3	34.4	82.1	69.8	80.8
Germany	5	6	5.4	43.9	84.1	57.0	73.7
Greece	15	9	35.4	44.5	69.7	49.6	78.7
Ireland	9	14	4.9	10.4	81.7	32.5	67.4
Italy	18	12	10.6	35.4	74.3	46.2	83.9[a]
Luxembourg	*	*	3.1	33.3	87.2	39.2	65.3
Netherlands	9	9	3.6	27.3	87.9	42.3	74.2[b]
Portugal	7	5	27.3	49.7	76.3	62.2	*
Spain	25	19	13.0	27.1	72.4	30.8	*
UK	9	11	1.2	22.7	80.9	69.8	69.5
EC	12	11	7.3	35.3	79.4	56.1	*

Source: [1] Millar (1992, 32); [2] Braithwaite (1994, 47); [3] Prondzynski (1989, 41, 60–61) Note: * no data; [a] 1985; [b] 1986.

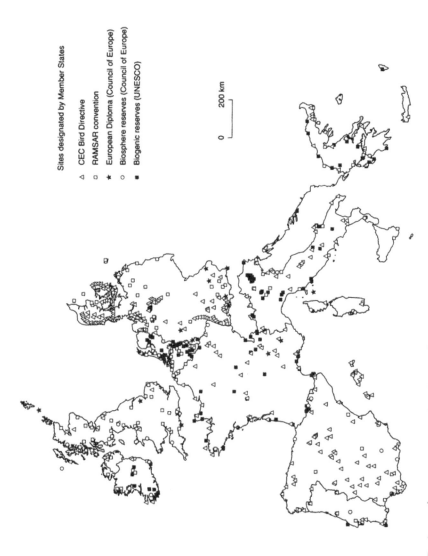

Sites designated by Member States

△ CEC Bird Directive
□ RAMSAR convention
★ European Diploma (Council of Europe)
○ Biosphere reserves (Council of Europe)
■ Biogenic reserves (UNESCO)

0   200 km

Figure 2.17 Designated nature conservation sites in the EC. Source: CEC DG-XVI (1991, 129)

None of this points to the geography of new investment or new production practices being easy to predict. Significantly, in contrast to the position in Britain, where the willingness of farmers to divert their land away from agricultural production seems to be linked to pre-existing forestry and conservation options (Potter and Gasson, 1988), set-aside in Italy is dominated by the Mezzogiorno (Istituto Nazionale Economia Agraria, 1992, 99). Here, afforestation and conservation possibilities have attracted little attention (Istituto Nazionale Economia Agraria, 1992, 110), so heightening concerns that set-aside is an ill-conceived stop-gap policy that will largely see the poorest land taken out of production, since yields on marginal land, relative to state payments for setting it aside, are low (Briggs and Kerrell, 1992; Gilg, 1992). The British-Italian contrast can hardly be explained by a differential ethos favouring conservation; or if it is, the relationship is the reverse of what we would expect. This is readily illustrated by OECD (1993b) figures suggesting that only 4.3% of the Italian land mass is covered by landscape protection policies, compared with 18.9% in Britain. Even accepting the official Italian figure of 6.5%, Britain still stands in a quite different position from Italy (Ministero dell'Agricoltura e delle Foreste, 1991). Yet, when we examine the incidence of internationally recognized conservation sites across Europe, a British-Italian disparity is less evident (see Figure 2.17, page 64). Put simply, international organizations might recognize dissimilar intensities of 'deserving' conservation measures, but these pale beside the disparities that exist in nationally-imposed conservation priorities. According to the same OECD figures, within the EC of 1995 the least protected national territories are in Ireland with 0.4% of its land mass so covered, Greece with 0.8% and Finland with 2.4%, while the highest protection ratios are in Austria with 19.0%, Britain, and then Germany with 13.9%.

The weight that is placed on landscape policies owes much to the power of environmental lobbies within nations, and the growing propensity for environmental activists to resist what they see as unwelcome 'developments' (e.g. Lowe *et al.* , 1986; Grande, 1989). And while questions need to be asked about the translation of legislation into action, with recent legal changes governing sex discrimination clearly revealing that hearts and minds require as much change as the law (Haugen, 1994), it is clear that legal provisions provide a fundamental basis on which change is promoted. This arises because, even post-Maastricht, EC policy-making relies heavily on agreement amongst the governments of member states, and occasionally their electorates (Nugent, 1991). Meanwhile, the role of states is reinforced since so much EC legislation is subject to national discretion and interpretation (e.g. over eligibility for fiscal aid).

In this section, we have seen that differential tendencies in the development of rural areas relate not only to existing socioeconomic disparities, measured by the level of unemployment or regional GDP, but also to past and future trajectories linked to the ageing of populations, new demands for shifts in employment, the role of women in the labour force and environmental

concerns. To a certain extent, differences have been highlighted here between European nations. But nations themselves are not uniform territories
and it might even be that more similarity exists between regions in different
nations than between places within the same country. In so far as this is true,
at the very least we might expect national governments to favour similar
policies toward their rural areas. Such an expectation rests on the assumption
that nations share common political cultures and socioeconomic trajectories.
If this is not the case, then their orientation towards even similar 'objective'
circumstances could be quite different. Assessing whether such basic socioeconomic similarities within cross-national regions do exist constitutes the
key issue for the next section.

Regional and national similarities?

The main question for this section is the extent to which cross-national
similarities in socioeconomic circumstances can be identified which point to
underlying consistencies in rural development prospects at a regional level. If
these exist in present socioeconomic circumstances, we need to ask whether
they result from common trends in the historical evolution of economies and
societies, as well as asking whether cross-border regularities are informed by
nation-specific influences. The fact that some transnational consistencies
exist is self-evident. If only in terms of physical conditions, there are clear
commonalities within, say, northern and southern European regions in terms
of climate and these incline agrarian economies toward particular product
mixes. But even if we restrict our attention to agricultural communities, we
should not expect an inevitable consistency in farm structures or incomes to
result from climatic similarities. Climate imposes restraints on the economic
feasibility of producing certain farm products, but the manner in which this
feeds into the socioeconomic standing of farm owners and labourers is more
significantly influenced by landholding structures and tenure arrangements
(Marsden *et al*., 1993), state support for farm activities (e.g. Just, 1994),
prospects for alternative jobs (Arkleton Trust (Research) Ltd., 1992) and,
within households, by established practices governing inheritance and gender
relations (e.g. Greenwood, 1976; Whatmore, 1991). Illustrating this point is
not difficult. The upland areas of the European Community, for instance,
carry a special legislative position, as farm operations in these areas are
subject to special payments under EC Directive 75/268 and Regulation 797/
85. Yet there is enormous variety in these areas, even within single nations.
Thus, in the upland areas of France, the percentage of employment in the
primary sector ranges from 5.5% in Vosges to 32% in Corse, with population
change for 1961–1981 falling between –7.7% in Corse to +8.1% in the
northern Alpes (Economic and Social Consultative Assembly, 1990). When

we look across national borders the picture is more complex, for the definition of an 'upland area' is subject to national designation (despite the resulting eligibility for extra EC farm payments). As our start, therefore, we take neither physical conditions nor nationhood as a focal point for assessing EC-wide similarities in socioeconomic standing. At the same time, owing to the subjective nature of such an enterprise and its dependence on the geographical scale of analysis, we do not begin our examination with the intention of producing a definitive categorization of places based on their shared socioeconomic position. Rather we seek to review existing work that highlights the varying socioeconomic circumstances of regions within the European Community to see if, based on a variety of classificatory devices, there is a common view on the existence of a European regionalization.

Logically, the starting place for any socioeconomic definition should be the EC's own scheme of regional divisions, which, for our purposes, is most visibly seen in the list of places that are eligible for regional policy assistance (such as Objective 1, 2 and 5(b) areas). However, as Wishlade (1994) points out, the criteria that are used to define such areas are highly subjective, while the process of agreeing which places fall under these area designations is highly political. The end-result is that the designation of places with poor development prospects varies across EC programmes. For example, if we compare Objective 1 areas that are eligible for regional policy assistance with the less favoured areas under EC competition policy, we find that regions like Asturias, Corse and Valencia qualify for special assistance as poor areas under one scheme but not under the other. In addition, as Objective 1 designation is decided principally by per capita GDP (falling below 75% of the EC average according to purchasing power parity), this class often overrides Objective 5(b) designations, which relate more directly to agricultural and rural development problems. As such, the map of EC regional policy designations tends to place many areas with severe rural development problems under the broader Objective 1 class (compare Figure 2.14 with Figure 2.10).

A further problem with EC regional designations comes from their immediate policy objectives, for their areal categorization is based on a slice in time, rather than development trajectories. As such, the areas that receive EC aid vary (at the margin at least) over relatively short periods of time. If we adopt a longer term view of development trajectories, a rather simplistic division of 'problem regions' can be made between 'traditional' economies, whose low levels of development result largely from them occupying a minor position in industrial growth processes in the last century, and places that were caught up in nineteenth century growth processes but have since been stranded by shifts in the economic tide. For Molle (1990), the former are well illustrated by the Mezzogiorno and Ireland, while the latter are characterized by north east England and Saarland. This rather straightforward dichotomy captures basic distinctions between European regions but the breadth of these areal types results in places with dissimilar present-day development problems and future growth prospects being grouped together. A finer tuned classification has been presented by Kersten (1990), who distinguished six regional types (backward regions, agricultural regions, peripheral regions, declining industrial areas, urban problem zones and frontier areas). Although it only focuses on economically disadvantaged areas, this typology goes some way toward identifying different categories of rural zone. Significantly, however, these classes overlap. At one level, this can be seen as a theoretically sound proposition; for it points to the multidimensional nature of development processes and warns that mutually exclusive regional designations run the risk of 'forcing' socioeconomic trends into a unilinear model that disguises the complexity and inter-connectedness of development trends. But this creditable feature is counteracted by the scale of Kersten's classes, with his peripheral areas alone covering 55% of the land area of the EC12. Such broad groupings collapse great variety into a single area type, even if we acknowledge that Kersten did recognize sub-classes, such as the division of agricultural regions into the sub-categories 'Mediterranean' and 'mountain/ hill areas' (Economic and Social Consultative Assembly, 1990; Figure 2.3). Perhaps more appropriately, therefore, we need to examine differences between places in terms of their standing along single dimensions of socio-economic differentiation.

Most evident in this regard is the division between places whose economies are tightly bound to nearby urban centres through processes of commuting and urban decentralization and areas with economic and social orientations that are less centred on single places and might even have few direct ties to urban centres. From the perspective of the highly 'urbanized' rurality of Britain, such a divide might seem to do little more than distinguish highly peripheral areas, like the northern lands of Scandinavia or the Highlands and Islands of Scotland, from the mainstream of rural Europe. But in reality most European nations contain sizeable, though often dispersed, areas that are 'isolated' from systematic city-centred influences (e.g. Centro de Estudios de Ordenación del Territorio y Medio Ambiente, 1981). Unlike diffusionist models that portray rural development as dependent on city economic growth (Hoggart and Buller, 1987), we do not believe that any unidirectional development consequence can be linked to a division of the rural world into city-centred and 'semi-autonomous' area types. This point is made obvious not simply by the inability of city attributes to account for the spatial expression of economic growth (Hall and Hay, 1980), but also in the emergence of localized growth centres in rural regions that lack a significant industrial or urban tradition. In some cases, these latter processes are aligned with spill-over or deconcentration trends from cities (e.g. Cooke and Rosa Pires, 1985; Hall *et al*., 1987; Holmes, 1989). In others they relate to an indigenous growth impetus (Lewis and Williams, 1987; Pecqueur and Silva, 1992; Vàzquez-Barquero, 1992b) or multinational corporations favouring new sites for their investment in the face of market globalization (Netherlands Economic Institute, 1993; Peck and Stone, 1994). Nevertheless, rural areas whose economic and social life is deeply interwoven into nearby urban centres find that their development prospects are predicated upon the ongoing economic successes of this proximate city, or more accurately perhaps on the region as a whole, with its fortunes being symbolized and, through its pattern of internal coherence, driven by the competitiveness of its central city. Hence, the very dynamism of urban centres along Europe's 'vital axis' lies at the heart of economic growth in this region and brings in its wake significant economic gains for surrounding rural areas (Dunford and Perrons, 1994). Significantly, as Cheshire and Hay (1989) point out, it is the lack of urban dynamism in (particularly southern European) areas beyond this axis that results in differential rural fortunes. Indeed, while more recent years have seen a new vigour in the economies of some southern European cities (Dunford, 1993), the apparent vitality of many of these places, as for example in southern Italy, appears to have resulted more from a lack of options for rural residents than their innate economic dynamism. Rather than spreading growth benefits into the countryside, many southern European urban centres have been weighed down by having to cope with the consequences of rural impoverishment, so that centres with the worst urban problems find their plight reinforced by pressures trending toward further decline (Cheshire and Hay, 1989). Of course, it is facile to place too much weight on a division of city-regions between those located within the vital axis and those beyond, just as it is to see a key division simply in terms of northern versus southern Europe. So, while there is some comfort in such broad clusters, we should not emphasize them too strongly, given that research shows how difficult it is to predict urban growth trajectories (Hall and Hay, 1980). Nonetheless, while acknowledging that some cities beyond its bounds have vibrant economies (e.g. Barcelona, Lisboa or Stockholm), it is evident that in the long-term urban centres within the 'vital axis' are well-placed to induce socioeconomic change that is beneficial to their surrounding rural zones (Figure 2.5).

While cities are said to be advantaged as growth centres owing to their concentration of activities, ease of communication and propensity to act as a catalyst for new ideas, this should not distract us from countervailing forces. Even some supposed city advantages turn out to be less clear-cut on examination. As evidence on the registration of patents for new products reveals, smaller centres are often associated with the generation of more new ideas than larger cities (e.g. Fischer *et al*., 1994). In addition, much economic activity in rural areas is not tied in a direct way to local settlement structures. Most evident in this regard is agriculture. There is a degree of overlap between areas of urban dynamism and places with more vibrant agricultural economies. But these two are not tight fits, as the map of agricultural fortunes commonly groups places that stretch well beyond core urban regions. Nevertheless, some basic covariations can be observed. Take, for example, the classification of agricultural areas by Mykolenko and associates (1987), which is based on measures like labour productivity and economic and labour market situations. Here, the list of areas with the most vulnerable agrarian structures, which have an unfavourable local economic climate and a large farm population with low rates of productivity, contains a disproportionate array of places beyond the 'vital axis' (the north and west of Ireland, Northern Ireland, Aragon, Castilla-La Mancha, Extremadura, Portugal, the Mezzogiorno, and most of Greece). By contrast this classification identifies the most favoured farming areas as lying in the Netherlands, Flanders, the Paris Basin, northern Germany, Denmark, all England except the South West, and Emilia-Romagna (e.g. Figure 2.14).

Caution is required in interpreting such regionalization schemes, for they have the potential to lull us into too simplistic an understanding of divergences in economic wellbeing. In particular it is critical to recognize that such classifications depend significantly on the measures they use to distinguish places. Thus, while Denmark has long been viewed as having a strong agricultural economy (Tracy, 1989), which its identification as one of the least vulnerable farming areas in Europe seems to confirm (Mykolenko *et al* ., 1987), effective farm incomes, and hence farmer socioeconomic wellbeing, are not that favourable. In fact, although high farm productivity results in large gross farm incomes (Table 2.5), farm debt repayments are also far higher than elsewhere in the EC, owing to the peculiarities of Danish inheritance practices which result in many farmers purchasing their property from their parents or buying-out their siblings' share. The end result is that net farm incomes are relatively low in all sectors except dairying and pigs/ poultry (CEC, 1993a). In terms of spending power, then, farm incomes in Denmark place their farmers alongside those of Greece, Portugal and Spain, and well below the norm for northern European nations (CEC, 1992a). At one level, this clearly puts Danish farmers at the periphery of the European league, despite their good productivity performance. In this, they have something in common with the three nations that joined the EC in 1995. Each of these reveal farm structures and incomes that are less strong than those of the Benelux countries, Britain or France (OECD, 1987, 1988, 1989b; CEC, 1993b, 1993c). Again, though, we can see problems with too much reliance on a single indicator of area performance. By themselves measures of farm activity are far from sufficient criteria for assessing rural fortunes, given the manner in which extra work, both on and off the farm, is a central feature of rural economies in much of Europe (Arkleton Trust (Research) Ltd., 1992). In Austria, for one, the ready availability of nonfarm work (e.g. tourism) has a major effect on the income of farm households (Niessler, 1987), given that only 36% of the nation's farms are defined as full-time (CEC, 1993b, 20).

If this points to the uncertainty that can arise from focusing on one economic measure, it is not difficult to heighten this uncertainty by examining other indicators. Thus, a core-periphery divide, with its resemblance to an accessibility surface (Figure 2.2), has associations not only with urban fortunes and agricultural wealth, but also potentially with manufacturing performance. At a national level we get some feel for this in the commitment of nations to R&D (Table 2.6). A finer grained analysis by the Netherlands

Economic Institute (1993, 113) points to south east England (London), the Ile de France (Paris) and Lombardia (Milan) standing apart as centres of high employment for engineers and technicians. Yet a simple core-periphery interpretation of manufacturing activity is difficult to maintain, for areas beyond the 'vital axis' such as Cataluña (Barcelona), Madrid, Lazio (Roma), central Scotland and north west England also score strongly on this latter measure (Netherlands Economic Institute, 1993, 113). Adding further confusion is the fact that small companies, so often praised for their positive growth impressions (Hart and Gudgin, 1994; Keeble and Walker, 1994), are identified as raising problems for economic progress in southern Europe (Lewis and Williams, 1988). Likewise, in so far as a positive aura attaches itself to moves toward a post-industrial society, the attractions of harbouring a recreational core to a regional economy should be viewed with some favour. In this regard, 'common sense' might lead us to expect that southern European nations will lie in an advantageous position. In reality, though, their earning capacity from tourism is not as deep or widespread as might be imagined (e.g. see Table 2.9). And even when national economies gain major injections of income from tourism, these do not come without attached problems of social disruption, economic dependency and a price sensitivity that generally favours low cost (and so low wage) labour markets (Nash,

Table 2.9 Main national tourism earners in the EC

Nation	Tourism receipts in 1992 ($US millions)[1]	World rank as a tourism earner in 1992[1]	Tourism receipts as percent GNP in 1990[2]	Average annual growth rate 1980–1992[1], percent	World rank as tourism earner in 1980[1]
France	22 190	2	0.02	8.61	2
Spain	21 035	3	4.36	9.64	4
Italy	20 013	4	1.39	7.70	3
Austria	14 804	5	7.44	7.18	7
UK	13 600	6	1.34	5.83	5
Germany	11 100	7	0.72	4.47	6
Netherlands	4967	12	1.36	9.52	14
Portugal	3885	17	6.94	10.70	18
Belgium	3750	18	1.98	6.26	11
Denmark	3500	20	2.30	8.35	16
Sweden	2800	23	1.37	9.31	21
Greece	2600	25	3.73	3.43	13

Source: [1] WTO (1993, 29); [2] CEC DG-XXIII Tourism Unit (1994, 148).

1978; Cohen, 1988; Gannon, 1994). While tourism is a key vehicle for economic advancement in a limited array of areas, therefore, it offers less compensation for peripheral locations in the European continent (e.g. note the relatively poor performance of the Greek tourist industry over the last decade; Table 2.9).

Taking these points in aggregate, we are left to ask about their combined effect on regional growth. The geographical distribution of GDP indicators very obviously points to the low standing of southern Europe (Figure 2.5), although recent trends show an improved position for Spain, even if southern Italy and Greece, and to a lesser extent Portugal, continue to lag behind (Figure 2.6; Table 2.6). But does this point to comparable regional standings across nations? In part, the answer lies in the eye of the beholder. At one level, we can identify places in different countries that appear to have a common socioeconomic standing and trajectory. The so-called 'vital axis' provides one example of this; at least in so far as it identifies regions that are regarded as having a shared prospect of positive growth in the future (Figure 2.9). Yet not all areas in this axis have common expectations about future advancement (compare south east England with places like Burgundy, for instance). For less advantaged European regions, such divisions are even more evident. Thus one scheme for regionalization devised by Nam and Reuter (1991) suggests the need to distinguish EC Objective 1 areas according to their likely prospects for economic improvement. For Abruzzo, Molise, Murcia and Valencia, Nam and Reuter see a future of good development prospects. For other regions in Italy (like Basilicata and Campania) and Spain (Asturias, Castilla-La Mancha, Castilla-León, Extremadura and Galicia), as well as for almost the whole of Greece (all except the Athens and Saloniki areas) and Portugal (all except Lisboa), their portrayal of the future development outlook is bleak. Positioned between these two are places with specific development problems that have the potential for achieving significant improvements in living standards, provided they can overcome these handicaps. With Athens, Saloniki and Lisboa (Lisbon) in this last group, we find that the whole of Greece and Portugal are conceptualized as having significant development problems, whether or not places are within city-regions. The positioning of Ireland in this last group further points in this direction. However, Nam and Reuter foresee that Piedmont, Roma, Trentino, Toscana (Tuscany) and Veneto in Italy, and Islas Baleares and Madrid in Spain, all have good prospects for economic expansion, with Puglia (Apulia), Andalucía and Islas Canarias having open development prospects if they can overcome their particular problems.

Based on this categorization, there is a confusion of development trajectories for both Italy and Spain (even amongst Objective 1 areas). Perhaps the size of these nations has something to do with this difference, but to rely heavily on this explanation would be misplaced. For one, it is notable that France does not have so uneven an array of development prospects as either Italy or Spain (Nam and Reuter, 1991). In some measure we might trace these differences to the long-enduring strength of regional economic distinctions in these two nations (e.g. Brenan, 1950; Graziano, 1978; Davis, 1979; Harrison, 1985). Yet the recent history of Europe has also shown some remarkable adjustments to the geographical surface of economic activity, which has seen traditionally weak regional economies experience significant economic growth. The new growth axis of southern Europe offers some obvious examples (Figure 2.9). It is perhaps more likely therefore that one of the key features that distinguishes France from Italy and Spain is its occupation of a central geographical position within the European Community. The importance of this is made evident by the difficulties of peripheral places in attracting mobile capital investment (Netherlands Economic Institute, 1993). The appearance of Islas Baleares and Madrid amongst the list of areas that are believed to have positive development prospects also indicates that peripherality alone is no explanation for uneven development prospects. Quite apart from indigenous growth that favours places almost irrespective of location (Andrikopoulou, 1987; Vàzquez-Barquero, 1992b), growth potential is influenced by the organization and structure of economic activities in a region.

We owe a debt to Pompili (1994) for providing an analysis of this kind. Admittedly, this is solely for the less developed regions of the EC. But examination of regional development areas computed by Pompili indicates that these distinguish widely recognized divisions within European nations, such as Britain's north-south divide, Germany's eastern and western *länder* (the former nations of East and West Germany), Paris and the French 'desert', the isolation of Jutland and the Danish islands, and the distinctiveness of the South in Italy (Parker, 1979; Ørum, 1985; Clout, 1986; CEC, 1987; Lewis and Townsend, 1989). Moreover, consideration of places that are excluded from EC regional development assistance reveals few areas that are not located within or at the margins of the continent's 'vital axis'. The contribution of Pompili is to ask whether areas beyond this economic core are distinctive in their economic structure and performance. The answer provided is that they are, with five regional types identified:

- *strong urban regions*, with the lowest values for agriculture and the highest for consumer services, R&D commitments, scientific output and infrastructural endowments. The east and southwest of Ireland, Northern Ireland and Attiki in Greece are placed in this class.
- *weak urban regions*, with high productive values in manufacturing and producer services, high levels of urbanization, but low scores for secondary schooling, graduates and scientific output. The Portuguese regions of Lisboa, Madeira and Norte fall under this heading.
- *Gaelic regions*, which are characterized by high levels of external control of economic enterprises, diffused entrepreneurship and high levels of consumer service provision. These are accompanied by low levels of producer services, slight infrastructural provision and few urban centres.

The remainder of Ireland occupies this class, and is seen to be 'well-positioned to start a take-off process ... maybe of the Third Italy type, but still missing an autonomous development strategy' (Pompili, 1994, 686).

- *Latin regions*, with high manufacturing and infrastructural endowments, good secondary schooling and low labour costs, but also with low values for producer services, R&D expenditure and scientific output, so they lack the human capital resources that are a key for new firm formation. This class includes all the Italian Objective 1, 2 and 5(b) areas except Molise, all Spanish areas and the Portuguese Centro region.

- *marginal areas*, which are the most agricultural in orientation, with the smallest manufacturing base, few consumer services, little external control of economic enterprises, much diffuse entrepreneurship, little secondary schooling, few people living in urban centres and little spending on R&D. Such marginal areas include all of Greece except Attiki, the Portuguese regions of Alentejo, Algarve and Açores, and Molise in Italy. These regions are argued to have missed out on process of productive decentralization, and lack the conditions for attracting either inward investment or indigenous-led growth, although it is ironic that Pompili's own figures show that at least in the 1970s these regions had higher average growth rates than either the Gaelic or Latin regions that supposedly have better conditions for growth. This pattern perhaps owes much to the role of tourism, whose potential was not directly examined in Pompili's study.

The fact that Pompili (1994) only examined Objective 1 areas perhaps explains some of the dissimilarities between his classification and the messages of Nam and Reuter (1991). After all, within Italy Objective 1 only covers the South (Figure 2.10), whose distinctiveness in development terms has been long-recognized (e.g. Davis, 1979; Wade, 1979; Ørum, 1985). Nevertheless, with the exception of Greece, which both studies identify as possessing development problems on a major scale throughout the land, it is notable that these two studies effectively reverse the findings they present for Portugal and Spain. On the one hand, Portugal comes across as a highly diverse nation in Pompili's portrayal, but only Lisboa emerges as being distinct from the rest of the country in the classification of Nam and Reuter. On the other hand, for Pompili Spain has a high coincidence in regional socioeconomic circumstances, whereas Nam and Reuter find substantial variety in regional prospects.

Our intention is not to argue that one of these studies is superior to the other. But there are important points in their conclusions that should be grasped. First, as Pompili (1994) argues, it is instructive to note that regional clusters do emerge which point to the importance of national effects. Certainly these might be most pronounced for the poorer regions of the EC, with some sharp distinctions persisting between richer and poorer territories within nations (as for Italy). Yet this does not detract from the observation that places of seemingly similar socioeconomic standing do display some distinctive features that separate them from their counterparts in other nations. Although there will be dispute over the precise form of these national effects, some of this national 'boundedness' occurs despite a background of similarity in physical environments (as with much of southern Europe), in economic traditions and even in national political circumstances (Giner, 1985). What this points to is the emergence of dissimilar responses to common historical conditions. In the next chapter we will explore whether we can begin to identify a basis for such national specificities in the values, interpretations and behaviour norms that have come to attach themselves to rural places in European countries.

|3|

Rural traditions in Europe

Despite the diversity in their physical environments, in the human activities that occupy their spaces and in their pace of socioeconomic and demographic transformation, European rural areas share two common, longstanding features: they continue to support relatively low densities of population and their landscape is dominated by extensive land-use forms that are linked, to a significant extent, to the physical environment. This geographical fix is about as far as we can go in providing a consistent image of *rurality* that encapsulates core elements in divergent national administrative classifications, as well as capturing popular images of the 'rural'. Nevertheless, the enduring predominance of these two features endows rural areas with a distinctive thematic homogeneity (Clout, 1972). As we noted in an earlier volume:

> ... even recognizing that the term 'rural' disguises a wide variety of socioeconomic and political conditions, it is believed [by many] that the spatial structures of such areas impose restraints and offer possibilities for action that justify their being analyzed in their own right (Hoggart and Buller, 1987, 27–28).

Even now, the tendency to group European non-urban space under a single nomenclature remains attractive both to policy-makers (CEC, 1988; OECD, 1993a) and to academics, as this current contribution bears witness. Such a tendency for a long time incorporated an inherent assumption that rural areas exhibit largely analogous economic, political and social characteristics. These varied in intensity (on a continuum) or differed entirely from those found in non-rural zones. Thus, the notion of a distinct rural society has a long-held currency in Europe (Friedmann, 1953: Mendras, 1976), and survived with few challenges until the 1960s when both the socioeconomic and physical extension of urban areas brought the notion of a distinct rurality into question (Pahl, 1966; Bauer and Roux, 1976; Koch, 1980). The recognition that residential expansion for urban commuters was inducing socioeconomic change in villages (Anderson and Anderson, 1964, 1965; Pahl, 1965; Spindler, 1973) brought a new emphasis into rural studies that shifted the focus away from investigations of rural depopulation, the demise of rural economies emanating from agricultural restructuring and the social structure of isolated rural communities (Saville, 1957; Littlejohn, 1963; Franklin, 1969). From then there was increasing recognition that rural areas were components of a broader socioeconomic transformation which many held to be part of societal modernization. Included amongst these influential trends were population and capital concentration (O'Brien and Keyder, 1978), urbanization (Rambaud, 1969), agricultural intensification (Tracy, 1989), economic diversification (Keeble *et al* ., 1983) and later rural repopulation (Fielding, 1982). More recently, causal associations between trends in rural areas and wider society have been recognized in the linking of local socioeconomic change to international processes such as global capital restructuring (Marsden *et al* ., 1993), international residential mobility (Buller and Hoggart, 1994a), geopolitical reorganization (Dolfus, 1979) and the manipulation of consumer demand (Cloke and Goodwin, 1992).

Not all European states, or indeed regions, have experienced these processes or felt their impact at the same time or in the same way, in part because dissimilarities in national traditions impose themselves between already varied socioeconomic characteristics (see Chapter 2) and evolving developmental trajectories. In working towards an understanding of whether such national distinctions are deeply-rooted in cultural, political or socioeconomic practices, or originate more from short-term expediency, we start our analysis by focusing in this chapter on rural 'traditions' within Europe. Our understanding of these 'traditions' combines both the importance of rural space in national identities and the dissimilar images that are conjured by the concept 'rural'. As such, we are interested not simply in whether 'rural' means different things to people in different member states of the EC but also whether significance is attached to that rurality which has a bearing on socioeconomic development.

Historical evolution of European rural landscapes

The strength of previous tendencies to group rural areas into broadly homogeneous wholes stemmed in part from an underlying assumption that agriculture was the quintessential, even the definitive, activity of the rural landscape (e.g. Gervais *et al* ., 1977; Hervieu, 1994). This led to functional classifications of rural areas in terms of the attributes of their agricultural base (Clout, 1991). Associated with this fundamental linkage was a set of economic, political and social assumptions about the character of social organization in agricultural communities (Michel, 1986). The frailty of placing such assumptions at the centre of a research tradition was readily exposed when the importance of agriculture in rural economies diminished (see Table 2.2, page 39), while a theoretical vacuum was created when the notion of a rural-urban continuum came to be dismantled in the 1960s (Newby, 1980). One reason why the demise of this conceptualization created such a significant theoretical gap was due to the neglect of rural issues in mainstream social science theorizing. As Newby (1980, 9) explained: 'Rural sociology [which we would extend to other branches of rural studies] has been very badly served by the classical writers in the history of sociological thought ... The rural has frequently been regarded as residual ...'. Recognition that socioeconomic change in the countryside was significantly informed by non-local forces also proved to offer little theoretical cheer. Such forces too readily came to be called 'urban', and this urban-centred view of space became associated with a 'white land' conception of rurality, in which the countryside was seen primarily as the bits between towns which waited to be acted upon by town dwellers (whether for urban growth or recreation). As Klatzmann (1971, 23) put it: 'Rural areas are from where twentieth century man [sic] gets his space' (our translation). In practice, neither an agricultural nor an urban-centred vision of the countryside takes us far, for they both rely on an essentially univariate conception of economic, political and social causation. But even in past centuries, when a certain legitimacy could be claimed for rural areas experiencing more unifying than divisive economic, political and social practices, the mosaic of contemporary rural Europe was being created through a complex interaction of emerging nationalisms and greater integration into a developing global political economy.

If we take the early years of this emerging world-system, it is not unreasonable to view the period leading up to the sixteenth century as a time of medieval unity in Christian Europe. In much of northern Europe, this was associated with a model of rural social organization founded on a feudal landowning nobility and more or less free peasants (Koenigsberger and Mosse, 1971; Dodgshon, 1987). In modern terms, this feudal model was remarkably widespread and consistent, with similarities in climate aligning with a low level of agrarian technology to promote broadly common agricultural landscapes in much of northern Europe (Dion, 1934). Distinguished by the later enclosure of its common land and the development of new agricultural techniques which enabled marginal, formerly unused land to be brought into production, the second great landscape type, the *bocage*, extended down the continent's western flank, from Norway in the north to parts of Portugal and Spain in the south (Lebeau, 1991). Completing a threefold division of Europe into broad clusters, the landscape form which emerged in much of the southern Mediterranean lands was distinguished both by its climate and the physical discontinuities that resulted from extensive upland zones. Perhaps even more notably, vast landholdings accrued through 'empire' building and then the expulsion of the Moors from Portugal and Spain which left a distinctive landscape mark (Brenan, 1950; Ribeiro, 1986).

But whereas pre-modern rural Europe was arguably characterized by a certain organizational homogeneity, this began to alter with the strengthening of nation-states. In seeking to promote or define national coherence, these sought to integrate their rural parts into a common nation. In many cases, this was achieved through changes to land ownership structures that strengthened the hand of local leaders who supported the dominant regime; although it also saw expression in deliberate efforts to humble regionally-based counter-cultures (e.g. Reece, 1979; Agnew, 1981). The British Parliamentary enclosure movement of the eighteenth century, alternatively presented as either a triumph of state modernism over local pastoralism or the deathknell of traditional rural England, was essentially a process of creating capitalist economic and social relations within the hitherto semi-feudal countryside (Williams, 1973; Mingay, 1977). It thus initially facilitated and later complemented the emergence of Britain as a world industrial power in the nineteenth century. In stark contrast to the British experience, the emphasis on equal property divisions after death in French post-revolutionary land reforms led to the pastoralization of the French rural economy (Crouzet, 1962). This promoted the development of small, fragmented landholdings both in France and in certain of the French territorial conquests of the Napoleonic era, such as southern Germany (Lambert, 1963; Mellor, 1978), and the north, but not the south, of Iberia (O'Flanagan, 1980; Black, 1992). Indeed, the establishment of a free landowning peasantry became the backbone of the French republican idea (Barral, 1968). In turn, the perpetuation of this peasant farm economy did not occur by chance, but was nurtured. In this we see the clear hand of the Church, particularly the Roman Catholic Church, whose leaders, even after the Second World War, adhered to a mythological contrast of rural virtue and city vice. The result was that they commonly pressed for governmental policies or even introduced measures of their own accord that helped maintain the rural population (such as cooperatives). Across much of Catholic Europe, small peasant farms were a favoured mechanism for achieving this outcome (Brenan, 1950; Hendriks, 1994).

The existence of a large peasantry, associated as this was with widespread property access, was not in itself sufficient to capture a place for rurality in a national psyche. It is true that the creation of a landowning peasantry played a critical role in defining post-unification national identities and in incorporating rural areas into the post-Byzantine sense of Greek nationhood. Indeed, in the second half of the nineteenth and early twentieth century, this was founded largely upon a belief in the ideological homogeneity and national solidarity that the peasantry appeared to enshrine (Kyriakidou-Nestoros, 1985). It is also the case that late industrializing nations, like Finland and France, still find themselves in a position where the bulk of urban dwellers have recent family associations with peasant farming. The state has needed to bestow such farming systems with virtues of nationhood in order to strengthen communal ties, given the slow rate at which integration into the national

economy followed political incorporation (S. Berger, 1972; Alapuro, 1982;
Kielstra, 1985; Abrahams, 1991). However, no such unity exists in nations
like Germany, Italy, Portugal or Spain, where peasant farming systems have
a long history but where peasant farm structures existed alongside large-scale
manorial systems. These weakened the sense of rural homogeneity and so
diluted any prospect of a rural imagery capturing a core place in national
identity (see Table 3.1). Thus, in southern Italy, the creation of numerous
smallholdings is a recent phenomenon that resulted largely from a desire to
quell farm labourer unrest, rather than originating in a sense of national
purpose (Ørum, 1985; King, 1987). Meanwhile, in Germany, smallholdings
have essentially been abandoned by policy-makers, who have focused their
largesse on larger farms, leaving those who retain a semblance of peasant
motivation (viz. that are not ruthlessly commercial) to turn to nonfarm work
for much of their living wage (Heinze and Voelzkow, 1993).

It is important to be careful about terminology when thinking in terms of
smallholdings. For one thing, we should not confuse smallholdings with
family farms. The mythology surrounding the notion of a family farm is one
that farm organizations have long played with, calling up images of democracy, enterprise and independence to ennoble their claims for further cash
support. But the concept of a family farm is a slippery one (Hill, 1993; von
Cramon-Taubadel, 1993). While in the past its essence might have been
captured by small peasant holdings, in the Europe of today these units are
unlikely to provide a sufficient income to sustain a family, unless other, non-
agricultural income prospects can be relied on (Arkleton Trust (Research)
Ltd., 1992). The notion of a peasantry is also problematic. In much of the
literature on rural Europe, the notions of 'peasant' and 'small family farm'
appear to be used interchangeably, but more precise definitions are available.
Ellis (1988, 4), for instance, attempts an economic definition, which
'... centres on the idea that peasants are only partially integrated into

Table 3.1 Dominant forms of land occupation and style of farming, c1800

Commercial, mercantile estates and farms	Peasant small-holdings (owned or tenanted)	Mixed peasant smallholding and manorial	Manorial
England	Belgium	Austria (alpine)	Austria (not alpine)
Scotland	Germany (western)	Denmark	Germany (eastern)
	Italy (northern)	France	Italy (southern)
	Netherlands	Ireland	Portugal (southern)
	Northern Ireland	Portugal (northern)	Spain (southern)
	Norway	Spain (northern)	
	Sweden		

Source: Urwin (1980, 32).

incomplete markets'. Key to this notion is the idea that although peasants may produce goods for sale, this is not their primary activity or rationale; whilst the markets they are involved in are not fully commercial. This distinguishes peasant farms from small commercial family farms where the primary purpose of production is the response to commercial market opportunities. There is also a social side to the definition of peasants, which relates to the fact that a 'peasant unit' is both an enterprise and a household, which '... simultaneously engages in both production and consumption' (Ellis, 1988, 7). This dual function means that on peasant farms, decisions are governed both by (commercial) economic criteria and by social and subsistence criteria relating to the internal organization of the household. As Pina-Cabral (1986, 10) puts it for northern Portugal, peasant farmers have a '... justified suspicion of the market system'; it is this suspicion that leads them to respond to commercial stimuli in different ways to fully commercial farms. While the question of the definition and survival of distinctively 'peasant' forms of production is returned to in Chapter 6, for present purposes it should be noted that the regional dominance of smallholdings cannot be equated with similar cultural traditions across agricultural regions.

Distinctive regional trajectories

To understand the present-day picture of Europe's rural economies and societies, we need to explore the association of emerging nation-states and technological advances. For Sivignon (1993) the eighteenth and nineteenth century saw widespread agricultural revolutions in northern Europe, even if the adoption of new technologies was slow in coming in some areas, as with many Breton farmers recognizing the merits of farm mechanization only through working as prisoners of war in Germany (Segalen, 1991). In contrast, in southern Europe, even in the twentieth century, the local power of large landowners combined with politically isolated authoritarian regimes to keep wages low and lessen the spread of 'outside' innovations which might have produced social change. Indeed, as these economies began to open-up, mechanization was often forced on landowners due to a shortage of labour that resulted from farm workers fleeing their poverty-stricken lot for higher wages overseas or in tourist outlets (Manganara, 1977; Mansvelt Beck, 1988), rather than from a desire to introduce technological improvement for economic gain (Black, 1992). These very basic differences lie at the core of what became emergent 'northern' and 'southern' rural trajectories.

Northern European models

Where the impetus for technological advance favoured a more progressive vision for agriculture, two alternative development models for rural areas can be identified (Sivignon, 1993). The first, originating in Britain and remaining largely confined to that nation, drew upon a longstanding concentration of agricultural holdings and the early introduction of an unprotected and rigourous agricultural capitalism. Here, with the industrial revolution taking hold early, the battle lines were soon drawn between a *nouveau riche* manufacturing class and an aristocratic landowner class. This was a conflict in which the aristocrats were not united, for while only a few turned their land over to manufacturing and mining, many more had strong and longstanding connections with commerce, which manufacturing did so much to make highly profitable (Mingay, 1976; Beckett, 1986). In addition, the prevailing ethos was supportive of economic progress, which inevitably led to the economic pendulum favouring industrialists, given that higher rates of return accrued from investments in finance and manufacturing.[1] In fact, rather than being reticent about supporting the development of overseas food suppliers, British investors were instrumental in promoting their productive capacity (Edelstein, 1982; Denoon, 1983). Ironically this increased British farm exposure to global market competition (Peet, 1969; Schedvin, 1990), and induced an agricultural depression that required commercial responses if survival was to be achieved (Grigg, 1974).

Complementing this competitive thrust was a companion trend, which, by way of contrast, was founded upon the creation, partly by state initiative, of a family farming peasant agricultural structure. Originating in Denmark, this model had much in common with improvements in agriculture in Britain prior to the Industrial Revolution. Both relied on the adoption of more proficient agricultural technologies and a more systematic utilization of principles from the animal and plant sciences. Similarly, both were founded upon substantial changes in land ownership patterns and rights. Yet while the British process led to a concentration in land ownership and the emptying of the countryside, by forcing agricultural labourers and tenants off the land into the new factories of nineteenth century industrialization (Mingay, 1990; see Table 3.2, page 83), the latter, both in Denmark and Sweden, contributed to a general emancipation of the farm labour force (Cabouret, 1984; Jones, 1986). The reason for this important divergence lies in the political and cultural climate of the late eighteenth and early nineteenth centuries in the two nations. The seeds of the Danish model saw their early fruition in the

---

1 One illustration of this change is provided in 88% of British millionaires coming from the landowning classes between 1809 and 1879, but just 33% doing so for 1880–1914 (Cannadine, 1990, 91).

Table 3.2 Percentage of national population dependent on agriculture, c1900

Country	Percentage	Country	Percentage
Austria	62	Ireland	45
Belgium	25	Italy	59
Denmark	46	Netherlands	34
Finland	71	Portugal	65
France	40	Spain	72
Germany	43	Sweden	55
Greece	46	United Kingdom	12

Source: Urwin (1980, 65).

1748 palace revolution by the Prince Regent, which heralded possibly the most progressive vision of national development, and of the role of ordinary citizens within it, in Europe at that time (Rying, 1988). The peasant land reforms of the late eighteenth century were born in a strongly Lutheran culture, and drew their roots from a combination of philanthropy, an enlightened monarchy and Rousseau-like idealism (Jones, 1986). A later stimulus came from the desire to restore national pride after the loss of Norway and Schleswig-Holstein in the nineteenth century (Webster, 1973). Together these led to the painstaking transformation of former heathlands into productive farmland (Kamp, 1975; Rying, 1988). It is no doubt stretching the point to see the Danish model as the product of '... a typically protestant world view' (Sivignon, 1993, 137) but the ethic of protestantism unquestionably provided significant cues for the behaviour of Danish leaders. Again revealing links between protestantism and capitalism, the publication in Danish of Adam Smith's *The wealth of nations* in 1779 was to produce a 'revolution' in economic thinking that made free-enterprise part of the conventional wisdom (Miller, 1991). This was accompanied by an early sight of what became the Scandinavian welfare model;[2] with such innovations as liberal tariff reforms in 1797, compulsory education for all children aged 7-14 in 1814, old-age pensions in 1891 and health insurance in 1892 (Johansen, 1987; Miller, 1991). With a more direct bearing on agriculture, the Danes encouraged tenant purchases of landholdings to strengthen farmer independence and provide a strong foundation for a property-owning democracy. Later in the century, extensive cooperative systems for the production and marketing of farm produce were introduced, which helped raise farm efficiency, increased the income that farmers received from their product sales and strengthened rural communities (Jones, 1986; Rying, 1988).

2 Without labouring the point, we should emphasize that Denmark and Sweden still share many similarities in their commitment to social welfare, even if this has come under strain in recent years (Persson and Westholm, 1993). A current extension of this sense of communal responsibility is seen in a strong commitment to environmental protection.

With manufacturing poorly developed, Denmark was set on a course that demanded international competitiveness in its farm sector. To retain this position it needed to adapt quickly to changing market circumstances, with its capacity to do so revealed in a rapid switch to animal products once the challenge of New World grain production was appreciated in the late nineteenth century (Johansen, 1987). It is important to recognize that one reason why the Danes were able to achieve this was because Britain purchased so much of its surplus production, as well as becoming a key supplier of needed materials for the Danish economy, with 80% of coal needs coming from Britain as late as 1932 (Miller, 1991). Thus, in the 1870s and 1880s, some 70% of farm exports went to the UK (Just, 1994), with a figure of 62% in 1913 (Nash and Attwood, 1961). The two world wars and the intervening years of economic uncertainty and depression disrupted market possibilities, but the commitment to free-market competition was confirmed when Denmark became the only European nation to dismantle its war-time state controls after 1945 (Tracy, 1989). None of this substantially lessened dependence on the British market, which made accession to the EC a 'natural' step once Britain sought entry. In fact, although attempts to broaden the market base bore fruit, with exports to Germany rising to 23.4% of the total in 1955/58, when the CAP became operational, principles of Community preference combined with import tariffs to reduce access to the German market.

Given the Danish model's emphasis upon the values of owner-occupying farmers, it is not surprising that its underlying principles appeared attractive to those states for whom the peasantry formed a major demographic and political organ in the nation's corpus. As a consequence, principles from the Danish model found their way into the agricultural systems of Belgium, Germany, the Netherlands and, somewhat later in a modified form, in France, where tariff barriers assured both the immediate short-term survival of farming and its ultimate recession (Gervais *et al* ., 1977). Of these, the Netherlands constitutes the purest transposition, with '...benefits from flat terrain, deep soils, rich pastures, ample water supplies, and a temperate climate' (OECD, 1994b, 173) providing what is probably a more favourable physical basis than Denmark for a rich agricultural economy. However, like Denmark, the Netherlands was short of the raw materials that were demanded by nineteenth century industrial growth, which placed considerable pressure on agriculture to produce commodities for export (Riley and Ashworth, 1975). As Tracy (1989) makes clear, the combined impact of ideas from the Reformation, the struggle to free themselves from Spanish rule (Parker, 1977), the age-long combat against the sea, plus the longstanding importance of trade to this small country, combined to produce an independent, energetic and commercial-minded business class. This is well reflected in the Dutch vision of family farms as institutions that depend upon rational entrepreneurial skills, rather than manifestations of a peculiarly agrarian ethic (de Haan, 1993). Already an exporter of agricultural produce in the nineteenth century, the leaders of this relatively small nation did not find it difficult to see the benefits that were likely to accrue from free-trade (Strijker, 1993). Moreover, as there was scarcity in factors of production, farmers were prompted to welcome technological change in order to maintain their international competitiveness. The end-product, as Hommes (1990, 45) described it, is that '... farms have come a long way towards looking like factories', with the volume of pollutants they produce revealing some disturbing analogies with traditional manufacturing plants (e.g. Figure 2.16, page 62).

A southern European model

Such an impetus favouring agricultural modernization came to the Mediterranean states much more recently, and has been particularly relevant only in the postwar era. Resistance by landed estates (Brenan, 1950), the diversity of agricultural systems (Sáenz Lorite, 1988), the marginalization of agriculture in national development policies (King, 1987) and the growing acquisition of rural lands by the urban bourgeoisie, both in the past (Brenan, 1950) and following agrarian reform (Prost and Vandenbrouke, 1993; Sivignon, 1993), have all been offered as explanations for the late entry of the Danish model into the Mediterranean. Whatever the diversity of reasons involved, we can see the clear hand of dominant power structures in the trajectories that rural areas have followed. For the Mediterranean nations of the EC, this has often meant retarded development options. The mood engendered has been well captured by Bell (1979, 1): 'Because so much of the past was beyond human understanding, because the present was uncontrollable, and because the future was unpredictable except for those with magical powers, fatalism (*fortuna*) dominated the peasants world view'. As a result, '... harsh present realities and a pervasive sense that the future was immutable beckoned society to live in the past' (Bell, 1979, 21). This did not come about through any 'natural' or environmentally engineered process, but resulted from deliberate efforts by power elites to maintain a particular social structure. Mediterranean nations were not alone in finding that their rural areas were being utilized to perpetuate social stability or to extend the power base of particular leaders. For example, nineteenth century attempts to alleviate possible social tension led to encouragement of industrial workers to live in Belgium's villages (Mormont, 1990) and to Nazi portrayals of the peasant farmer as the nation's 'life source' in Germany (Thieme, 1983). Nevertheless, it is in southern Europe that we see this effect most sharply.

Transparently, Italy illustrates the point, with the creation of the nation in 1861 drawing its roots from a compromise between northern industrialists and southern landowners (Graziano, 1978; Davis, 1979). This compromise strengthened the resource base of southern landowners, so they could call on a stronger state structure, and especially its military and police, to enforce their local dominance and sustain a little changing rural society. This subjugation of landless agrarian workers was not simply enacted through elite compromises at the national level, but also saw expression in local pacts i8n both northern and southern Italy (e.g. MacDonald, 1963; Cardoza, 1979). Unlike the British situation of large landowners being driven by capitalist imperatives, in the South in particular propertied interests prevented development occurring through their efforts to perpetuate a semi-feudal society. Understandably, for Graziano (1978, 291): 'The truth of the matter seems to be that the South has suffered less from capitalist exploitation than from the lack of it'. Yet it is also the case that agriculture has been plundered in Italy. The state has been the primary agent in this process of seemingly deliberate undevelopment, in which manufacturing and commerce benefited from investments and incentives that were funded by high taxation and low investment in agriculture (e.g. Snowden, 1979). This pattern was also not of short duration. Even after the Second World War, although southern landowners lost their power base through land reform, the dominance of manufacturing interests continued, leaving agriculture in a subservient, retarded state. Thus King (1987, 138) remarks: 'Italy's partial catching-up of other European countries in the fields of industry and services has been achieved by plundering the resources of agriculture, depriving it of its youngest, most able workers, usurping much good land close to urban markets and appropriating many of the most plentiful sources of water'.

Perhaps the earlier formation of their present states removes the potential for a directly comparable trend in Portugal and Spain, yet a large landowning class was dominant in certain regions of these nations and produced a similar pattern of retarded development (Brenan, 1950; Cutileiro, 1971). Elsewhere, as in Greece, there was the problem of adverse physical conditions, whether from the weather or the rugged terrain (Clout, 1986; Figure 2.3, page 34). Greece does differ from Portugal and Spain, in that its large landholdings were largely eradicated following military defeat by Turkey in 1922. This brought to an end the long-held aim of expanding the nation's land area to what was considered the 'real Greece' (Ioakimidis, 1984). Defeat saw more than a million and a half overseas Greeks 'returning' to the Greek national territory, which provoked the government to introduce a package of land reforms that commentators have credited with producing an egalitarian peasantry (Burgel, 1981). What the break-up of these large landholdings visibly demonstrates is the dominant role of the Greek state in the nation's economic and social affairs. Possibly in a more extreme form, but sharing features with Italy (Silverman, 1965; Littlewood, 1981), and even parts of Portugal and Spain (e.g. Kenny, 1960), we can see the continuance of principles from the Hellenic feudal system that promote 'clientelism' as a major factor in the functioning of the Greek state, such that there is no clear separation in concept or practice between the state and civil society (Wenturis, 1994). In effect: 'The Greek state appears to constitute in itself the main mechanism through which decisions about the functioning of the economy are made' (Tsoulvouvis, 1987, 516). For rural areas this has had particularly adverse effects, for state functionaries have regarded the countryside with disfavour. Thus rural areas came to be seen as a remnant of the past, and the prospects of modernization were strongly linked to urbanity (Kyriakidou-Nestoros, 1985). In turn, economic activity become concentrated almost exclusively in a few large cities (Nikolinakos, 1985; Clout, 1986). Not that the pace of development has been rapid, for Greece shares with Portugal and Spain the experience of recent spells of authoritarian rule which were marked by an unwillingness to promote the free exchange of people, products and ideas that a dynamic economy requires (Jouganatos, 1992). And while effective economic bankruptcy led to a freeing-up of the Spanish economy after 1959, bringing in its wake huge increases in foreign investment and substantial growth through tourism (Harrison, 1985), Greece has passed through periods in which foreign investment has been limited (Jouganatos, 1992). Thus opposition from wealthy landowners (Loukissas, 1977), plus poor management, weak marketing and inadequate service quality (Briassoulis, 1993), have all limited the potential for tourist-based expansion (see Table 2.9, page 71). In addition, the long-existing pattern of state dominance over decision structures has had a major effect on the potential for development initiatives coming from other national sources (Wenturis, 1994). As a result, development programmes that rely on some input from local leaders and enterprises have commonly floundered (Arkleton Trust (Research) Ltd., 1992; Georgiou, 1994). A similar situation can be seen in Portugal (Syrett, 1994).

The role of political change

We could proceed with detailed consideration of each of the member states of the EC, drawing out their peculiarities and particularities. But the main purpose of this section has been served if the reader has appreciated that there is variety in the historical trajectories through which areas in rural Europe have travelled. In the twentieth century, of course, such paths have become somewhat confused, with two world wars leading to the redrawing of national boundaries, so that Austria and Germany lost important territorial units that severely disrupted images of 'their' rurality. In Germany's case, the loss of territory after 1945 deprived it of most of its large farm sector (Mellor, 1978; Tracy, 1989), whilst in the case of Austria, an extended Allied (four power) occupation after World War Two proved influential. Here, the fear that their nation would be subjected to an east-west division led economic and political leaders to emphasize mutual tolerance and understanding

(Scully, 1990). This resulted in political compromises that favoured a significant governmental role in economic affairs and led to early governmental
efforts to assist rural areas through broadening their infrastructural base. In
turn, small farmers, who are particularly dominant in the alpine zones, were
able to benefit from conservation measures and job openings from tourism
and allied activities (OECD, 1987). As in Germany, therefore, many part-
time farmers in Austria owe their retention of a farm interest to off-farm work
opportunities, although in Austria the positive hand of government is more
evident in promoting this situation.

The recent nature of the political disruptions that have penetrated the
mores of rural life in Austria and Germany distinguish these places from
Finland and Ireland, for whom political uncertainty and/or antagonism was
a long-enduring phenomenon. In both cases a key issue in their sociopolitical
history has been the late arrival of effective national sovereignty. For Finland
this arose from being part of Sweden and then Russia, with independence
only coming as a consequence of the Russian Revolution of 1917. Even then
the Finnish vista was dominated by the shadow of the Russia bear, so that
arguably it has only been since the disintegration of the Soviet Union in 1991
that Finland has had an open hand to play. One consequence of dependence
on neighbouring powers is that Finland is an 'unhistoric nation' (Alapuro,
1982), whose nationalism is of recent origin. As regards the role its rural areas
have in national sentiments, these can be traced to bouts of hunger that have
emerged after markets were cut-off due to conflict or political change, along
with a desire to build and strengthen social stability by broadening the
property owning base (Singleton, 1986). As a consequence, land reform came
early to Finland, with the establishment of small-scale property ownership
being encouraged and a strong emphasis being placed on national farm self-
sufficiency (OECD, 1989b). A significant consequence, as Abrahams (1991,
170) notes, is that: 'The idea of a fundamentally agrarian Finnish spirit is a
much more realistic one, despite industrial development and settlement
patterns, than would be the case for Britain'.

A somewhat similar situation exists for Ireland, where independence from
Britain did not come until the 1920s. Since that time, a confusing picture has
been generated by continuing economic ties to the UK, coupled with a strong
political reaction to British rule. Indeed, economic dependence on the UK has
only lessened significantly since accession of both countries to the EC (Brunt,
1988). Strong sentiments favouring smallholdings developed as a counterbalance to the prior dominance of large estates that were controlled by English
absentee-owners (Clark, 1975). With the arrival of nationhood, then, land
reform quickly followed (Leeuwis, 1989), and the search for national identity
became rooted in an idealized traditional rurality of owner-occupying small-
holders (Breen *et al* ., 1990). However, the lingering impact of colonial rule,
with its emphasis on restricting local leadership potential, interacted with the
political kudos that emanated from the combination of post-independence
sentiment and gratitude for land reform benefits, to produce a centralized

and patronage-driven political structure (Bax, 1976; Sacks, 1976). The existence of a large property-owning population in rural Ireland did not thereby provide the kind of political independence that one associates with nations like Britain, Denmark and Finland, but rather led to a leadership structure that more closely resembles Greece.

The reason for our lengthy attention to these shorter and longer term trends in the historical setting of rural Europe has been justified if we have provided some understanding of the way in which processes of evolution have created, from a situation of relative homogeneity, a set of persistent divisions and specializations which both derive from and reinforce rural traditions in individual nations (e.g. see Table 3.3). Thus, whereas we might find common trends in the agricultural development trajectories of European states, and can trace the origins of some of these back to the Danish model of the late eighteenth century, rural organization reflects, above all, the cultural, political and socioeconomic diversity of European nations. And there are significant contemporary elements to this historical parade of initially unifying and later divisive forces which impinge on cross-national rural traditions. The establishment of the Common Agricultural Policy in 1957 and its gradual extension throughout Europe represents not only the onset of a genuine transnational agricultural policy but also the re-emergence of a European agricultural space (Mykolenko and Calmes, 1985). Within this space broad

Table 3.3 Rural land-use types in the European Community, 1988

	Total surface 1000 km²	Arable percent	Grassland percent	Permanent crops percent	Forests percent	Water percent	Other percent
Belgium	30.5	24.2	20.3	0.5	20.2	0.9	33.9
Denmark	43.1	59.8	5.1	0.3	11.4	1.6	21.7
France	549.1	32.0	21.4	2.3	26.9	1.1	16.3
Germany	248.7	29.2	17.9	0.7	29.6	1.8	20.8
Greece	132.0	22.2	13.6	*	43.6	2.4	*
Ireland	70.3	14.6	66.4	*	4.7	2.0	12.3
Italy	301.3	29.9	16.1	11.1	21.0	2.4	19.4
Luxembourg	2.6	21.4	26.7	0.6	34.3	0.4	16.6
Netherlands	41.5	21.5	26.0	0.9	7.9	9.1	24.0
Portugal	92.1	31.6	8.3	9.4	32.2	0.5	18.1
Spain	504.8	30.8	13.2	9.7	24.8	1.1	20.4
UK	244.1	28.1	45.4	0.2	9.4	1.3	15.5
EC	2258.9	29.8	21.3	*	23.8	1.6	*

Source: Eurostat (1992). Note: * missing data.

agricultural regions of production specialization might well lead to the re-establishment of common cross-frontier landscapes (Bowler, 1986; Hervieu, 1993; Neveau, 1993). It is significant that the basic principles of the Danish eighteenth century approach underlie the CAP: a belief in the family-owned farm as the central agricultural unit, an emphasis on agricultural production and a belief in the farmer as a central actor within rural space (CEC, 1991b). At a superficial level, we have come full circle. The relative universality of medieval agricultural (and by association, rural) economic and social organiza-tion, with its parallel linguistic forms (for example, outfield, *utmark*, *essen*, *méjous*) and its broadly common agrarian structures, was gradually replaced by the emergence of independent national identities. This led to the fragmentation, not only of rural organization but also, and crucially, of the relative importance attached both to rural economic activities and to rural populations. In the more recent past, the CAP and other EC policies, notably regional policy, have contributed to reducing the distinctiveness of national initiatives for the countryside by establishing a European framework for rural policy. Today, the member states have a new rural lexicon (for example, set-aside, *gel de terres*). Nonetheless, as the recent history of the CAP and international agreements such as GATT so amply demonstrate, adherence to national frameworks for rural policy-making and appeals to the particulari-ties of national circumstances remain a strong force in EC decision-making (Kjeldahl and Tracy, 1994).

Diverse national images

It is certainly the case that all national self-images, whether they be the 'official' representations of tourist boards and pageantry, the brochures of industry or the visions of artists and writers, are elaborate composites. Over the last century or so, rural areas have come to play a major part in the iconography of many European nations (Keith, 1975; Green, 1990); with portrayals that align themselves with the Christian virtues of a peasant life (e.g. Millet's *l'Angélus*, painted in 1858; see Caille, 1986), natural beauty that is touched by the divine (Blake's *Jerusalem*) or the legitimacy, or otherwise, of social organization (Bertolucci's film *1990*). Within such representations, we find evidence of national political ideology (J. Berger, 1972). Paradoxi-cally, it is the most urbanized nations within Europe that have incorporated their rural imagery most vigorously. If the vision of a 'green and pleasant land' characterizes a longstanding protectionist image of the Britain's coun-tryside, and indeed goes a long way toward conveying the very identity of the nation (Newby, 1987), it derives not so much from any unique physical features of the British Isles as from the particular sociopolitical circumstances of eighteenth and nineteenth century urban culture (Lowe, 1989; Daniels,

1993). Alternatively, Greece, in demographic terms one of the most rural nations in Europe, has shied away from rural images, for rurality has been associated with economic and social backwardness (Papadopoulus, 1985).

Certainly, in Britain and France, the countryside has become a powerful cultural symbol of nationhood. Yet this has occurred for markedly differing reasons. These reflect, above all, the perceived role and function of rural space within each nation. For Britain, an essentially aesthetic and preservationist conception of the countryside has developed which has long dominated both public and political representation. In this vision, natural beauty is emphasized and portrayed as an alternative amenity environment to the town (Howkins, 1986; Lowe, 1989). In France, where the 'eternal order of the fields' has traditionally provided a sense of national continuity that reflects the nation's particular view on its *campagne* (Braudel, 1990), rural areas are largely seen as natural resources to be tamed and exploited by human activity (Gavignaud, 1990). These two nations provide the classic example of varying rural traditions within Europe.

> Up to the mid-century and in their respective national contexts, the peasantist movement in France and the preservation movement in Britain were the most prominent sociopolitical forces seeking to define the cultural significance of rurality in the modern world. However tinged they were with retrospective regret, both helped to forge distinctive and potent rural ideologies which have outlived their progenitors (Lowe and Buller, 1990, 14).

Similar instances of dominant rural symbolism can be found that underlie the very notion of nationhood (e.g. MOPU, 1987; Volker, 1988; Bodenstedt, 1990; Abrahams, 1991). The German *heimat* and the French *pays* represent fundamental territorial reference points for national populations (Rovan, 1994). The common theme in such symbolism is not so much the content of rural space as the importance given to that content; it is how nations perceive their own rurality.

Clearly, rural areas in the 1990s possess a wide range of production and consumption functions that challenge the viability of any simple unifunctional definition of rural space. This functional pluralism, in which different activities overlap in both time and space, makes rural areas less and less distinct from their supposedly urban counterparts (e.g. Convery and Blackwell, 1988). Consequently, the last 20 or so years have seen increasing challenges to the pertinence of the rural-urban distinction both in terms of definable rural functions (Bonnamour, 1993), identifiable rural sociological and behavioural distinctiveness (Pahl, 1966; Association des Ruralists Français, 1986) and indeed as a distinct field of social science investigation (Grignon and Weber, 1994). Nonetheless, the essential paradox of rurality within developed nations is that while the existence of a specific and exclusive rurality is increasingly contested at one level (Hoggart, 1990), the traditional components of rural areas (farmers, peasants, rural landscapes, fields and forests, etc.), continue to occupy a central and growing place in many national, regional, local and personal representations (Halfacree, 1993).

This central place nonetheless draws on different elements and infuses rurality with different emphases from one country to another. Critical differences have been said to exist not only between and within northern and southern Europe (Baker *et al* ., 1994), but also between western and eastern Europe (Szucs, 1985). From the multiplicity of rural histories and ideologies that exist, four broad images stand out in terms of their embodiment of distinctive rural characteristics. These are seen: first, when rural traditions retain an independent pertinence, that is often attached to agriculture or to agricultural-forestry dominance; second, in settings where the ethos of rurality is essentially defined and maintained by an urban-centred population (and arguably *for* that population as well); third, in places in which urban and rural distinctions have little ideological or cultural value in terms of identities; and, finally, in environments that are more hostile to human settlement, but where the wilderness or mountain habitat carry symbolic weight as emblems of national identity. We might construe from this initial distinction, four pseudo-independent rural traditions: an *agrarian tradition*, wherein the conception of rural areas is essentially one of productive surfaces on which the human exploitation of natural resources has led to the development of a distinct rural culture; a *naturalist tradition*, which derives more overtly from a conception of rural space as an arena for consumption, which often but not exclusively is an urban-centred notion of rurality; a *Mediterranean tradition*, in which either a close integration of city and countryside was achieved early (as in Greece; Urwin, 1980) or where political instability under repressive regimes have provided few opportunities for independent urban-rural distinctions to emerge; and, what might be termed *traditions of the margins*, where constraints on human activity arising from the character of the physical environment have generated a conception of the rural that is strongly linked to the sustainable use of natural resources. In slightly different guise we recognize this last tradition in the Nordic countries, in nations with a mountainous physical framework, such as Austria and Switzerland, as well as on the Atlantic Rim (Ribeiro, 1986; Scully, 1990; Abrahams, 1991).

We do not suggest that these four traditions are mutually exclusive or all-embracing. They co-exist within some nations and vie with each other in political and social representations. Neither do we suggest that these are static conceptions. On the contrary, increasing social modernization has clearly provided a base for shifts in emphasis from agrarian to naturalist notions of rurality, with such adjustments being associated with socioeconomic transformations and conflict between different 'rural' interests (Marsden *et al* ., 1993). What we do maintain, however, is that broadly common rural traditions are discernible in Europe, with some important associations between these traditions and national identities. Having understood the character of these traditions we can then appreciate their material impact on socioeconomic trends in rural development and rural policy within the European Community.

Agrarian traditions

For the bulk of Europe, rural space is dominated by agriculture and forestry (Table 3.3). Although primary production is limited as a basis for defining rural areas, with rural economic diversification gnawing away at overlaps between rural economies and farming (Clout, 1991), agriculture is the dominant rural land-use in all European states. Furthermore, it has generally been from agricultural policy that rural development policy has grown, both at the national level (e.g. Chosson, 1990) and within the European Community (CEC, 1988). Also, agriculture, along with forestry in some upland and climatically marginal areas (Table 3.3), is to all intents and purposes the primary contributor to the rural landscape. Hence, broad geographical areas become defined by their dominant agriculture. This is particularly evident in much of Mediterranean Europe, where a longstanding combination of low density pastoralism, physical constraints, traditional permanent crop production (like olives and vines) and, in many cases, semi-feudal social organization, are considered central elements of the Mediterranean identity itself (Le Coz, 1990).

The role and function of agriculture within national economies and the place of farmers within national socio-professional and political structures are basic to national representations of rural areas. Indeed, for some, the agricultural sector is the bedrock of European political identity (Hervieu and Lagrave, 1992) and the mainstay of any sustainable rural future (Kayser *et al* ., 1994). However, although agrarian traditions are present in virtually all European states, only in certain nations do they retain a dominant position in national representations of rural space. What might explain this? Agrarian traditions derive from a combination of elements including the role of farming within the national economy, although this has declined significantly over time (e.g. Table 2.2, page 39), and the demographic significance of the agricultural population, which has also been sharply reduced in this century (comparison of Table 3.2 with Table 2.7, page 59, provides some indication of this). However, more important than both of these is the symbolic importance that agriculture has occupied in political and cultural organization, which has placed the rural at the centre of national consciousness. Hence, the share of the 1960 GNP that came from agriculture was 30% in Greece, 26% in both Portugal and Spain, and 17% in Italy, but only 14% in Denmark and 9% in France (Urwin, 1980, 83). And yet the latter two can be said to have much stronger agrarian traditions, since in these latter states agriculture still retains a key position in terms of the nation's overall economic image. Thus French agriculture is frequently hailed as the nation's *petrole vert*, whilst in the Danish case, core values for democracy have also derived from the early introduction of laws that supported farm owner-occupation, restricted the growth of large landholdings and promoted cooperative organizations as a means of sustaining farm viability (Jones,

Table 3.4 Percentage of farm area owner-occupied, c1930

Country	Percentage	Country	Percentage
Austria	93	Italy	67
Belgium	41	Luxembourg	81[d]
Denmark	94	Netherlands	49
Finland	98[b]	Portugal	73*
France	60	Spain	64[b]
Germany	86[a]	Sweden	73
Greece	92[b]	United Kingdom	38[c]
Ireland	98		

Source: Urwin (1980, 76). Note: [a] 1907; [b] 1950; [c] 1958; [d] 1961; [e] figure for 1950 and for England and Wales only (the percentage for Scotland was 42); * this figure is for the percentage of farms, not for the percentage of the land area

1986; Johansen, 1987). Significantly, such cooperative movements were not only important for the economic gains they provided, but also for the manner in which they could strengthen rural community solidarity and agrarian traditions. As Susan Berger (1972, 71) noted for Bretagne: 'In order to shield the countryside from the impact of the city and the state, the founders of syndicalism began to create institutions that could meet all those needs which traditional rural society could not satisfy. For this purpose, the syndicate was to be the coordinating centre of all rural services and associations ...'. As we have already noted, in Ireland, agricultural improvement, the extension of farm owner-occupation and efforts to expel the British likewise forged a strong link between agrarianism and national identity (e.g. Curtin and Varley, 1991).[3]

This cross-national pattern provides a cautionary message about associating 'objective' measures with the subjective sentiments that are expressed in agrarian traditions. In the same way that the relative weight of agrarian (or rural) populations is not an adequate proxy for this ideological vision (Table 3.2), so too the link between levels of owner-occupation and agrarian traditions is problematic. This applies particularly to periods after the First World War, when the threat of peasant revolt produced significant land reallocations in European nations like Austria and Finland (Urwin, 1980), with a similar process in Greece following military defeat by Turkey in 1922 (Ioakimidis, 1984). As a result, 1930 land ownership structures had little connection with underlying rural traditions (see Table 3.4). The significance of cooperative organization in agriculture certainly provides a more valuable

3 This association was also one that was common to many nationalist movements in eastern Europe, although not in Greece, which had seen an early economic integration of its rural and urban spheres, so breaking the imagery that aligned nationalism to an agrarian democracy based on widespread land ownership (Urwin, 1980).

index (e.g. Table 3.5), although its coincidence with national sentiments is dependent upon the relative importance of production sectors across nations (and any coincidence of cooperative organization and agrarian values that holds within the EC will not automatically hold beyond its boundaries). Indeed, whatever the organizational form of agricultural endeavours, a key feature of agrarian traditions is a traditional peasant cornerstone to national stability and identity, along with a clear sense that agricultural issues have had a central place in national politics (Gervais *et al.*, 1977; Fabiani, 1986; Cleary, 1989; Tracy, 1989; Danish Farmers' Union, 1990).

Where states like Denmark, France and Ireland differ from other countries, especially those in southern Europe, is not simply because a family farming agricultural economy has been viewed as an effective development model (Sivignon, 1993), but also because it has been taken as a key embodiment of national identity. As Thompson (1970, 127) explained for rural France, inertia in government policy '... was permitted by the tariff protection and encouraged by the political mystique of the traditional rural way of life as a symbol of the nation's strength and solidarity'. Capturing the centrality of this impetus, the notion of a nation without peasants has invoked a remarkable sense of national crisis in France (*Le Monde*, 1992). Similarly, the relative absence of inequalities in landholding size in Ireland lent itself to a remarkably uniform rural imagery based upon the belief that the 'real Ireland' lay in small family-owned farms (Tovey, 1994), so that: 'In Ireland, rural culture is inextricably linked to the identity national' (Reid, 1992, 4).

Table 3.5 Percentage of agricultural products sold through cooperatives, 1991

	Pig meat	Beef/ veal	Poultry meat	Eggs	Milk	Sugar beets	Cereals	Fruit	Vegetables
Belgium	15	1	0	0	65	0	25–30	60–65	70–75
Denmark	97	53	0	60	92	0	50	90	90
France	80	30	30	25	50	16	70	45	35
Germany	23	35	*	*	56	*	0	20–40	55–65
Greece	3	2	20	3	20	0	49	51	12
Ireland	55	9	20	0	98	0	26	14	8
Italy	15	6	0	5	32	0	35	31	10
Luxem.	35	25	0	0	81	0	79	10	0
Nether.	24	16	21	18	84	63	65	78	69
Portugal	*	*	*	*	*	*	*	*	*
Spain	5	6	8	18	16	20	16	30	15
UK	20	5	0	18	4	0	21	30	19

Source: CEC (1994f, T132). Note: * no data.

Yet it is important to note the dynamism of traditions, for Tovey (1994) claims that contemporary leaders in Irish society are trying to distance a new vision of a 'European Ireland' from what they see as the backward agrarian tradition of the past.

Where democratic forms of government were associated with a large agrarian populations, agrarian values were likely to make themselves manifest in ways ranging from national political discourse, through the organization of space, to the formulation of rural policy and into political representation. Each of these then contribute to the manner in which nations respond to international pressures relating to rural areas. In recent years, the classic example of this has been France's response to threats to its agricultural sector, whether these be from the enlargement of the European Community to admit Portugal and Spain (Tacet, 1992), the 1992 CAP reforms (Robinson, 1993) or the Uruguay Round of GATT negotiations (Ingersent *et al.*, 1994b). In all three debates, the French have claimed a special interest in the protection of agriculture on economic, social and 'nationalist' grounds; arguing, in particular, that as the biggest national producer of EC farm commodities the Community's interests are best served by maintaining a healthy French farming sector. Thus, the accession of Spain was initially resisted because French officials feared increased competition from within the Community (Tacet, 1992). Similarly, during debates over reform of the CAP in 1992 the extension of mandatory set-aside was fought, as anathema both to the government-inspired productivist spirit and peasant pride in agricultural enterprise (Deverre, 1995). Underlying these responses has been a deep-rooted sense of linkage between agriculture and nationalism. Thus, when Brice Lalonde, the French minister of the environment, attacked modern farm practices that create environmental damage, and criticized the relative freedom from environmental regulation that farmers enjoy, this drew a fierce rebuke both from fellow ministers and from farm leaders, who described the minister's statement as a 'disgrace to the Republic' (quoted in Berlan-Darqué and Kalaora, 1992, 112).

More evident in the landscape, persistent agrarian traditions are woven into geographical structures in rural areas. Thus, despite extensive reforms in the recent past (Garner, 1975), administrative divisions in Denmark largely distinguish urban and agricultural districts, with the latter being characterized by a planning presumption in favour of agricultural activities (Therkildsen *et al.*, 1990). Belgium, the Netherlands and Sweden also reveal tendencies of geographical specialization in which agriculture holds a quasi-monopoly over other rural activities, except in specific recreational or settlement zones (Christians and Daels, 1988; Thissen, 1992; Persson and Westholm, 1994). In these nations, in place of the holistic approach to rural planning that some argue is characteristic of Britain, a more spatially and functionally sectoral approach has developed which permits increasing specialization of rural zones (van de Berg, 1989).

A pervading agrarian tradition is also reflected in the ways rural and agricultural interests are represented in national politics, as well as in the influence that representation has on rural policies. There is no doubt that agricultural lobbies gain strength from their nation's agrarian traditions, in addition to developing a sense of the justified need for farmer self-preservation in the face of urban, industrial interests (Hervieu and Lagrave, 1992). This has given rise, almost universally within northern Europe, to some form of agricultural corporatism (Andrlik, 1981; Muller, 1984; Cox *et al.*, 1985; von Doninck, 1992; Just, 1994), which possibly reaches its apogee in the Dutch 'Green Front' (Frouws, 1988). Providing another example of dissonance between the 'objective' circumstances of the farm population and its behavioural importance, it is worth noting that in few countries where such lobbies are effective is there a significant farm population, nor even a key electoral position for farm constituencies. As long ago as the the 1950s, for instance, only around one-tenth of Belgian parliamentary deputies had links with the farmers' organization the *Boerenbond*, with the figure for Germany at around the same level in the late 1950s (Milward, 1992). Such numbers can be contrasted with special bounties that governments have provided for their farm populations. In France, for instance, a distinct social security system exists for farmers and their families, while local bureau of the Ministry of Agriculture and indeed the broad rural development remit of the parent ministry are dominated by farm interests (Houée, 1979). Somewhat similar patterns are found in Germany (Tangermann, 1992), where the formation of the federal Ministry of Food, Agriculture and Forestry saw its most important officials being taken directly from the *Deutscher Bauernverband* (DBV), the German farmers' union (Milward, 1992).

To be clear, the strength of these groups does not rely on the existence of farm-oriented political parties, which exist in few European nations, and generally receive scant electoral support where they are found (e.g. Nooij, 1969). In many nations, as with France, political parties have failed to capture the farm vote, despite attempts to do so (Milward, 1992), and even where close links exist between one party and the farm lobby, as in Germany (von Cramon-Taubadel, 1993), the agricultural sector has not suffered from years of 'opposition' rule. As Condradt (1982, 108) pointed out, German politicians have been mindful of past agrarian support for right-wing political movements, including the Nazis, so all political parties have taken steps to placate farmer interests, including the Social Democrats, who receive very little electoral support from farmers. More significant than direct political representation has been the organizational ability of farm lobbies, which in some instances have had to demonstrate remarkable astuteness to maintain cohesion amongst farmers themselves. In France, for instance, agrarian unity has been preserved because landholders from the old nobility and the conservative upper class bourgeoisie were able to promote adherence to the myth of a rurality dominated by a single class *paysanne* (Tracy, 1989). This has been maintained despite the existence of sharp class distinctions in farming communities in parts of the nation, as in the sharecropping zones of the Central Massif and southern France (Tarrow, 1977). However, although drawing sustenance from the strength of agrarian sentiments, there is no direct association between this affection and the political effectiveness of farm lobbies. At one level this is evinced in the uneven efficacy of such groups in seemingly similar contexts, along with their changing political role over time (Just, 1994). More impressively, it is seen in the performance of the British National Farmers' Union, which has been described as '... arguably the best and organizationally strongest of western agricultural interest groups' (Wilson, 1978, 31), with the influence of this organization being furthered by a House of Lords that has traditionally provided a home for the scions of the rural gentry (Norton-Taylor, 1982).

It is pertinent to draw attention to the special position of Britain here, for within northern Europe it does stand apart as a nation with less intense agrarian sentiments (e.g. Table 3.2). As Abrahams (1991, 169) explained:

> In a country like Britain, where industrialization and urbanization have occurred on a grand scale relatively early, conceptions and misconceptions about rural life have commonly provided little more than a welcome sense that all is not lost ... the time when agriculture was the mainstay of our economic life is too far past for Kipling's sentimental verses about Sussex yeomen, or even Blake's genuinely emotive visions of a 'green and pleasant land', to provide an effective basis for most Britons' sense of their identity, except perhaps in time of crisis.

Indeed, with the early introduction of agrarian capitalism being followed by rapid urbanization and rural depopulation, the form of agricultural organization and its symbolic function changed at an early stage in Britain, as is well illustrated by the early removal of a population dependence on farming (Table 3.2). Early in the nineteenth century, agriculture became a relatively minor element in the economic and social fabric of the nation; with effective protection from foreign competition being denied in favour of cheaper prices for consumers and manufacturers. This contrasts sharply with most of Europe, save for nations like Denmark that depended heavily on the British market or pre-1870 Germany, which had yet to realize that its farmers could not compete with producers in the New World (Jones, 1986; Borchardt, 1991). Notwithstanding this, British countryside planning policy in the first half of the twentieth century went hand in glove with the belief that agriculture was the key rural activity, not only economically and socially, but also environmentally, as the 1942 Scott report so clearly stated (Buller, 1992). Thus, when today's land-use planning system was introduced in 1947 one of its major aims was to protect farmland, with farm activities largely exempt from planning regulation. As Marsden and associates (1993, 113) note: 'The situation owed much to an agricultural fundamentalism that still pervades the rural land-use debate, with agriculture being seen as having a pre-emptive claim on the use of rural land'. Illustrating the fragility of British agrarianism, since the 1950s this preferential treatment for farm enterprises has generated extensive political conflict. For Lowe and colleagues (1986, 11): 'Rural issues and conflicts have become a central feature of local and national politics and are hotly contested in public inquiries, council chambers, Parliament and the press'. One consequence is that British agricultural corporatism is in retreat (Cox *et al.*, 1986), with its agrarian ideological underpinning being increasingly forced back into the political domain of agriculture *per se*. By contrast, elsewhere in northern Europe this ideology is still a potent force in rural policy-making. In the Netherlands and France, for example, rural development policy is interpreted largely in terms of agricultural investment (Bauwens and Douw, 1986; Chosson, 1990), whereas in Britain there is a longstanding disassociation of agricultural and rural policy (Buller and Wright, 1990).

It follows that agrarian traditions have a determining impact on how images of rural change are perceived and presented. When agrarian sentiments are strong, agricultural modernization tends to be accepted only up to a certain point, after which it represents a threat to rural sustainability as marginal farms are forced out of operation (Lévy, 1994). Likewise, increased rural residence for urban-centred commuters is a double-edged sword, in that while it is commonly associated with economic diversification, it ultimately increases the economic gap between remote and intermediate rural zones (Béteille, 1981; Margaris, 1988; Thissen, 1992). In short, the rural 'crisis', the theme of a vast literature in various European nations, is essentially portrayed as an outcome of agriculture's declining demographic, economic and political hegemony in rural areas. The tide might have turned against agriculture, but its symbolic importance has far from declined in much of northern Europe. When French farmers demonstrated against GATT negotiations in 1993 they 'planted' fields of wheat along the Champs Elysées in Paris. This potent imagery was greeted by both public and political observers with disgust at the resurgence of old-formula peasantism but, significantly, also with pride over the bounty of French agricultural endeavour.

Naturalist traditions

That the countryside offers the basis for a distinct relationship, when held in comparison with the town, between human society and nature, has been a persistent theme of European cultural enterprise (Williams, 1973; Corbin, 1988; Green, 1990). Although social scientists are challenging the pertinence of urban-rural distinctiveness, the general 'proximity' of nature remains as one of the countryside's few remaining indigenous characteristics. As a strand in national orientations towards their rural areas, perceptions of this proximity are significant elements in representations of rural space. Indeed, the strength of 'naturalist traditions' reflects the importance that nations accord to 'non-productive' uses of rural space, particularly in terms of their aesthetic, public health, amenity, leisure and life-style qualities.

As with agrarian traditions, naturalist sentiments illustrate key variations in the way rural areas are conceived, defined, represented, managed and studied in different nations (Jollivet, 1994). Although the notion of rural space as 'preserved countryside' is an increasingly prevalent model in Europe (Marsden *et al.*, 1993), major distinctions can be drawn between nations, and even within nations, regarding both the strength of naturalist ideals and the impact that tradition has had upon rural policy. Significantly, in parallel with agrarian traditions, the strength and the impact of naturalist traditions cannot be said to be directly linked to landscape quality *per se*. All the European states have national parks and protected natural areas (Richez, 1991; Figure 2.17, page 64), yet only a few can be said to have a naturalist tradition that plays a major role in the conception of rural areas and the definition of rural policy.

As we interpret it, a naturalist tradition is a composite notion founded on the representation of the countryside first and foremost in terms of landscape and nature. This vision is generally accompanied by the belief that a (rather ill-defined) traditional rural way of life is superior both to urban and to contemporary rural life. Taken together, these arcadian and pastoral myths constitute the basis for a rural nostalgia that in Britain at least has become '... the basis of the role that the countryside plays in national identity' (Short, 1991, 34). Policies to protect the countryside have then reinforced this identity by enacting and generating definitions of Englishness (Jeans, 1990; Lowenthal, 1991; Daniels, 1993). The irony of this misplaced mythology is not lost on Short (1991, 67):

> It is ironic that the typical English countryside, the super-charged image of English environmental ideology, which can still conjure up notions of community, unchanging values and national sentiments, is in reality the imprint of a profit-based exercise which destroyed the English peasantry and replaced a moral economy of traditional rights and obligations with the cash nexus of commercial capitalism.

In tracing the development of naturalist traditions, we might initially distinguish the early industrialized nations (such as Britain) from those countries that came later to industrialization (such as Italy). The tendency to fall into a simple north-south discourse needs to be guarded against, for considerable opposition to this tradition exists in northern Europe (Pridham, 1994). Even so, the development of industrial and agrarian capitalism fundamentally altered the relationship of urban and rural areas (Williams, 1973; Mingay, 1990). In nations that experienced these changes early (notably Britain, but also the Netherlands and Germany), naturalist rural traditions first saw institutional expression in the formation of amenity groups, preservationist societies and natural history organizations (Howkins, 1986; Lowe, 1989). At least for the urban middle classes, rural areas soon became aesthetic and amenity havens from urban life, initially to visit and later to inhabit (Newby, 1987), as well as becoming the symbolic terrain of naturalism (Bruckmeier, 1994). Urbanization alone was not sufficient to develop such a naturalist tradition, as Italy ably demonstrates. But, having turned the centre of its economic vision so early toward manufacturing and commerce, the wealth that accrued to Britain had provoked a sentimental, consumption-driven image of 'the countryside' that originated in the upper classes but soon permeated all social strata (Williams, 1973). Certainly, those who came into new wealth through industrial or colonial endeavours revealed an obvious propensity to 'play' the landed aristocrat, by acquiring a manicured country residence (Wiener, 1981; Edwardes, 1991). In the longer term, as has been widely documented (Lowe and Goyder, 1983; Cox *et al.*, 1988), the juxta-position of land ownership structures, rural management traditions, population concentration and urban social movements in the nineteenth and early twentieth centuries brought widespread support for the protection of the countryside and its symbolic natural monuments from threats to its imagined character. This vision has now become a central element in national self-definition. Thus, in one of the central environmental policy documents of recent years, the British government defined the countryside in the following manner:

> Our countryside and our coasts are a central part of our heritage. Most of us, wherever we live, visit them from time to time for pleasure. Our landscapes have been an inspiration for centuries to writers, painters and nature lovers. They help form our sense of national identity (Secretary of State for the Environment and others, 1990, 96).

The significance of this statement is captured in the word 'visit', for in residential terms the British population is one of the most urban in Europe. Yet in no other nation in Europe are naturalist sentiments so unambiguously a cornerstone of rural policy (Newby, 1987; Shoard, 1987; Bodiguel *et al.*, 1990). Indeed, in no other nation has countryside protection become such a political rallying point for (commonly) middle class home-owners, united not so much as co-members in ecological or environmental movements as by a shared concern for the depreciating value of a consumer good; to whit, rural amenity and more specifically the sanctity of rural residential amenity.

Elsewhere, the emergence of a broadly naturalist rural tradition is rela-tively recent. This is clearly demonstrated by France, where a repositioning of 'rural' with respect to 'environment' has involved a recent shift from agrarian to naturalist conceptions (Mathieu and Jollivet, 1989). This late arrival is notable given that France had an early scientific and social elite-based rural protectionist movement that developed toward the end of the nineteenth century, and consequently was a broad contemporary with its albeit more widely-based English counterpart (Cadoret, 1985). In Belgium too, naturalist approaches to rural space have gained ground in recent years (Mormont, 1987); although here, the absence of a strong antecedent rural protectionist tradition reflects not so much the predominance of agrarian traditions as the particular form of peri-urbanization that has characterized the Belgian countryside since industrialization (Mormont, 1994a).

To a certain degree, neither agrarian nor naturalist traditions are sufficient as embodiments of rurality. Both are rooted in partial visions of rural reality and, as such, both are myths. Nonetheless, what these myths have in common is the fact that they endow rural areas with an identity that seeks to differentiate them from the town. Significantly, despite their innate differences, what unites these two rural visions is the high standing that is afforded to rural living and rural occupations. This imagery puts the central tenets of these traditions in a quite different footing from the rural traditions that have long dominated southern Europe.

Mediterranean traditions

Although one instinctively shies away from broad regional generalizations that group nations as culturally, historically and economically distinct as Greece, Italy, Portugal and Spain, there is some justification in claiming that these four countries display broadly common rural traditions, and that these traditions have emerged out of similar experiences. Thus all four have experienced, over the last two hundred years, deeply divisive political histories. Both Greece and Italy are relatively new states, while all four nations have experienced significant eras of anti-democratic dictatorial rule during critical periods in their socioeconomic development (Giner, 1985). Their processes of nation-building, which some see as still incomplete in Spain and Italy (Ritaine, 1994), have moreover largely taken place within a broad European context of economic, political and social modernization. The significance of this for rural areas is two-fold. First, these nations at various stages placed the agrarian question at the centre of their processes of national unification, democratization and/or nation-building (Kyriakidou-Nestoros, 1985; Carrière, 1989; Drain, 1990; Rivière, 1990). Second, recent processes of agricultural modernization have had to be predicated upon major structural reforms. The difficulties inherent in such reforms have meant that the economic modernization of the Mediterranean states has not, until recently, involved the bulk of the agricultural sector. Where agricultural modernization has been achieved, notably in coastal Spain and parts of northern Italy, which were often already relatively productive agricultural regions following the northern European agrarian tradition (Gay and Wagret, 1986), it has emphasized and reinforced the dual nature of both the agricultural sector (de Meo, 1991; Tricart, 1993) and general development paths within these nations (Benenati, 1981; Balbanian et al., 1991).

At the same time, all four Mediterranean countries contain archetypical urban civilizations in which spatial organization has been dictated by cities, ports and major towns (Braudel, 1973). The organization of rural space and rural activities has fallen under this basic geographical principle, which is reinforced by the physical structure of the Mediterranean region, which divides essentially into upland and coastal strips (Coccossis, 1991). Taken together, these common factors facilitate the collective treatment of these four states (e.g. King, 1982; Giner, 1985; Hadjimichalis, 1994), albeit with caveats over the Atlantic fringe of Portugal and Spain (Ribeiro, 1986), and partly justify the notion that a distinct rural tradition is identifiable across such a wide region. Certainly, one characteristic of this region is a resistance to sentiments allied to agrarian traditions. In good part this is because peasants and peasant traditions have been regarded as hindrances to the establishment of a modern national economy (Carrière, 1989; Sivignon, 1989; Black, 1992). As Papadopoulus (1985, 87) notes, it is significant that, in a nation whose employment structure is dominated by farmers, shepherds and fishing (e.g. Table 2.7, page 59; Table 3.2), Greece possesses no museums for agriculture, forestry, fishing or rural life, other than those concentrating on ancient Greek civilization. In addition, at the time when northern European nations were embracing a more consumption-based idea of 'the rural', Greece, Portugal and Spain were experiencing sociopolitical isolation as a result of their rule by authoritarian regimes (Giner, 1985; Wenturis, 1994). This meant that they have been slow to pick-up recent rural development trajectories. Illustrative of this is the late arrival of urban-to-rural migration flows (e.g. Fielding, 1982; Puyol Antolín, 1989).

The size of the farm workforce and a relatively poor farm performance also sets the four Mediterranean nations apart from other members of the EC (Table 2.4, page 41; Table 2.7, page 59). Yet these countries collectively contain 48% of the agricultural land surface of the EC12, as well as accounting for 66% of agricultural holdings within the Community. Although they dominate certain sectors of Community commodity production, their contribution to total EC agricultural output is just 36.6% (compared with 23.0% for France alone). Internally, the contribution of the agricultural sector to the GDP of Greece, Italy, Portugal and Spain is highly variable (11.6%, 3.0%, 4.1% and 4.1%, respectively, in 1990; Table 2.2, page 39). In short, much of the land area of these Mediterranean nations is occupied by low intensity farming on smallholdings, which support a relatively high density agricultural population. What is significant for the current discussion, and indeed is paradoxical in the light of traditions in other areas, is that the existence of a large agricultural population and the presence of a sizeable farm area has not given rise to a durable agrarian tradition; although alternative rural land-uses are not on the whole a major preoccupation. This is not to say that peasant traditions and ideologies have only been a minor component of the recent political landscape of the Mediterranean states. On the contrary, postwar agricultural policy in Italy, Portugal and Spain have all been formulated around principles of a peasant farm economy. Thus the encouragement of family farms, owner-occupation and a stress on independence in agricultural decision-making have all received prominence. In Italy, these principles were enshrined in the Sila and Stralcio laws of 1949 and 1950, which permitted the expropriation of some 8.5 million hectares of southern *latifundia* for redistribution (Gay and Wagret, 1986). Peasant ideals also permeated to the highest ranks of Christian Democrat governments and strongly influenced Italy's responses to the CAP during the 1960s and 1970s (Rivière, 1990). But while the peasant movement was crucial to the unification of both Greece and Italy, and coloured national politics in the early twentieth century (Avdelidis, 1981; Vermeulen, 1981), unlike Denmark, France, and the Netherlands, neither the egalitarian pretensions of peasant society (Burgel, 1981), nor the ideological predominance of peasantism in rural representation, survived late twentieth century economic modernization.

The tradition that has emerged in the southern states is therefore complex. At one and the same time it is a critical cornerstone to nationhood yet a hindrance to the development of a modern European profile. In the absence of rural nostalgia, the peasant tradition is the poor cousin of national imagery. As Hokwerda (1992) illustrates, in the novels of Dimitris Chatzis, one of Greece's most important postwar prose writers, the countryside is consistently portrayed as a backward, unwanted alternative to the city, with the latter presented as the future for a modern nation. Similarly, Douglass (1992) shows that the process of becoming a modern European nation has necessitated the denial of longstanding Spanish traditions that are seen as 'un-European'; notably bull fighting, which is felt to reinforce the image of Spain as a peripheral, backward state.

In the postwar period, Mediterranean rural traditions have been incorporated into, but fundamentally marginalized within, national developmental priorities whose primary aims have been industrialization and European integration. Although these rural traditions persist in local and regional contexts, they are not a major component of national identity in the way that agrarian and/or naturalist traditions are in other European states. One reason for this apparent paradox is the lateness of agrarian reforms and their subsequent limited impact on local (and often household) economies (Ørum, 1985; Cruz Villalón, 1987). The dominant position on the agricultural landscape of giant *latifundia* certainly provided a distinctive element in many Mediterranean regions well into the twentieth century; especially as the landed nobility had historically shown little interest in agricultural productivity (Brenan, 1950; Péchoux, 1975; Drain, 1985; Daumas, 1988). But while reform of such units was virtually ubiquitous in Greece, elsewhere farm structures are confused by regional diversity that results in a national farm sector comprised of both giant landholdings and smallholdings that are generally too small to support a household (King, 1987; Sáenz Lorite, 1988; Black, 1992). This duality has restricted reform attempts in the past, as well as promoting emigration as a development solution for low farm incomes

(Brandes, 1976; Manganara, 1977; Black, 1992). What it also illustrates is the complexity of the agrarian question in southern Europe and partially accounts for the relative absence of a unified agrarian identity.

Adding a further dimension to the incorporation and eventual marginalization of the agricultural sector have been national development trends which are essentially urban and industrial. Agrarian traditions and peasant ideologies, though present at the local and regional level, have served to reinforce rather than to challenge this commitment to urban and industrial growth, in nations that over much of their territory have historically typified urban civilization. A high density peasant population certainly facilitated rural depopulation and so provided the manpower for industrialization (Rivière, 1990). The effect has been to tie the profitable rural and urban sectors in extensive city-regions, while emphasizing the difference between these regions and the vast under-exploited and at times mountainous interior (Sivignon, 1990). For Drain (1990) it is only in post-revolutionary Portugal that the development of a peasant tradition escaped the controlling hands of an urban-based bourgeoisie, although even in a Mediterranean context, Portugal's agricultural sector is underdeveloped (Avillez, 1993).

It follows that the southern European states present a paradoxical notion of rurality. Rural areas are conceived and defined almost wholly as agricultural areas. Yet, this functional exclusivity has not yielded a coherent rural imagery or set of rural representations. For the reasons outlined above, it has given rise to a complex and highly variable agrarian profile which has not imposed itself on the national iconography. The Mediterranean notion of rurality remains deeply-rooted in the traditions of urban civilization, whether this is seen through the spatial organization of agricultural production (Sivignon, 1989) or through the flexible interaction of urban and rural labour forces and patterns of property ownership (Rivière, 1990). How long this distinctive position will be sustained is a matter of conjecture, for recent decades have seen significant economic changes in this region and entry into the European Community has stimulated a fundamental reappraisal of the role and position of rural zones in national economies (La Callé Dominguez, 1994).

Nowhere is this paradox more evident than in Italy. Here, the divided character of the Italian space-economy is apparent in urbanization trends in the Centre and South occurring alongside counterurbanization in the North as late as the 1970s (Fielding, 1982). This distinction alone draws attention to the rather peculiar position of Italy within a Mediterranean grouping, as well as warning against too ready an assumption of uniformity within the region. In particular we see this difference displayed in the very evident disparities in economic performance of the North and the South (e.g. Figure 2.5, page 44), with Italy standing as the only Mediterranean nation that has any of its territory falling in Europe's current vital economic axis (Figure 2.9, page 50). Of course, Italy is not the only nation that has marked inequities in socioeconomic and cultural-political traditions across its regions (e.g. Brenan, 1950). But on a European scale Italian regional divergences are more striking (Figure 2.5, page 44). Moreover, these differences are not of recent origin but date back to, and were even reinforced by, the process of national unification in the nineteenth century (Graziano, 1978; Davis, 1979). Capturing the magnitude of the chasm that separated the two halves of Italy, Furlong (1994, 35) noted how: 'The Mezzogiorno appeared to the northern liberals much like a foreign country, to which different solutions applied'. These solutions often meant that genuine development efforts were minimal, with emigration providing the historical safety valve for the rural masses (MacDonald, 1963). Land reform, when it eventually arrived after the Second World War, owed most to fears of social unrest and left the new peasant farmers with landholdings that were too small and insufficiently equipped for viable farm operations (Ørum, 1985; King, 1987). Effectively, in the South, as in much of Mediterranean Europe, the mainstream rural population lived for well into the twentieth century in a state of neo-serfdom (Bell, 1979; Schweizer, 1988). Meanwhile, in the North a pattern of indus-trialization and peasant farming developed that was not much different from that in many northern European nations. However, in terms of national orientations, the compromises that had been reached between northern industrialists and southern landowners at the time of national unity resulted in a southern, large landowner bias in rural policy formulation. Even today, more than any other European nation, Italy presents an agricultural system of two extremes: a geographically limited dynamic sector, that is currently threatened by the reform of the CAP, and an extensive but stagnant sector which, despite national and Community aid, shows little sign of emerging from its longstanding marginalization (De Meo, 1991; Mucci, 1992).

Traditions of the margins

Adding a final dimension to the character of European rural traditions, we can identify similarities in the rural images of three traditions that trace their roots to regions of economic marginality. The first of these is a Nordic tradition, which is associated with the states of Finland, Norway and Sweden, and is founded upon their particular physical and environmental profiles. In part, the shared nature of these images derives from the past political unity that these nations have experienced (e.g. Singleton, 1986), which have also produced some shared associations with Denmark (Rying, 1988). At the same time, in so far as a rural tradition has been forged that is distinctive from agrarian traditions, what separates this Nordic tradition is the relatively inhospitable nature of its physical environment. A measure of this is the much stronger association of farming with forestry than is usual for areas domi-nated by 'purer' agrarian traditions (e.g. Singleton, 1986; Abrahams, 1991).

At the same time, we can identify a strong attachment to rural places and a lauding of their distinctive sociocultural features (e.g. Aarebrot, 1982). The sense of distinction that this provides is readily apparent in Norwegian rural opposition to incorporation in the EC, despite the supposed agrarian core of its rural policy-making.

While we would not wish to stretch the point too far, given dissimilar political-cultural histories, there are some grounds for recognizing similarities in Nordic sentiments with the ruralities of nations with significant mountain zones, like Austria and Switzerland. For Wolf (1966) these countries are more accurately defined as having a transalpine tradition (if we take his ecotype as sharing many features with a rural tradition). But what distinguishes this ecotype are the constraints that the environment has imposed on the colonization of national territories and on the economic activities that can be undertaken in the countryside. Herein we find many original features of the Nordic conception of rural space. These elements have come to play a key symbolic role in the definition and presentation of national identity which illustrates the central importance of the natural environment in both transalpine and Nordic sentiments.

In the Nordic states this rural-natural environment relationship is manifest in four key ways: first, in the dominant place given to rural representation in national imagery and policy (Arter, 1993; Persson and Westholm, 1993; Abrahamsson, 1994); second, in a longstanding recognition of the importance of integrating environmental protection within local agricultural, forestry and fishing practices (Pettersson, 1993); third, in the close, and often nostalgic, links that urban dwellers retain with the countryside and rural life (Abrahams, 1991); and finally, in a strong regulatory tradition within national politics that is founded upon an implicit recognition of the importance of natural resource protection (Lavoux, 1991). These sentiments appear to favour the egalitarian treatment of citizens, with a specific interest in supporting those living in remoter areas (Aarebrot, 1982; Persson, 1992) and build upon an awareness of the potentially precarious nature of national food supplies (Peterson, 1990; Just, 1994).

Echoes of such sentiments are likewise found in both Austria and Switzerland, where national policy has long sought to integrate human activities within a broad rural-environmental framework. In these states, where the natural environment plays such a major part in national imagery, rural environmental protection and environmental risk management have long been central policy themes (Chabert, 1993; Sieper, 1993). The nature and extent of regulations on farm-based pollution in Switzerland, for example, are strongly dependent upon the altitude of the farm unit concerned (Knoepfel, 1990). Similarly, agri-environmental measures, which seek to retain a farm population more for environmental than for productive reasons, have longstanding precedents in these mountainous states (OECD, 1987; Chabert, 1993). We would not want to emphasize the distinctiveness of this tradition unduly. At heart, what we find in these countries is a strong tradition of agrarianism that has a great deal in common with most northern European states. What the peculiarities of their physical environments and the specifics of their political-cultural histories add to this is a particular twist that provides a somewhat different orientation in national traditions. In particular, a stronger concern for communal wellbeing, with its associated concerns for citizens living in remoter rural areas, as well as for the preservation of the character of the physical environment, set these nations in a slightly different mould from their northern European counterparts.

Allied to the Nordic and transalpine traditions is a third tradition associated with the Atlantic periphery of Europe. In the case of this Atlantic tradition, elements of the rural-natural environment relationship again exist, as does a sense of egalitarianism and communal support in traditional societies, in response to the role of agriculture in facing up to sometimes harsh climatic and physical circumstances. Thus the role of communal organization has been highlighted in areas of the Atlantic periphery ranging from northern Portugal (Dias, 1961; Bennema, 1978), and Galicia (O'Flanagan, 1978) to western Ireland (Arensberg, 1937; Messenger, 1969), even if the stability of these rural societies has recently been undermined by agricultural modernization, emigration and closer integration with the mainstream European economy. In such areas, however, the strength of this tradition has not been translated into a significant impact on national images of rural areas, being outweighed in national consciousness by other traditions, such as the 'agrarian' and 'naturalist' traditions of France and the UK, or the 'Mediterranean' tradition of Spain. An exception perhaps is the case of Portugal, where a classic text on the geography of Portugal by Orlando Ribeiro (1986) draws a specific contrast between what are seen as the distinctive worlds of 'Mediterranean' and 'Atlantic' Portugal, and questions Portugal's true 'vocation' as a product of these two worlds. Nonetheless, in the case of Galicia, as well as in areas such as Bretagne (Brittany) and the Highlands and Islands of Scotland (Penrose, 1993), the Atlantic tradition does arguably form part of an alternative 'nationalist' vision based on rejection of existing nation states, and an assertion of the distinctiveness and right to self-determination of these regions or nations.

Conclusion

One aspect of the material that is explored in the chapters that follow is whether the four rural traditions examined in this chapter find expression in socioeconomic changes in rural Europe. As readers will have already grasped, it is not expected that these traditions draw sharp lines between nations, for national orientations toward their rural areas are not mutually exclusive, nor are they consistent even within a single country. Moreover, our aim is not to impose an interpretation of socioeconomic change that draws out the particular role of rural traditions. Rather, it is to explore the character of rural transformation within the EC, asking whether we can identify nation-specific inputs to change processes. In doing this, rural traditions comprise one element of national effects, but they are just one element, that has to be put alongside short-term political expediency, existing socioeconomic conditions and the forces that arise from local and, more particularly, globalization trends.

|4|

Nations and transnational policies

Almost without exception, texts on the European economy or the development of the European Community devote at least one discrete chapter to agriculture or more specifically to the Common Agricultural Policy. This is not surprising, since the CAP is the leading and most comprehensive sectoral policy of the EC, as well as being the most highly visible example of sectoral integration over European space.

> It is a monument to the determination of politicians to work together for a united Community. Because the CAP is the first major common policy involved in the integration process it has become the symbol of cooperation. In this spirit of determined cooperation the politicians have nurtured the CAP in defence of economic logic and the long-term interests of the Community (Hill, 1984, 118).

The complexities of the CAP and the centrality of negotiations over its establishment and reform to the development of the EC merit this regular enumeration. However, for the purposes of this book, the importance of the CAP arises less from the various reforms that have been implemented to address issues like farm over-production or the high cost of subsidies, and more from the compromises that have been reached to achieve policy changes. As we will show below, the CAP began its life as a cross-national policy, requiring significant national cooperation in order to promote a uniformity of approach across the EC. Initially, within the agreements that EC member states made, there was considerable scope for national preferences to be introduced. The accession of more countries to the EC, and the strengthening economic cohesion of both the CAP and the EC over time enhanced the importance of this European policy as the EC has become a stronger player on the global economic scene. As such, while formulated as a transnational policy, and while retaining a cross-national tone amongst EC members (even if more muted than in the past), beyond the EC the CAP has acquired the aura of an internal policy (viz. intra-national). Most visibly, the changing position of the CAP has been seen in negotiations for the various GATT Rounds on world trade (Ingersent *et al.*, 1994b). In this regard the CAP stands apart from other EC initiatives. Policies for manufacturing and service industries, for example, have been slower to develop and, through their concentration on declining industries (Swann, 1984; Molle, 1988), have drawn less attention to themselves outside Europe. Quite apart from the 'natural' focus of CAP policies on rural areas, agricultural policy provides a microcosm of other inter-state debates over the direction of the European Community, albeit one in which special factors and unique actors intervene, and where agriculture offers the central focus for discussion in the absence of a coherent European rural policy (Clout, 1993b). Whether or not other sectors follow the path established by EC farm policies, agriculture stands alone in the insights it provides on European policy formulation.

The key objective of this chapter is to explore the nature of policy debates and compromises that have led to the emergence of EC initiatives. Our attention will not be restricted to the CAP, but the central place of this instrument of economic regulation in EC policy-making, along with its importance for rural areas *per se*, inevitably gives the CAP a heavy weight in our discussion. Of course, policy change has been independent of economic change of an institutional and technological kind, such as mechanization, the expansion of large corporations, economic globalization, and employment shifts from the primary to the secondary and tertiary sectors. However, this chapter focuses explicitly on policy initiatives in order to explore national impacts that should be readily identifiable, on account of the institutional channels through which they are expressed (e.g. through governments). Having established themes and trends from this analysis, we will be better placed to turn in Chapter 5 to the more diverse and less publicly negotiated (or visible) changes that result from economic globalization.

Before turning to explicit policy arenas, however, we should recognize that there are more holistic policies that have a significant determining effect on rural development. Most evident in this regard is the postwar 'consensus' on demand management and the welfare state, which has so characterized trends in European development since 1945. The breakdown of this consensus in the face of growing international economic competition and increased stresses on public expenditure has created significant pressures on national governments that forced them to introduce policy change, even if this was against their 'better judgement' (e.g. Peterson, 1990). These changes resulted from a long-term recognition that national economic management strategies needed reform (e.g. Maynard, 1988; Grande, 1989; Nannestad, 1991). Growing economic globalization and the lessened regulatory influence of US hegemonic power also played a key role in provoking the need for change (Tussie, 1991; Blake and Walters, 1992; Agnew, 1993). Whatever the precise causes of these changes, an understanding of European rural policy initiatives will not be adequate unless it is temporally located within the unfolding traumas of the unsettled, dynamic political economy of the last few decades.

The changing postwar international order

Immediately after the Second World War, there was a considerable uniformity amongst European governments about the direction both national and international economies should take, with free-trade as a replacement for pre-war nationalist protectionism at the top of the agenda. Such consensus is unsurprising, as war-torn European nations were keen to rebuild their economies after the physical ravages of war, not simply to erase the painful memory of the fascist era but also to provide a stable and successful economic base from which to resist the threat of communism and stave-off the potential for economic chaos that brought pre-war facist regimes to power (Conradt, 1982; Furlong, 1994). In addition, this era of physical and moral reconstruction was an ideal arena in which the victors of World War Two, and especially the United States, could establish hegemony over the political and economic direction of newly installed governments on the European mainland. This opportunity was reinforced by investments made under the Marshall Plan and subsequent financial transfers to rebuild the European economy (Block, 1977; Agnew, 1987; Gill and Law, 1988). Thus open government and open economies, allied to a 'New Deal' for Europe's citizens, were part of an externally-regulated agenda for Europe's growth. The dollar became the international reference currency, linked to the Gold standard, and the unit in which most or all commodities, including agricultural commodities, were priced.

By no means all of the European political system was created or recreated in a US image. One particular theme of postwar reconstruction was the significant role that would be played by the state and other public bodies in the economy, ranging from relatively widespread nationalizations of key industries by the newly-elected Labour government in Britain and the parallel consolidation of social democracy in much of Scandinavia, to more limited but nonetheless significant state involvement in industries in France, Germany and Italy (e.g. Winchester, 1993; Furlong, 1994). In some countries, this included part-state ownership of companies and state regulation of company affairs, especially in the field of industrial relations. In addition, the growing role of the state included the consolidation of comprehensive health and welfare programmes in many (especially northern European) countries, to the point at which the state was an inescapable feature of ordinary people's lives, whether they lived in towns or rural areas (Ashford, 1986). This higher state profile has been theorized by some as part of a new 'mode of regulation' (Offe, 1976), in which the state takes a mediating role between capital and organized labour. In its breadth and depth, this amounted in some cases to a radical socialization of everyday life, with governments becoming involved in what many on the right saw as ever-increasing domains of citizen and company behaviour (Barry, 1987).

One consequence of this intra-national trend was a growing dissonance between what was happening within nations and global processes of change. A clear sign of this growing disparity occurred in 1971, when the USA had a trade balance deficit for the first time in the twentieth century. More worrying for the USA than its deficit in commodity trade was the negative trade balance on invisibles, which had gone into the red in the 1950s. Much of this imbalance was a result of foreign policy initiatives. For one, after the Second World War the US government encouraged overseas investment by US-based corporations so as to help rebuild European economies and halt the spread of socialism and communism (Block, 1977). Overseas military intervention and support for 'friendly' governments and guerrilla movements all increased the outflow of dollars from the US economy (Cox, 1987). The US responded to this by increasing the national debt and printing more money (Castells, 1980; Cox, 1987). As dollar surpluses accumulated abroad, the costs of sustaining its position as regulator of the global economy proved too great. As Britain found to its cost in the nineteenth and early twentieth century, the role of hegemonic power at times requires the subjugation of national interests to what is seen as the greater good of the world-economy (Ingham, 1984). When confronted with this choice, US leaders put their national interests first; as when President Nixon ended the US commitment to a fixed value for the dollar in 1971, so introducing floating exchange rates that lowered the value of the dollar and made US exports cheaper. The disruption this caused to the world-economy was soon heightened by the fourfold increase in oil prices in 1973. This was followed by massive efforts by international banks to unload their surplus dollars (for oil was paid for in dollars), which, with the advanced economies in recession, led to sharply increased bank loans to the Third World (Holley, 1987). Many such investments were made in a hurried fashion, so that within 10 years the world-economy was further destabilized by fears that the financial system would collapse (Griffith-Jones, 1984). Meanwhile, with large corporations responding to the squeeze on their profits by seeking low cost production sites, observers began to foretell the deindustrialization of the advanced economies, as Third World and newly industrializing countries with less stringent health, labour, safety and welfare systems captured increasing shares of global manufacturing output (e.g. Rowthorn and Wells, 1987). Fears about economic decline, along with the expectation that social disruption might follow, were intensified by the seeming inability of governments to introduce policies that effectively addressed very evident economic weaknesses (e.g. Nannestad, 1991). But while calls for increased trade protection from Third World imports grew in volume (e.g. Tussie, 1991), the fact that European and US-based multinational corporations were central to the increasing globalization of economic markets cautioned governments against implementing such measures (Chase-Dunn, 1989). Indeed, right-wing administrations in government in various key economic nations ensured that, in intent at least, the emphasis was more likely to be on clawing-back the role of the state and 'releasing' the vitality of free-markets (e.g. Barry, 1987; Maynard, 1988).[1] Less powerful nations that depended significantly on agricultural commodity trade gave an early lead in this regard (e.g. Cloke, 1989b), with analysts predicting huge gains for the world-economy from freer trade (e.g. Stoeckel and Breckling, 1989). It is in this confused and uncertain environment that the major issues of agricultural policy, and more generally rural development options, have been debated and agonized over in the 1980s and 1990s. Before we can trace how these changes have penetrated contemporary mores on rural policy, we first have to grasp the nature and entrenched character of dominant rural policies that have had to be confronted.

National interests and the emergence of the CAP

Underlying European agricultural policies in the immediate postwar years were a set of relatively consistent objectives across nations. Principal amongst these was the desire to secure self-sufficiency in foodstuffs, thus attacking the twin problems of food shortages and balance of payments difficulties which affected most European nations at the end of the Second World War (Tracy, 1989; Milward, 1992). Specific measures were introduced to stabilize or control commodity markets, as well as increasing production and productivity either through the encouragement of higher prices or improvements in technology. Signifying the extent of sentiment favouring these goals, the immediate postwar years saw major policy instruments introduced that established new frameworks for agricultural production in Britain (1947), France (1947–51), Switzerland (1947 codified in law in 1951), and Italy (1948), while countries like Belgium and the Netherlands retained much of the support measures that were introduced during the war. Only Denmark explicitly dismantled its war-time agricultural controls and subsidies, seeking instead to win markets through free-trade in agricultural commodities and the diversification of its farm exports (Tracy, 1989). However, by 1958 even this policy had begun to be abandoned as the Grain Marketing Act set guaranteed producer prices for some cereals. Given the long association of Denmark with free-trade sentiments, this Act alone provides clear testimony to the difficulty of pursuing economic policies that were significantly out of line with the European norm. Epitomizing this norm (West) Germany's 1955 Agriculture Act codified principles designed to raise farm incomes and productivity, through the stabilization of agricultural prices and supplies.

1 Although it should be said that governments of all political persuasions were disoriented by economic crisis in the 1970s and early 1980s, with some left-wing administrations acting in a overtly right-wing manner (e.g. Lauber, 1989), while some governments of the right failed to implement free-market and pro-capital policies (e.g. Foppen, 1989).

This doctrine, with one or two modifications, set the tone for Article 39 of the Treaty of Rome, which established the principal aims of the Common Agricultural Policy just two years later.

The origin and initial form of the Common Agricultural Policy is a matter of public record and much academic writing (e.g. Hill, 1984; Bowler, 1985; Fennell, 1987). This makes it superfluous to reiterate the detail and mechanics of the policy in its original form. However, the principles on which the CAP was established are worth drawing out, for they have had a major effect on rural development since the European Community was created. By principles we do not refer to the stated aims that were written down in the Treaty of Rome to direct agricultural policy. These aims, and their well-known inconsistencies that allow them to be used to justify widely differing policy stances, were: (1) to increase farm productivity; (2) to provide for a decent standard of living for farmers; (3) to promote market stabilization for agricultural products; (4) to ensure security of food supplies; and, (5) to foster price structures that were 'fair' for the consumers of farm products (Averyt, 1977). Important though these have been for rural development, only some are what we regard as fundamental principles for the formation of the CAP (or, more broadly, for European integration). In particular, as Bowler (1985) points out, in the creation of the CAP no nation had to defend its *status quo* against change, since the key issues of food security, market stabilization and adequate farm incomes were already established in national policies for agricultural protection. This concern for food security extended beyond the bounds of World War Two participants, with notable similarities between the aims of Swedish postwar agricultural policy and the CAP (OECD, 1988). Indeed, it was the United States, despite its strong export performance and global economic ascendancy, that most vigorously objected to the liberalization of trade in farm commodities in the 1950s (Hillman, 1994; McMichael, 1994).

But the hopes that lay behind the creation of the CAP owed their roots to more than a desire for agricultural protectionism. This could have been achieved without recourse to common cross-national farm policies. Even before the end of the war we see signs of a growing appreciation of the benefits that would accrue from closer European integration. The Benelux countries provided an early sight of this when they agreed to a customs union in 1944, which led to shared commercial policies in 1948, and a Treaty of Economic Union in 1958 which was introduced after the Treaty of Rome in order to codify and further extend cross-national cooperation (de Schoutheete, 1990). This might be written-off as little more than a group of small nations seeking to limit the prospects that they were 'trampled on' by their large neighbours (France and Germany), but there is no doubt that what lay behind these measures was genuine support for European integration (de Schoutheete, 1992). The strength of this feeling might have been less intense in France, Germany and Italy, but its potency in these nations was also strong. Indeed, the old cliché that the CAP was a compromise between France and Germany, allowing France to reap agrarian benefits in return for German manufacturing gains, provides so simplified a picture as to distort the reality of this policy initiative (Milward, 1992). For one thing, manufacturing was always a more important sector in the eyes of negotiators than farming; for the enlarged common market that the Treaty of Rome introduced was much more important economically for manufacturing than for agriculture. This was especially true given that, for many food products, the new member states were still net importers, with markets readily available for nations like France and the Netherlands who had significant export sales. Moreover, if France did hope to replicate the postwar successes of the USA by using agriculture to promote export earnings, the compromise with Germany that resulted in higher cereal prices lessened the prospect of this outcome (Coulomb and Delorme, 1989). Thus, while an abundance of cereal surpluses could have provided a cheap source of animal feed that would have enabled the French livestock industry to compete with other northern European nations, in agreeing to high cereal prices France restricted the ability of its livestock producers to develop this new market (Le Heron, 1994).

In reality, the reason why agricultural policy stands out as a policy initiative has less to do with horse-trading over gains than it has to do with the structure of agricultural production. In mainstream manufacturing activity a relatively small number of companies tends to dominate sales in single market sectors, so producers are able to set their own prices within certain limits. Such an oligopolistic market structure differs significantly from the competitive markets of agricultural production, in which the large number of small producers ensures that farmers are price takers (O'Connor, 1973). In such competitive markets, the wellbeing of producers is determined in good part by their ability to secure governmental regulation (Tarrant, 1992). Without this, producers are subject to price vicissitudes, which can be due to many effects, including the weather or 'too many' farmers switching their interest to the high priced crops of the previous summer. As such, under present ownership structures, it is inevitable that governments in capitalist economies have a more visible involvement in agriculture than in other economic sectors, especially at a time when governmental involvement is driven by considerations of food security.[2] Effectively, while the needs of many European manufacturing sectors could be partly met by open access to foreign markets, for agriculture such openings needed to be accompanied by visible, shared policies across states.

Hence, while France was keen to see agriculture as part of the new common market, it was not the only nation to be so inclined. Indeed,

2 We add one caveat to this. Agriculture is not the only competitive market sector in European economies. The fact that it is subject to so much state regulation owes much to three things: its strategic importance to states, especially in times of war; the strength of favourable agrarian sentiments in many countries, which is associated with the historical evolution of national economies; and, the fact that agriculture help satisfy basic needs for human sustenance, which sectors like tourism do not.

although the French might be said to have ensured that agriculture was part of the final cooperative agreement, it was the Dutch that largely gave the CAP its form (Duchêne *et al.*, 1985). This form overtly demonstrated the heavy hand of compromise. This was inevitable if agreement was to be reached, since national policies had shared aims but their operating principles were different. Belgium and Germany, for instance, had highly protected farm sectors, whereas the Netherlands had a long-held liberal policy for its farm sector, and was well established as an exporter of agricultural produce (Tracy, 1989; Hommes, 1990). Price levels were also different, with France having relatively low farm product prices, whereas Germany's were high even by European standards (Averyt, 1977). Thus its grain prices were some 30% greater than those in France (Hendriks, 1994). In sum, France gained from CAP price increases, as did the Netherlands (dairy) and Italy (fruit and vegetables), but it might also have scored highly with lower farm prices since this could have strengthened its export performance (Coulomb and Delorme, 1989). To have pushed for such an outcome would hardly have been in the spirit of promoting a new European cooperation. If there was a difference in attitudes towards the CAP on the part of France and Germany, this arose more from a German fear that common commodity prices would lead to a politically embarrassing lowering of farmgate recipients for its farmers (George, 1985; Hendriks, 1994).

Agricultural lobbying for the nation

In reality, compared with the entity we know today, the CAP began life as a loose cooperation that allowed national governments considerable scope to influence farm policy. At a superficial level, this is seen in the time it took for common prices to be used across the Community. For cereals this occurred in 1967, followed by milk products, beef, rice, sugar and olive oil in 1968, and then tobacco and wine in 1970 (Averyt, 1977). Reflecting the persistence of national differences, the common grain price stood some 8% above the French level and 11% below the German level in 1967. Not that the introduction of common prices heralded a genuine common market. Anyone who held such an illusion did not wait long for this misunderstanding to be exposed. Thus, in 1969 the French franc was devalued by 11.1%, which should have raised farmgate prices in France (to maintain a common EC price). But, at a time of high inflation, the French were able to persuade the Council of Ministers that new prices should be phased in over two years (Fennell, 1987). This break from parity between real exchange rates effectively led to the so-called green exchange rate, with Monetary Compensation Amounts (MCAs) introduced as a border 'tax' to prevent farm products from unfairly competing in other national markets as a result of the disparity

Table 4.1 Farm commodity prices in EC nations (ECU per 100 kilograms), 1992

	Common wheat	Barley	Oats	Potatoes	Onions[1]	Desert apples	Pigs[2]	Butter
Belgium	15.59	14.44	16.75	3.69	10.39	28.78	162.35	314.78
Denmark	16.40	16.35	17.14	18.21	14.95	38.21	137.39	321.40
France	14.58	12.81	16.11	5.42	31.10	44.83	*	408.83
Germany	16.63	14.92	16.01	11.84	11.32	49.09	142.55	347.03
Greece	16.95	16.60	19.32	19.37	22.90	29.28	183.31	386.77
Ireland	13.61	13.36	13.94	*	*	*	*	*
Italy	19.69	18.19	23.00	23.24	38.13	42.08	211.05	343.44
Netherlands	15.76	17.08	16.84	5.72	*	35.17	*	313.42
Portugal	22.78	18.35	20.01	11.79	17.51	52.70	196.86	416.85
Spain	20.19	16.51	17.48	11.59	12.85	29.42	186.68	308.65
UK	16.50	16.03	16.00	10.49	16.30	56.97	130.81	286.31

Source: CEC (1994f, T86). Note: * no data; [1] all qualities; [2] carcass weight.

between their real currency exchange value and that used for farm commodities (Hill, 1984). The need to ward against this tendency was confirmed two months later, when the German mark rose in value by 9.3%. A confusing, and increasingly complex system of green exchange rates resulted, with governments utilizing disparities between real and green rates for political purposes. As late as 1978, for instance, the divergence in farm prices resulting from this disparity was 25% for France and Germany (Delorme, 1991). Then, from 1974-1985, the politically expedient quest to keep food prices and so inflation artificially low led successive British governments to manipulate the green exchange rate (Hill, 1984; Phillips, 1990). As a result *real* food prices differed by as much as 50% from those in Germany (Phillips, 1990, 57). Even today, with the EMS in principle offering more price stability, and MCAs having been abolished,[3] there are still substantial farm price differentials across EC nations (see Table 4.1).

An ascendancy of politics over economics is easy to appreciate when strategic interests dictate that food self-sufficiency is a key state goal. Indeed, when most member states (and the EC as a whole) were net importers of

3 From 1 February 1995 the EC changed the relationship between currency and green exchange rates. From this date, producers have less of a cushion against the impact of currency changes, for a degree of currency fluctuation is now allowed without any revaluation of green conversion rates. In effect, up to a 20% increase in ECU prices for farm commodities will be offset by a 20% lowering in green rates (this step is expected to save the CAP budget six billion ECU a year; CEC London, 1995).

agricultural produce, there was a certain logic to the argument that increases in farm output were beneficial if a country was not self-sufficient, for they reduced imports and so helped the balance of payments (Milward, 1992). Today, however, there is a high degree of self-sufficiency within the Community (Table 1.1, page 12). [4] But maintenance of food supplies is not the only argument that is put forward for CAP procedures. Even arbitrary price setting mechanisms have been justified by the CEC, as when contending that world prices are an inadequate yardstick for (some) EC products, since in '... beef and veal, wine and tobacco, a world market is virtually non-existent and prices differ according to the destination of exports' (CEC, 1989b, 23). However, given the long-established beef trade of Argentina and the USA with Europe, and recent assertions that beef production (amongst other products) is becoming ever more globalized (Sanderson, 1986; Le Heron, 1994), this assertion strikes a strange chord. Perhaps there is a complex reasoning involved here; one that is linked to the dominance of the EC in certain commodities, such as wine (Niederbacher, 1988), along with a lagged reaction to the former need of the EC to import other products, such as beef and veal, even though CAP policies have promoted so substantial an output increase that the EC is now self-sufficient in many of the products it once imported (CEC, 1994d).

Another interpretation is that the power of agricultural lobbies in the EC led to an ethos within the Commission that enveloped the agricultural sector with a protective shield. In part, the stimulus for this protectionist mode came from national governments, who have used CAP provisions to further their own national interests. Indeed, some see this as a critical weakness of the CAP, as it dampened efforts to institute the long-recognized need for reform (Hill, 1984; Bowler, 1985). That the promotion of national interests has compromised the CAP might seem a peculiar assertion to make, given the criticisms of European agricultural policy that have come from Britain (e.g. Body, 1991), or for that matter from elsewhere. Yet we must be careful when interpreting charges made for and against the CAP, given that they often have ulterior motives. As Duchêne and colleagues (1985, 8) recognized: '... the CAP is far from being only an agricultural policy. It is the more or less muted political battleground of all the forces for and against European federation, and attitudes towards it often have little to do with agriculture'. Even when they do have a clear agricultural focus: 'There is a strong sense that each country is still inclined to point out how its neighbours might mend their ways the better to retain their own' (Duchêne *et al*., 1985, 132). And CAP procedures provide plenty of opportunity for such actions, or at least for single nations to perpetuate the policies they favour. At the national level, this arises because the CAP has provided no objective methodology for setting

4 Providing one indication of the changes that have occurred in productive capacity, whereas France was 97% self-sufficient in food in 1974 by 1989 this percentage had risen to 126 (Pivot, 1993).

commodity prices, with most of its history also seeing no effective mechanism for relating product supply to consumer demand (Bowler, 1985). Thus national governments have been able to adjust prices, both internally and through the green exchange rate, to suit their own purposes. Indeed, when the EC has not provided the bounties national governments either promise or desire for their farmers, there has been little to stop national measures being introduced with scant regard for Community norms (Duchêne *et al.*, 1985). One example of this was the French response to 1993 protests over EC farm price cuts which allowed French farmers to benefit from a progressive reduction of land taxes (and their removal by 1996), a doubling of the limit on farm investments that can be set against taxes, and continued government responsibility for the social security contributions of farmers in economic difficulty (Naylor, 1994). This flexibility in CAP provisions is not unexpected, given the difficulties of achieving agreement for fundamental change. As Philip (1990, 73) explained: 'When you have 12 member states, several of them with important regional and provincial elections to defend, there are very few opportune periods of time for difficult decisions to be reasonably taken and for political candour to be tolerated'.

Beside which, most states are net beneficiaries of CAP largesse. Admittedly, the magnitude of gains and losses are subject to debate, and appear to depend substantially on the precise assumptions and estimation procedures of calculations. Nonetheless, there seems to be widespread agreement that only Britain and Germany are unequivocal net contributors (Brown, 1988; Croci-Angelini, 1992). Comparing the strong economic performance of Germany with the weak growth record of Britain (e.g. Table 2.6, page 55), and placing this alongside widely recognized German manufacturing gains from the enlarged EC market (e.g. Dunford and Perrons, 1994), it is easy to point to why Germany seemingly accepts its CAP contributions as a price worth paying for EC membership, whereas Britain is a strong critic of the CAP (Brown, 1988; de Schoutheete, 1990). But pressure from Britain for reductions in CAP spending, and more particularly in Britain's contribution to the EC budget (of which the CAP takes the lion's share), cannot be divorced from the politicking of British Conservative governments since 1979. British administrations have relied overtly on the attractions of tax cuts as vote winners and yet have proved singularly unsuccessful in cutting public expenditure sufficiently for longer term ('real'?) tax cuts to be made (Maynard, 1988). Add the weak electoral base of the farm community and the British position is clearer. Effectively, like Germany, agricultural costs were accepted as a necessary consequence of the gains Britain expected for its other economic sectors on joining the EC (Wijkman, 1990; Dunford and Perrons, 1994). In reality, the accumulation of poor national economic performances, rather than the cost of the CAP, lies at the heart of Britain's antagonism to EC agricultural policies.

What adds to Britain's concerns, given their financial basis, is the difficulty of controlling CAP expenditure under recent EC policy-making procedures.

Phillips (1990) explains these difficulties in the following terms:

1 the fact that the EC budget ceiling is artificial, as its tax base is an arbitrary percentage of VAT contributions, and despite Britain's opposition this percentage has shown an upward trend in the past
2 an imbalance in the timing of financial years, which sees the EC budget year only overlapping with six months of the CAP budget period, so making control over farm spending more difficult
3 the fact that CAP has a legal basis in the Treaty of Rome, which requires that its spending commitments be met
4 the inability of the European Parliament to control individual budget items, as its powers are restricted to rejecting a whole budget (as it did in 1979) rather than parts of it.

Of course, it takes little thought to recognize that many of these problems could be erased, given the political will. But for most of the years in which the CAP has been operational this will has not been present. In the early years of the EC this was understandable, as the CAP was the premier policy of this supranational organization, so that its acceptance, establishment and progressive improvement symbolized the vitality of European cooperation (Hill, 1984). But as EC membership increased, the prospects for reform seemed to lessen. The accession of Denmark and Ireland in 1973, for example, introduced two states that almost every study unequivocally identifies as major net beneficiaries of CAP spending (Croci-Angelini, 1992). Indeed, as Conway (1991) pointed out, Ireland has been keen to raise CAP spending, as it costs the Irish exchequer little and brings considerable economic benefits. Thus some 5% to 10% of the Irish GNP is estimated to come from CAP transfers. Add Greece to the mixture in 1981 and the impetus for no fundamental change increases. Not only does 78% of this nation receive extra EC payments owing to its designation as a less favoured area (Clout *et al.*, 1993), but there is also high economic dependency on farming in much of Greece (Figure 2.4, page 38). The PASOK government was able to heighten these returns by negotiating extra structural assistance on accession that was largely channelled to rural areas (Williams, 1991). As for the Iberian enlargement in 1986, here farmers in Portugal and Spain were obvious beneficiaries, given that their agricultural receipts under the CAP were much higher than prior support systems (Mykolenko *et al.*, 1987). Not that membership has been trouble free, for the potential for heightened regional inequity is problematical (Arroyo Ilera, 1990; Avillez, 1993), whilst in its first year of membership at least, Spain ended up as a net contributor to the EC budget, a situation only addressed when the structure of CAP funding had been shifted more towards Mediterranean products and structural support.

If the Iberian extension created any dilemmas, these were not simply important for Portugal and Spain. In the other Mediterranean countries considerable reservations were expressed over the impact that (particularly) Spanish agriculture would have on their farm sectors (Williams, 1991;

Tacet, 1992). The result was a drawn-out negotiation, lengthy transition periods for different (farm and nonfarm) products and, in the Integrated Mediterranean Programmes (e.g. Plaskovitis, 1994), the introduction of further measures to assist rural development prospects in France, Greece and Italy (with 40% of funds going directly to agriculture; Konsalas, 1992). Moreover, as the Mediterranean contingent has grown, the oft-criticized northern European product bias in CAP support has had more chance of being redressed, so offering the prospect of more positive policy impacts for southern Europe (e.g. Franzmeyer *et al.*, 1991).

It is clear that most member states have tended to see agricultural policy as an arena in which they are net direct beneficiaries from the EC. Viewed in terms of both direct and indirect costs, however, most outside observers estimate that the CAP has had a negative impact on general income levels (e.g. Stoeckel and Breckling, 1989). For Germany, for instance, estimates suggest that the abolition of the CAP would reduce consumer prices by 1.75%, increase aggregate employment by 5.5% and give a 3.5% boost to the GDP (Rosenblatt *et al.*, 1988). Such figures are hypothetical estimates, whose validity depends heavily on the assumptions that underlie their calculation. Moreover, to produce an effective input into policy-making, their message has had to overcome counter proposals for yet more price support from established, deeply-entrenched power interests (e.g. Urwin, 1980). Until the 1980s at least, any realistic prospect that such abstract arguments could overcome inbuilt tendencies favouring protection and production increases was remote.

To understand this we must recognize the central position of the Franco-German axis within EC policy-making. As de Schoutheete (1990) indicates, Britain was in a position to break into this key group, but its early years of membership were occasioned both by uncertainties over renegotiating its terms of entry and by economic disruption. These were then followed by the self-inflicted wound of the 'hateful climate' that the Thatcher governments provoked in seeking to lower British EC budget contributions (see also Arter, 1993). Moreover, at least as far as agriculture was concerned, a precondition for France not exerting its veto on British entry was acceptance of the CAP and, with the fate of its agricultural industry having been abandoned long before in favour of manufacturing (Tracy, 1989), Britain was neither in the position nor had the inclination to show leadership in the battle for agrarian policy reform. Its pressures for change were driven in significant measure by fiscal concerns, but its arguments found little support until an accumulation of fears over foreign trade and the environment added to fiscal worries to persuade other members states to alter their views.

Prior to these relatively recent changes in opinion, the CAP was nurtured significantly by strong support from the CEC, the French government and the German government. With EC spending so dominated by the CAP, officials within DG-VI, the Directorate responsible for agriculture, were in a strong position to manage policy. In the 1960s and 1970s, they did this with a political astuteness that saw a close working relationship with national governments, national farm organizations and the cross-national farm lobby group COPA (Committee of Professional Agricultural Organizations). In order to bolster the position of this Directorate in farm policy-making, DG-VI was assisted by what Philip (1990) refers to as the 'stunning complexity' of the CAP, which made understanding and best utilizing its provisions far from easy.

> The system that has resulted is comprehensible only to a few aficionados, policy-makers and lobbyists, who are thus able to operate the policy for most of the time in a secret garden which is largely inaccessible to non-agricultural interests and beyond the comprehension of most journalists and interested outsiders (Philip, 1990, 70).

Added to this, senior officials in DG-VI were drawn heavily from the French civil service, which helped fuse a strong bond with one of the key players in farm policy decision-making (Phillips, 1990).

The importance of personalities should not be played down here, although it has to be recognized that opportunities for making a personal impression on a policy process are partly decided by the position a human agent occupies in prevailing power structures. Nevertheless, it is difficult to envisage the bilateral Franco-German policy relationship developing into so central a power nexus without the initial complicity of Monnet and Hallstein, followed as it was by fruitful personal associations between de Gaulle and Adenauer, Giscard d'Estaing and Schmidt and, most recently, Mitterand and Kohl (de Schoutheete, 1990). The stabilizing effect on policy of this Franco-German axis did not stop here, but was strengthened by the longevity of German ministers of agriculture, which enhanced the influence of Germany in the Council of Ministers. Thus just two Germans held the ministerial post for agriculture in the first 20 years of the CAP, in contrast with a rapid turnover of agricultural ministers from most other nations (Philip, 1990).

What vested this Franco-German interchange with a general aura of acceptability was its ability to appease divergent national interests within the Community. On one side we see this in the pro-market, productivist orientation of much French lobbying. The tone in the British mass media, although also in academic reports of previous decades (e.g. Thompson, 1970), is of a French agricultural sector dominated by small-scale producers, who have significant political sway over national policy-makers. This is seen to stem from farmers' voting power (Wright, 1955), widespread public support for the rural way of life (Tracy, 1989) and the renowned propensity for French farmers to engage in violent demonstration (Naylor, 1994). Such a vision misrepresents the reality of French agriculture: 'The outsiders' image of France remains a rural romantic one rather than that of a highly successful economic power' (Winchester, 1993, 225). In fact, even prior to the Second World War France was an important exporter of cereals (Phillips, 1990) and, while catastrophic harvests in the early postwar years undoubtedly led to calls for improved farm output (Tracy, 1989), soon after the Second World War it was large-scale farms and younger farmers that established primacy over the principal agricultural lobbying organization, the *Fédération Nationale des Syndicats d'Exploitants Agricoles* (FNSEA; Milward, 1992; Naylor, 1994). This organization not only established a position of privileged access to government policy-makers but also supported government programmes favouring farm modernization and an increased export orientation in farm production (Delorme, 1991; Naylor, 1994), both of which have come to characterize French farm policy since the Second World War (Pivot, 1993). Of course, the effectiveness of political actions by its small-scale farmers cannot be discounted (Naylor, 1994), for the political ethos of France has embodied strong support for peasant farmers (Chapter 3). This was even though large landowners had been the dominant force in agricultural policy-making for a long time (Tarrow, 1977; Tracy, 1989). Moreover, the French desire to promote farm improvement and enhance farm production (and so promote agricultural exports) found sympathy with policy-makers in the Netherlands, Luxembourg and later in Denmark (Johansen, 1987; Strijker, 1993). The headlines of national newspapers perhaps dwell on sensational French actions against foreign imports; the burning of British lamb carcasses, the destruction of Spanish fruit, conflict with Spanish wine producers and disputes with German livestock producers are all notable (Winchester, 1993). But in reality the French Government was persuaded by its large-scale farmers to realign agricultural commodity prices to world levels almost 20 years ago (Phillips, 1990).

This policy stance certainly caused concern in other nations, and particularly in Germany. If we are lulled into believing that Germany's interest in a European single market is dominated by considerations of manufactured exports, then it is easy to sympathize with the view that: '... the system of Community support that has made the CAP almost unbearably expensive has fallen on the Germans heavily as major contributors, and they see the French as major beneficiaries making far less attempt to rectify weaknesses than themselves' (Mellor, 1978, 304). But this vision flies in the face of real world events, for, as Hendriks (1994, 156) ably demonstrates: 'Bonn's role in the agricultural reform debate so far has been consistent: in order to protect farm incomes, the FRG has not only agreed to, but has pressed for, measures which while curbing over-production, would leave scope for price increases during the annual farm price reviews'. And if the Germans have been unable to win their case by persuasion, they have not been averse to rejecting price-cutting measures by obstruction or vetoing EC legislation (the most renowned example being in 1985, although its renown comes from the CEC finding the will and the means to circumvent the German veto).

While it might be imagined that the highly fragmented and general small-scale of German farm production lay behind this policy stance (Thieme, 1983), the actual position is more complex. For one, the postwar loss of productive growing areas in what became East Germany coincided with severe economic hardship and rekindled memories of the circumstances leading to the Nazi rise to power, thereby fully justifying measures to promote agrarian production (Mellor, 1978; Tracy, 1989). In addition, in a competitive party political environment, in which attaining and retaining political office often depends upon sustaining the support of delicate coalitions, the fact that the farm lobby represented an organized voting bloc which closely identified with the CDU and CSU parties gave the main farm organization, the *Deutscher Bauernverband* (DBV), significant political influence (78.3% of the farm vote since 1953 is reported to have gone to these two parties; von Cramon-Taubadel, 1993). This was seen not only in a 'special' lobbying arrangement with government officials, but even in its officials holding key positions at the very formation of the (West) German agricultural ministry (Milward, 1992). So, while the DBV has been criticized for its almost exclusive support for large-farm operations, to the detriment of part-time and small-scale producers (Andrlik, 1981; von Cramon-Taubadel, 1993), regular evaluations of its role in German policy-making are summarized by the sentiment, 'agricultural policy is made by farmers for farmers' (Phillips, 1990; von Cramon-Taubadel, 1993). And while any favouritism toward large-scale farms might seem to orient German policy-makers in favour of more commercial, export-oriented producers (as in France), the comparatively small-scale of German full-time farmers worked against this tendency (Table 2.3, page 40). To achieve income levels equivalent to manufacturing workers, German farmers need high prices for their produce, to make up for relatively low productivity and output (Table 2.4, page 41, Table 2.5, page 42). It is no surprise, then, that the mainstay of German agricultural policy has been to raise farm commodity prices. When this has not been feasible, a variety of mechanisms have been used to increase farmer benefits, such as the (1988) payment of DM4.8 billion toward the social security contributions of farmers. This figure was equal to 28% of the total net value added by German agriculture, and meant that a German farmer paid one-quarter of the contribution of a manufacturing worker for an equivalent social security benefit (Tangermann, 1992). That internal politics was central to this process is evident in the two-tracked subsidy system that was introduced in the mid-1970s, which gave greater benefits to the large-scale farm operators who formed the backbone of DBV support (Heinze and Voelzkow, 1993).

The rationale for this policy stance was of less concern to those in other nations than its effects. Germany was the primary driving force behind the development of the EC (Williams, 1991), and its strong bonds with France usually produced compromise policies that brought benefits for farmers in other nations. Thus Italy has reportedly been a net recipient of more the one billion ECU in CAP funding (Brown, 1988). Meanwhile, up to the entry of Britain, and perhaps only excluding Britain and Greece since then, it is probably accurate to state that all member states supported the CAP (the position of Greece is questionable, although it has experienced economic

Table 4.2 Percentage of agricultural holdings by area class size for EC nations, 1989

	1–5 ha	5–10 ha	10–20 ha	20–50 ha	Over 50 ha
Belgium	27.5	17.4	22.9	25.5	6.7
Denmark	1.6	15.1	25.0	38.9	19.4
France	20.3	12.3	17.6	31.2	18.6
Germany	29.0	17.2	21.3	25.3	7.2
Greece	70.0	19.0	7.9	2.6	0.5
Ireland	10.5	14.3	28.8	34.8	11.6
Italy	67.4	16.4	9.0	5.0	2.2
Luxembourg	19.1	9.7	11.6	29.4	30.2
Netherlands	24.8	19.1	22.6	28.1	5.4
Portugal	74.4	13.5	6.6	3.3	2.2
Spain	54.7	17.8	12.1	9.1	6.3
UK	11.3	12.9	15.8	25.7	34.3

Source: CEC (1994f, T128–T131).

problems after accession as it has become a net food importer; Gianntsis, 1994). Particularly for nations with unfavourable farm structures (see Table 4.2) or farm lobbies whose influence easily exceeded their contribution to the economy (Philip, 1990; van Doninck, 1992), it was easy to agree to CAP price increases that bought political support and cost comparatively little to national exchequers. This does not of course mean that all aspects of that policy were favoured. The Netherlands, for one, while supporting the policy's productivist overtones were less inclined to accept guaranteed price supports, since just 45% of Dutch farm products were underwritten by guaranteed EC minimum prices, compared with 60% for the EC as a whole (Hommes, 1990). More generally, what emerges is a picture of the Franco-German axis having a key determining effect on Community agricultural policy. This does not mean, however, that the French and German governments had similar motivating forces behind their agricultural policy stances. In the early days of the CAP there was some clear coincidence of interest, for there was an inner drive to address strategic issues of food security. But by the time member states felt comfortable with their degree of farm self-sufficiency, not only had global pressures for agricultural policy reform intensified but new pressures began to impose themselves on farm policy decisions.

The GATT, fiscal pressure and environmental concern

Given the amount of publicity and controversy it generated, it might be tempting to interpret the recent agreement for the Uruguay Round of the GATT as causing a fundamental change in EC agricultural policy. Indeed, the agreement did coincide with a significant lowering of EC guaranteed prices and a redirection of support. But while the GATT negotiations undoubtedly made an impression on EC policy-making, there is a risk of misrepresenting trends within the EC if the GATT is portrayed in too forceful a light. Rather than highlighting one causal circumstance, it is more appropriate to recognize that the conjunction of three specific forces had increased pressure for change in the CAP as the 1980s progressed, with the principal prompts consistently indicating that support for an expansionary production regime for the CAP needed to be drastically curtailed.

A first and very significant stimulant for reform of the CAP was provided by alterations in German attitudes towards the policy. Neither German policy-makers nor the DBV were immune to change in the economic environment and the need this created for adjustments to the farm economy. Thus, although strongly opposed to structural change in agriculture, the DBV recognized in the mid-1960s that such changes were inevitable and argued that structural adjustment should be accompanied by social policy initiatives. In the 1980s a similar attitudinal transition occurred, when opposition to environmental restrictions on farm production was supplanted with support for such measures so long as its costs were borne by taxpayers or consumers (Heinze and Voelzkow, 1993). Underlying concerns about environmental degradation were particularly advanced in Germany at this time, with the electoral share going to the Green Party (*Die Grünen*) providing one statistical index of this (Grande, 1989). Fears in the Netherlands that it could become the 'dustbin of Europe' were also influential (CEC, 1990a), with the waste produced by its intensive agricultural system providing a particular point of concern (e.g. Figure 2.16, page 62). However, over much of Europe (Last, 1988; Young, 1988), and despite growing scientific evidence on its damaging effects (Pitman, 1992), associations of environmental degradation were slow to attach themselves to farmers' actions. Even had they, policy in nations like Germany revealed a well-established tendency to subordinate environmental concerns to the logic of economic growth (Grande, 1989). So, while influencing public consciousness, environmental concerns alone had insufficient force to draw EC policy-makers into calling for major revisions to CAP principles.

A second stimulant for policy change came from the fact that the CAP increasingly became a victim of its own successes. As production mounted and overtook contemporary demand, the cost of expensive support programmes became more politically sensitive (as late as 1985 agricultural guidance payments accounted for 70% of total EC spending; General

Secretariat of the Council of the European Communities, annual). The charge could easily be made that farmers were being paid to produce commodities in a manner that not only raised serious questions about long-term environmental damage, but also raised the spectre of health problems (as with the 'eggs scare' that led to Edwina Currie losing her British cabinet office or the fears about 'mad cow disease', the latter of which created intra-Community tension as Germany threatened to ban British sales of beef).[5] Set against these concerns was an increasing public awareness that other production techniques (like organic farming) might provide more 'wholesome' food, without generating huge lakes or mountains of surplus food. Some of those who argued this case were not necessarily calling for reduced subsidies for the farm sector, but their arguments did favour farming practices that would lower production levels (Peterson, 1990; Heinze and Voelzkow, 1993; Clunies-Ross and Cox, 1994). At the same time there was considerable anxiety over the cost to taxpayers of farm subsidies; with this apprehension being heightened by reports of substantial fraud in the reporting of farm production (as with reports that the Mafia had netted £150 million from false claims about 1977–1979 Italian tomato production or the 1994 discovery of inflated reports of the subsidized suckler cow population in Corse).

For other reasons, sentiments in favour of reducing farm output found support in other economic sectors within the EC (von Cramon-Taubadel, 1993), as well as in other agricultural trading nations (Ingersent et *al.*, 1994b), so providing a third force for change in the CAP. Thus, although the EC is self-sufficient for many farm products (Table 1.1, page 12), it is still the largest importer of agricultural commodities in the world (Phillips, 1990). Yet agricultural policy in the EC, both at the national level and as pursued by the Commission, has aggressively promoted reductions in farm imports and also increased agricultural exports through raising farm output and subsidizing sales overseas. As regards restrictions on imports, the EC position is well illustrated in litigation to tighten the naming of food products in order to lessen market competition for prestige EC products (Moran, 1993). EC producers, particularly in France, have regularly used the courts to tighten the use of *appelations d'origine*. As one example, stricter regulations are now imposed on wine produced outside the original geographical centre of production, so that an original name can only be used if the end-product demonstrates the special characteristics of the initial growing, such as meeting conditions associated with grape type, method of cultivation, mechanisms for making wine, alcohol content, grape yield per hectare and the organoleptic characteristics of the product. Illustrating how this process can be used to restrict the usage of a product name, the 7500 producers of gruyère de Comté

5 Although in no sense simply a product of farm activities, one indication of these underlying health fears is provided by the increased incidence of food poisoning. According to the UK Office of Population Censuses and Surveys, for instance, food poisoning cases during the first 50 weeks of each year rose from an annual average of 18 837 for 1982/1985 to 55 555 for 1989/1992 (Erlichman, 1993).

cheese have agreed to use only one breed of cow (*pie rouge*) on their farms, not to permit silage from feed and to use little refrigeration for storing their milk on the farm, with each of these specifications taking on the character of production in the core region (Moran, 1993). Any failure of producers outside Franche-Comté to meet these stipulations will mean that they are not making their cheese in the style of the traditional producer region, so the use of the gruyère name can be challenged. Such protectionist measures hold out the prospect of reduced EC imports and, in so far as other producers find restrictions on their use of a 'fashionable' name, additionally raises the possibility of increased EC exports of some products.

This form of protectionism is increasing, though largely for higher quality products, so its impact is still restricted. However, for potential trading partners with the EC, another form of protectionism involving quantity quotas and tax levies on imported goods into the EC has received more vigorous criticism. In addition to basic restrictions, an overt policy developed of responding to the over-production of EC farmers by promoting subsidized exports of surplus commodities. In combination, these two trends have resulted in the EC capturing a larger stake in world agricultural commodity exports, at the same time as it has seen its share of global agricultural imports decline (see Table 4.3, page 130). As a result, distortions to global trade emanating from EC policies have been seen not only in subsidized exports but also in a willingness to 'unload' surpluses onto poorer nations in the form of food aid (Tarrant, 1982). Thus between 1969 and 1978, 43% of the total EC surplus was used in this way (Phillips, 1990, 157). As a multitude of critics have pointed out (e.g. Brown, 1988; Rosenblatt *et al.*, 1988), this destabilized farm commodity markets by decreasing trading prices (through subsidies), reducing export outlets for other nations, and increasing market unpredictability (since the disposal of surpluses was not predictable).

Turning to the policy context of global trade negotiations, the problem that critics of EC policies have faced is that agricultural products have always been treated as something of a cinderella item in trade negotiations. This effectively excluded the farm sector from the discipline of other commodity markets. As Hillman (1992) makes clear, for the origins of this situation we need look no further that the USA. It was here that governmental intervention in the agricultural sector started in earnest, with the tendency for internal agricultural policy to show little regard for international trade consequences being established in the USA well before it came to Europe. Indeed, it was the US that argued for and achieved the 1951 Torquay Protocol and the 1955 waiver (under GATT Article XXV: 5) that effectively gave the USA complete freedom from GATT trading discipline for its agricultural sector. Indeed, in pursuit of its own trading self-interest, the US wrote illiberal agricultural trade practices like quantitative import restrictions and export subsidies into the GATT from the outset (Grant, 1993). Initially, therefore, the fact that the EC followed the same practices was not politically troubling. Indeed, when the distorting effect of EC actions on

Table 4.3 Changes in EC trade in agricultural products, 1985–1992

	1985	1986	1987	1988	1989	1990	1991	1992
Percent total world exports that were agricultural products	13.7	13.4	13.1	13.3	13.1	11.9	11.6	11.5
Percent total EC exports from agricultural products	9.0	8.5	8.4	8.4	8.7	8.5	8.5	8.9
Percent total EC imports that were agricultural products	15.1	15.9	15.0	14.1	12.9	12.1	11.5	11.7
Index of the changing volume of world exports (all products)	100.0	108.4	124.4	140.0	151.3	161.4	163.7	173.3
Index of the volume of world agricultural exports	100.0	105.8	118.6	136.1	144.2	139.9	138.4	145.4
Index of the volume of EC agricultural exports	100.0	110.7	125.7	137.9	152.0	171.6	170.9	192.7
Index of the volume of EC agricultural imports	100.0	113.1	125.7	138.3	136.0	152.5	151.0	158.5

Source: CEC (1994f, T146).

world markets first began to raise concern, the US was not inclined to raise objections as the European Community was believed to have an important geopolitical role in the Cold War (Moyer and Josling, 1990). However, with the accession of Britain, and more especially Spain, trade options for major non-European agricultural producers began to close down, while increased production within the EC lowered demand for agricultural imports.[6] At the same time, the trade deficit of the USA was rising significantly (Cox, 1987; Gill and Law, 1988), with agricultural sales being one of the few bright spots (Nagle, 1976; Hillman, 1985). Certainly, the US did not reveal a noteworthy inclination to reform its own agricultural policies, which revealed notable similarities to those of the 1930s (Hillman, 1992) and shared many essential features with the EC and Japanese policies for agriculture (Grant, 1993). But under pressure to improve its trade balance, the US looked increasingly to

6 As Byman (1990) points out, with new biotechnologies (and especially improved European plant varieties) it has become more feasible for the EC to grow products it used to import from the USA. The significance of the accession of Portugal and Spain to the EC in this context arises because it broadened the climatic range of the EC, thereby raising the prospect of more crop varieties being produced, with a concomitant enhanced threat to US agricultural exports.

raise its farm exports, with the consequence that a series of trade 'wars' broke out with the European Community, as in 1962 over poultry production, in 1981 over flour, poultry, sugar and wheat, and in 1983 over citrus fruits and pasta. Over time, therefore, we find the US moving toward a position that contradicted its earlier demand that agriculture be excluded from GATT trade negotiations. Increasingly, with its need to gain foreign earnings from agricultural exports, the US argued that agriculture should no longer be treated any differently from other economic sectors (Guyomard *et al.*, 1993).

Certainly, when pressure was exerted on the EC to liberalize its agricultural trade policies this came late. In the early GATT rounds (like Dillon and Kennedy), EC officials had been able to utilize negotiations to push member states toward strengthening the CAP in order to protect European agriculture from outside competition (Phillips, 1990). In fact, agriculture only appeared as an issue in GATT negotiations during the Dillon Round (see Table 4.4), and did not appear as a separate agenda item until the Tokyo Round (Grant, 1993). As a consequence, the EC had considerable breathing space before international calls for policy change became irresistible. With widespread recognition that the removal of agricultural trade barriers would raise economic performance in almost all nations of the world, it would be unrealistic to expect Europe's highly protectionist trade policies to have a long-term future. Yet EC negotiators spent most of the Tokyo Round defending the CAP so that it did not have to change. Even if elements within state institutions accepted that the global economy would benefit from deregulating agricultural trade, this did not mean that governments felt comfortable about the consequences of such adjustments. For even if their national economies as a whole were expected to gain, it was inevitable that some economic sectors in each nation would lose out.

Table 4.4 The GATT negotiating rounds

Round	Dates	Number of countries in negotiations
Geneva	1947	23
Annecy	1949	33
Geneva	1956	22
Dillon	1961–1962	45
Kennedy	1964–1967	48
Tokyo	1973–1979	99
Uruguay	1986–1993	105

Source: Ingersent *et al.* (1994b, 1–7, 10).

Here we have the classic dilemma for the EC. Even if national governments accepted the hypothetical argument that the CAP had some negative
effects on growth performance, mostly they were faced with what they saw as
the reality of a positive EC budgetary position resulting from the distribution
of CAP payments (Brown, 1988; Croci-Angelini, 1992). Britain would not
have been swayed by this argument, but the start of the Tokyo Round
coincided with its entry into the EC, which had necessitated that it accept
CAP principles and policies. Otherwise, only Germany can be clearly identified as a net contributor to the EC budget. But here the chances of reform
were at their lowest, as a high priority was given to increasing farm commodity prices.[7] For Italy reductions in protective barriers were likewise anathema,
as they could be expected to heighten a negative agricultural trade balance
(see Table 4.5, page 133), given the rather mediocre performance of the
nation's farm sector (e.g. Table 2.4, page 141). Even in France there was
reluctance to see EC commodity markets open to free competition from
outside the Community. For although France had developed a significant
export market (Table 4.5), after the oil price hikes of the 1970s French farm
productivity did not rise as rapidly as its EC counterparts, and it increasingly
relied on subsidized sales to the Third World for its exports (Delorme, 1991;
Pivot, 1993). Hence, while there was growing recognition that change had to
come, even in the 1980s the deregulation of farm commodity markets was
not strongly supported in EC circles.

So, from the beginning of negotiations for the Uruguay Round (Table 4.4), EC negotiators adopted a defensive position on agricultural trade. [8]
They continued to do so when the deadline for completion of the Round
passed in 1990, with the strength of this sentiment being seen right up to the
eventual signing of an agreement late in 1993. In truth, in adopting a self-
interested, pro-protectionist stance, the EC was acting no differently from
the USA (McMichael, 1994), whose position was accurately portrayed by
Moyer and Josling (1990, 189): 'The US proposal for an elimination of all
trade-distorting support is premised on a belief that US agriculture would be
competitive in a free-trade environment. Given a "level playing-field", the
American [sic] farmer can use superior technology, large farm size, climatic
advantage and sophisticated management skills to sell on world markets'. As
a result, when US calls for the complete abolition of protectionist measures

---

7 In addition Germany has very high surpluses in certain farm commodities. According to
  George (1985, 121), for instance, 73% of the EC butter mountain and 61% of its
  powdered milk surpluses came from West Germany in 1977, with reductions in these
  excess supplies being assisted by such highly subsidized actions as the 1973 sale of butter
  to the USSR for one-quarter of its EC price.
8 This defensive posture gained sustenance from the protectionist measures of other nations
  (such as the US reintroduction of export subsidies for farm products in 1985). But, with
  nations like Australia and New Zealand making savage cuts in their protectionist armoury
  (Cloke, 1989b; Le Heron, 1994), and with the USA making loud inquiries about opening
  the European door to foreign farm products, there is no doubt that the EC (as well as
  Japan) was most resistance to change (Ingersent *et al* ., 1994b).

Table 4.5 Net exports of agricultural products for EC member states, 1990–1992

	With other EC members (in millions of ECU)			With non–EC members (in millions of ECU)		
	1990	1991	1992	1990	1991	1992
Belgium/ Luxembourg	163	–52	1	–1640	–1424	–1183
Denmark	3491	3655	3576	955	1086	1036
France	7043	5853	6320	607	193	1,141
Germany[1]	–9619	–11 214	–11 923	–6205	–6461	–6376
Greece	–525	–380	–425	–244	–143	82
Ireland	1783	1966	2638	649	569	738
Italy	–8468	–9174	–8742	–5395	–5322	–4352
Netherlands	10 847	11 683	11 576	–1061	–1235	–2426
Portugal	–445	–911	–1075	–1275	–1162	–1060
Spain	984	1073	723	–2501	–2733	–2742
UK	–6275	–5207	–5366	–4594	–4211	–3896

Source: CEC (1994f, T159) Note: [1] figures for 1990 are only for the western *länder* of Germany (viz. the former West Germany).

were denied, US attention shifted to securing the greatest percentage reduction in protection volumes and the most favourable base year against which to calculate this percentage, in order to win maximum benefits for US producers (McCorriston, 1993). But while blatant self-interest, as well as mutual misunderstanding (Hillman, 1992), were hallmarks of the Uruguay Round, negotiators' positions were not static over time. For the USA, which acted as the main protagonist for tariff reductions, such changes might seem like little more than the usual compromises that are struck in negotiating processes. But for the EC the adjustments that were made were predicated on genuine shifts in policy priorities.[9]

A variety of factors account for this change. Certainly there was a fear that failure to reach agreement on agriculture could result in retaliatory measures against EC producers in other industrial sectors. This could have had

---

9 This should not be read as implying that free-trade would open the floodgates to US (or other) farm imports into Europe, with EC producers inevitably losing out. Past compromises in trade negotiations invalidated such an assumption, as they provided some access for external products that free-trade would probably remove. For example, without the GATT obligation it had previously signed, the EC would probably import much less soya and cereal substitutes from the USA (these ran to some 30 million tonnes a year in the 1980s). In France in particular it was felt that the importation of these substitutes was a primary reason for weak domestic demand for home-produced wheat, which could be used as animal feed (Phillips, 1990, 159).

particularly significant consequences for high-tech manufacturing exports. Pressure to ensure that this did not happen came especially from Germany, where fears were growing over manufactured exports. Thus Germany recorded its first trade deficit for 10 years in 1991, when manufactured export orders were 10% down on the previous year (Gow, 1991). The importance of this change in circumstances was heightened by the deadline for the completion of the Uruguay Round having passed in 1990, so raising fears that an agreement would not be reached at all. With the impetus for less agricultural protection coming from a higher priority being placed on the prospect of securing a bigger global market for European manufacturing, the hallmark of self-interest is again plain to see. German manufacturers were expected to benefit most from such expansion, while Germany's farmers could be bought-off, as had happened so often in the past, by substituting EC funding with German government payments (von Cramon-Taubadel, 1993). Support also came from the British government, since it was clear that the Treasury would be the recipient of a large budget reduction if CAP payments fell.

However, factors favouring reform went beyond mere national self-interest. One element of 'communal self-interest' was the fear that the economies of Australia and New Zealand could be sufficiently undermined by CAP restrictions on their exports that they would be unable to play an effective military role in the Pacific region (Phillips, 1990). And with US and other negotiators so insistent that nothing would be concluded from the Uruguay Round unless agreement was reached on agricultural trade, compromises had to be reached within the EC or else it risked losing the gains it had fought hard to secure in other trade spheres. These included reductions in special trade concessions for Asia's newly industrializing countries, which were seen to be essential to promote the competitiveness of EC manufacturers. Added to this, in the drive to improve its international market position, the 1992 Single European Market initiative had seen significant reductions in cross-border restrictions within the EC, that fostered a more uniform market for manufacturers. But 1992 was expected to increase inter-regional economic disparities substantially, as manufacturers could serve the whole EC from core locations (Cheshire, 1993; Curbelo and Alburquerque, 1993). One consequence was an agreed extension of EC regional policy, which would increase demand for EC spending at a time when increases in agricultural production had already brought considerable strain on the Community budget. The enlargement of the EC, first to Greece and then to Portugal and Spain, had also increased demands on EC spending.[10] With agricultural over-production placing further stress on the tax burden, steps had already been taken to reduce farm production, as with the imposition of milk quotas in 1984, encouragement for production extensification and

10 As one indication of this, 1991 per capita payments to Greece were 72 ECU for the ERDF, 33 ECU for the ESF, and 26 ECU for EAGGF funds, when the average figures for the EC were 17, 11 and 6, respectively (Plaskovitis, 1994).

reductions in commodity prices (Briggs and Kerrell, 1992; Robinson, 1993; Ingersent *et al.*, 1994a). In combination, these had already begun to reduce the agricultural share of total EC spending (see Table 4.6).

Thus while the 1980s saw Britain as a lone protagonist for sweeping cuts in CAP expenditure, the 1990s heralded broader claims for the same. Particularly influential in this regard was the reunification of Germany in 1990. This led to large increments in public spending, and hence in taxation demands, which put considerable strain on Germany's previous almost open CAP spending commitment (von Cramon-Taubadel, 1993). Now, with the French government believing that its (large-scale) farmers could compete effectively even in a more open market, as its counterparts in Denmark and the Netherlands had long believed for their own farmers, support for a continuance of old CAP policies began to fade. In the case of Denmark and the Netherlands, a long history of significant farm export volumes had made farmers particularly aware that they needed to compete internationally. As Webster (1973) explained, Danish farmers have long been aware that the world does not owe them a living. In addition, as Phillips (1990, 99) put it, it was in the national interest of Britain (as a recipient of relatively few producer benefits), Germany (with the highest taxpayer costs) and France (with few 'deadweight' producers) to end the CAP. Faced with 'excessive' taxation demands, which threatened electoral prospects, particularly for

Table 4.6 Percentage share of EC spending under different headings, 1973–1993

	EAGGF guarantee payments	EAGGF guidance payments	Food aid	Research, energy, industry	Regional policy	Social Fund
1973	70.8	6.3	0.9	0.1	0.0	5.5
1981	62.8	2.9	*	1.7	10.6	4.0
1985	70.3	2.2	0.5	2.5	5.8	5.6
1986	62.9	2.1	1.6	2.2	7.3	7.4
1987	63.5	2.5	1.6	2.7	7.5	7.2
1988	62.8	2.6	0.8	2.6	7.2	6.2
1990	56.8	4.0[a]	*	3.8	10.8	7.7
1991	56.7	4.2[a]	1.0	3.8	11.9	7.3
1992	55.7	Δ	0.9	3.8	Δ	Δ
1993	51.1	Δ	0.9	4.2	Δ	Δ

Source: General Secretariat of the Council of the European Communities (annual). Note: Owing to changes in the manner in which data have been reported over time, some small budget items might have be included under these headings in one year but not in others. Too much weight should not be given to the figure after the decimal point. * no data; Δ these headings were not distinguished for this year; [a] these figures include set-aside payments. Without these the percentages were 3.5(1990) and 3.6 (1991)

right-wing governments in Britain and Germany, the claims of environmental conservation and the need to address potentially greater regional inequalities after 1992 gathered momentum.[11] These saw fruition in the 1992 reforms of the CAP which reduced market prices, decoupled income support from price support and introduced a range of agri-environmental measures (Robinson, 1993; Whitby, 1994).

But for rural areas beyond the 'vital axis' of the EC (Figure 2.9, page 50), there must be fear that reductions in agricultural support will significantly worsen development prospects or alternatively make the adoption of grant-aided environmentally sensitive farming practices more attractive (Whitby, 1994). If these fears arise simply because past payments for regional assist-ance have been meagre compared with their agricultural counterparts (Plaskovitis, 1994), then projections for the reallocation of the EC budget should provide reassurance. While CAP expenditure was 32.7 billion ECU in 1987 compared to 9.1 billion for EC Structural Fund allocations, the comparable figures for 1997 are expected to be 35.6 billion and 29.3 billion (Bandarra, 1993). However, much more than this is involved. In nations such as Greece, estimates point to more than a third of the agricultural product resulting from the protection afforded by CAP policies (Maraveyas, 1994, 66), with the seeming surety of this income contrasting markedly with the limited impact existing regional initiatives have had in the nation (Georgiou, 1994; Plaskovitis, 1994). At the same time, CAP payments can be seen to have placed money directly into the hands of small producers in the form of price subsidies; whereas social and regional funds pass through a complex of intermediaries such as entrepreneurs, training schemes and infrastructure projects, with a less direct and certain effect on rural populations. Hence, while increases in regional policy commitments might be welcomed, it is not guaranteed that they will assuage fears about economic decline in peripheral areas.

Regional inequality, 1992 and all that

Compared with its agricultural counterpart, debate and controversy over regional policy has been muted at the international level. In fact, for the original EC6, regional inequalities were regarded as a major policy issue in Italy alone, for other member states were felt to have relatively homogeneous regional economic structures (Molle, 1990; Tsoukalis, 1993). With the accession of Britain in 1973 attention to the alleviation of regional economic disparities increased; for Britain initially saw regional assistance as a counter-

---

11 The view that environmental concerns have endorsed rather than driven policy changes prompted by economic and political considerations is supported by the lack of clear integration of production limitation and environmental policy (Briggs and Kerrell, 1992). Although recently this situation has begun to change (Chapter 7).

weight to CAP spending (Cheshire *et al.*, 1991). Shortly after this, in 1975, the European Regional Development Fund (ERDF) was created, along with the less favoured areas (LFA) programme, that saw additional agricultural 'guidance' payments going to farmers in difficult agricultural areas such as mountain regions (Clout, 1984). In 1979 EC support for rural areas was broadened by the introduction of integrated development programmes, which formally recognized that the entire economic fabric of less favoured areas, and not simply their farm sectors, were economically vulnerable (CEC, 1989a). These schemes were extended with the start of the Integrated Mediterranean Programmes (IMPs) in 1985, which were introduced partly as a result of Italian pressure concerning biases in CAP funding in favour of northern European products and partly in response to the fears of France, Greece and Italy over the economic impact of the accession of Portugal and Spain (Konsalas, 1992). However, although each of these measures took the Community one step closer to a serious commitment to addressing regional inequities, it has only been since the 1987 Single European Act that legislative attention and monetary pledges have given regional initiatives a real sense of bite.

Up to this point, despite the array of projects, schemes and programmes that have been available, the overall commitment of the EC to the alleviation of regional inequalities has been weak. In the agricultural sector, which has received the vast bulk of EC expenditure, measures to improve the structural position of European farmers in poorer areas have palled beside the huge commitments that have been made to support commodity prices. Indeed, this price support has itself tended to favour those who produce more, in the more advantaged agricultural regions in the north (Bowers and Cheshire, 1983; Franzmeyer *et al.*, 1991). As the CEC (1989a) has admitted, only about 5% of CAP expenditure has been devoted to structural policy, although EC stated intentions are for around one-third of agricultural budgetary allocations to go for structural improvements such as farm consolidation, training young farmers, and enhanced farm investment. However, even if more funding was allocated to structural policies for agriculture, the manner in which EC programmes are fashioned does not necessarily assist problem regions, as schemes tend to operate uniformly across the Community. As a result, more economically developed and agriculturally advanced regions are better placed to make use of such funds (Strijker and de Veer, 1988). At least under the ERDF, Integrated Mediterranean Programmes and LFAs, and more recently under specific programmes such as those for the development of Portuguese agriculture and industry, specific budget headings have been devoted to poorer nations and regions (e.g. Table 4.7, page 138). But these have received slight fiscal allocations compared with other EC policies (Table 4.6), as well as frequently being handicapped as cross-national policy instruments by national quotas on expenditure levels (O'Donnell, 1991). Indeed, with national criteria being used for expenditure commitments and area designations (Wishlade, 1994), EC regional policy instruments have been little more than complements to national policy, wherein EC fiscal

Table 4.7 Budget allocations of EC structural funds by objective and country, 1989–1993

	Objective 1		Objective 2		Objective 5b	
	Million ECU	percent	Million ECU	percent	Million ECU	percent
Belgium	–	–	290	4.3	32.5	1.3
Denmark	–	–	37	0.6	23.0	0.4
France	88	2.5	1227	18.3	960.0	36.8
Germany	–	–	616	9.2	525.0	20.1
Greece	6667	18.4	–	–	–	–
Ireland	3672	10.1	–	–	–	–
Italy	7443	20.6	427	6.4	385.0	14.8
Luxembourg	–	–	21	0.3	2.5	0.1
Netherlands	–	–	174	2.6	44.0	1.7
Portugal	6958	19.2	–	–	–	–
Spain	9779	27.0	1361	20.3	285.0	10.9
UK	793	2.2	2545	38.0	350.0	13.4
Total	36 200	100.0	6698	100.0	2607.0	100.0

Source: CEC DG-XVI (1992).

allocations have been made as additional to national contributions (Molle, 1990). For poorer nations, difficulties in raising their share of the required matching funds has proved problematic in the past (Konsalas, 1992), although it should be noted that the percentage of total costs that come from national governments does now vary for some policies, as with France making larger contributions to initiatives under the Integrated Mediterranean Programme (Winchester, 1993).

In addition, political expediency or ideology have dictated that other nations have not taken up their share of EC funding. Thus according to one report only 60% of budgetary commitments for regional policy over the 1981-1987 period were actually paid out (Franzmeyer *et al.*, 1991, 12), an example being Britain's initially refusal to pass EC funds on to British coal mining areas (Beavis and Halsall, 1992) and its rejection of EC funding for the retraining of workers (Wintour, 1994). As far as problem regions are concerned, an additional weakness in EC funding is the lack of coordination in Structural Fund expenditure (O'Donnell, 1991). There is also the question of the 'undue' weight that has been given to infrastructural investment, which accounts for around half Structural Fund commitments to Italy and Spain, and about a third to Greece (e.g. Table 4.8, page 139). Arguably, manufacturing and services are more likely to promote economic advancement

Table 4.8 Percentage of national Objective 1 allocations by economic sector, 1989–1993

	France	Greece	Ireland	Italy	Portugal	Spain	UK	Total
Business investment and support infrastructure	9	14	28	22	18	11	18	17
Human resources (training and infrastructure)	41	21	24	18	33	24	33	24
Communications infrastructure	16	21	20	11	18	32	20	21
Agriculture and rural development	18	20	16	9	17	14	13	15
Environment	2	2	6	8	2	7	7	5
Tourism	5	3	5	10	3	2	7	5
Energy infrastructure	1	8	–	12	3	2	1	5
Water infrastructure	7	4	1	10	4	7	–	6
Miscellaneous	–	6	–	–	1	1	1	2
Technical assistance and implementation	1	1	–	–	1	–	–	1

Source: CEC DG-XVI (1992, 2).

(Bandarra, 1993), but these have received little direct funding. Moreover, with payments of just 14 billion ECU for regional programmes in 1993, the EC commitment to alleviating regional inequities is meagre compared with the benefits of 216 billion ECU that are expected to result from the Single European Market. Once again, many analysts see these gains going primarily to core economic regions (Grahl and Teague, 1990; Nam and Reuter, 1991; Dunford and Perrons, 1994; Hadjimichalis, 1994).

The fact that this economic core constitutes a small element of the total EC land area has been emphasized by the addition of Austria, Finland and Sweden to the Community (Figure 2.9, page 50). Even before these nations joined, 55% of the EC territory was eligible for LFA additional farm payments (CEC, 1994f) and more than 40% was eligible for regional policy incentives (Wishlade, 1994), that were designed to counter a poor general economic standing (Objective 1), decline in manufacturing (Objective 2) or difficult rural development prospects (Objective 5b; Figure 2.10, page 51). With such variety in area designations, it might be thought that all member states should have a commitment to regional policy in one or other of its guises. But while this might be the case, member state concerns are often national in scope rather than cross-national. One reason for this is that the policies of national governments in the wealthier EC nations have been formulated, and in some instances become entrenched, to address regional disparities which pale beside those that exist across the Community as a whole. This point is readily demonstrated by the accession of Austria, Finland and Sweden. Here, broad ranging policies of regional economic assistance already exist (CEC, 1993b, 1993c), but few of the beneficiary areas will be eligible for EC assistance. As the Commission itself recognized: 'Regional policy, especially in the Nordic countries had a strategic dimension of maintaining the population in remote northern and border regions and the applicant countries wished to continue their main elements of regional policy and enjoy "Objective 1" status. However, only Austria's state of Burgenland satisfies the economic criteria for designated "Objective 1" regions' (CEC London, 1994, 3). Although three extra counties (regions) will be eligible for other forms of structural assistance (Objective 2 or 5b), these new EC members will receive spartan assistance for what they consider to be economic problem zones, a fact that has already led to some public protest in Sweden, which only narrowly approved EC membership in its referendum on the subject. Perhaps the approval of a new Objective 6 will address some of these issues, as this designation is for areas with a population density of less than eight inhabitants per square kilometre. But the past points to EC performance falling well below expectations for northern states. As Strijker and Deinum (1992) indicate, while there might be many areas in the more healthy (north) European economies that do not technically qualify for Objective 1 status, across a variety of measures of economic wellbeing these regions are in a relatively poor position. Indeed, if alternative measures are used to assess regional deprivation, such as value added rather than the Objective 1 measure of GDP per capita (Pompili, 1994), somewhat different maps of economic disadvantage are created (Strijker and Deinum, 1992). The same is true if we compare different EC measures of economic wellbeing (Wishlade, 1994).

Why then has regional policy not become a more central feature of EC policy-making? One reason is that, during the early years of the Community, with the exception of southern Italy, which had begun to receive major infusions of Italian governmental investment in the 1950s (Wade, 1979; King, 1987), the member states were relatively homogeneous in economic standing (Tsoukalis, 1993). Both then and later, Germany, as the likely main contributor to any regional programme, was reluctant to countenance large-scale schemes (Williams, 1991). Moreover, given sensitivities to cost and potential electoral repercussions, nations were reluctant to lose control over this policy instrument. Even after Britain's accession, it proved difficult to agree to Community-wide criteria for area designations (Cheshire *et al.*, 1991). In the early years of the Community it is also possible that some governments were swayed by a belief that the benefits of a European common market would result in a lessening of regional inequalities (Dunford and Perrons, 1994; Hadjimichalis, 1994). Whatever hopes did exist along these lines were quickly dashed by economic recession consequent upon oil price rises and the US decision to come off the gold standard. From this time regional inequalities rose and the time seemed ripe for a strong policy response. However, as the main advocate of enhanced regional assistance, Britain was not inclined to push forcefully for this outcome. At first this was on account of its own dire national economic problems. Then, under the Thatcher governments, the governmental desire to let markets take their course, accompanied by a preference for low taxation and minimal public spending, put pressure on all forms of state expenditure (Gamble, 1988). Where regional policy was maintained, moreover, this tended to take on a much more urban orientation. Similar economic worries underlined the responses of the southern European states that joined the EC in the 1980s. Here, national policies had already been rewritten to mirror EC measures,.in an attempt to extract as much income from the CEC as possible (Arkleton Trust (Research) Ltd., 1992; Cabral, 1993). Additionally, on accession, the CAP gave the rural areas of these nations a substantial income boost (Mykolenko *et al.*, 1987; Economou, 1993). By the time of the Iberian enlargement, the prospects of the Single European Market (SEM) beckoned, and pressure for reduced CAP expenditure offered an opening for improved regional assistance, which was welcomed by many who expected regional inequities to be aggravated by the SEM (Cheshire, 1993; Curbelo and Alburquerque, 1993; Dunford, 1993).

Nationalism and internationalism

In this chapter we have focused on two primary policy fields; agriculture and regional policy. Clearly these are not alone in having an impact on rural development. However, in terms of institutionalized cross-national policy instruments, these are the two fields that have the clearest and most developed initiatives. With respect to manufacturing, EC policy has tended to centre on declining sectors in traditional industrial heartlands (Swann, 1984; Molle, 1988), with the absence of a clear EC industrial policy being apparent for most of the Community's life (Williams, 1991); although fears over competition from newly industrializing countries and Japan seem to have stirred the CEC and the Council of Ministers to appreciate this lacuna (Swann, 1984; Netherlands Economic Institute, 1993). As far as fields like the fisheries, social affairs, tourism and transport are concerned, the EC role is likewise somewhat restricted or else affects so small a percentage of the rural population that their consideration has inevitably had to be played down for a broad overview such as ours. Certainly we should have liked to have considered social policy in more depth but, as with transport issues, available research and policy directions provide an insufficiently distinctive rural focus or their applicability finds some expression in other policy spheres (like social security payments for farmers or infrastructural investment under regional policy). Very likely we will see more attention to tourism and social affairs in future texts on rural Europe, and each of these issues is considered to a certain extent in the following two chapters.

Further development of these themes will be a welcome step forward, particularly for those interested in national imprints on cross-national initiatives. Already we see that national governments are adding a distinctive tone to these fields, with the British government's shrill rejections of the Social Chapter providing the most overt national proclamation; which, for many non-British observers, must be a particularly incomprehensible assertion of national prerogative. Yet at one level this expression of national primacy finds little sustenance in British sociopolitical traditions, save for the easy recourse to xenophobia that characterizes other nations as well as Britain (e.g. Wenturis, 1994). But if dissonance exists between longer term traditions and recent policy enactments in Britain, then this nation is not alone in this regard. From the outset, the nations that came together to create the European Community were hardly united in their political vision. For France and Germany proximity and the possible repetition of global conflagration provided a stimulus for cooperation that was based on a shared need to revive economic fortunes. Perhaps this coming together did offer the prospect that the Benelux countries would be able to stand on a more equal footing with their larger neighbours, yet it also fused nations that had traditionally adopted distinctive rural policies. This is seen in Belgium's (and Germany's) emphasis on a highly protected farm sector compared with the more open-market orientation of the Netherlands (Tracy, 1989). For the Netherlands this vision of agricultural organization still remains strong, yet the Dutch are perhaps in a minority in retaining central tenets of their past traditions in their responses to EC initiatives. For Greece, Portugal and Spain we can perhaps see distinctions from the past expressing themselves in a desire to reject memories of previous authoritarian regimes by embracing a European vision. But acceptance of a strong European orientation was in no sense automatic and in some cases appears to have taken time to develop (Bacalhau, 1993; Wenturis, 1994). Indicative of the lack of association between EC membership and commitments to European integration, we see clear disparities between Danish and Irish views (Table 1.2, page 14), although both took their cues for joining the Community from their close economic integration with Britain. Perhaps in this regard we are seeing some traces of distinctiveness, for aspects of the Nordic tradition that favour an enhanced welfare state and more attention to environmental quality have arguably been less comprehensively addressed by EC policies than agricultural issues. In a similar vein, the more naturalist traditions of British public sentiments have provided a relatively fertile ground for criticism of EC farm policies. This is not withstanding the fact that in some circles these criticisms are reinforced by sentiments of regret over the loss of Empire, which membership of the EC brings into sharp focus. Similarly, the declining significance of the 'special relationship' with the USA is important here. EC agricultural policies point-edly signify the disjuncture of US and European priorities; and Britain is less strongly bonded to agricultural policies on account of a long-term downgrad-ing of agriculture in favour of manufacturing and services.

Taken together, we do see in the evolution of the EC a manifestation of (northern) European agrarian traditions in the aims and practices of the CAP. Perhaps had there been more Mediterranean input in the early years of the EC this policy might have taken a different form. However, by the time nations with different rural traditions joined the Community the ground rules had already been established. Moreover, as these new nations took their place in the Community they were each confronted with particular problematics that restricted their freedom of movement over policy change, even assuming they could accumulate sufficient support amongst other nations for such change. Added to this, at least amongst those nations with relatively weak economies (Table 2.6, page 55), membership of the Commu-nity did bring substantial inflows of EC aid. In a context where (particularly southern European) nations lacked a coherent rural policy, this not only provided sectoral programmes but also a high level of funding to go with them (Arkleton Trust (Research) Ltd., 1992). Effectively, the reservations of Britain and Denmark, along with the ready adoption of EC initiatives following the Mediterranean enlargements, suggest that cross-national policy initiatives have emerged more from (short- or medium-term) political expe-diency than from the infusion of rural traditions into institutional actions. Even within the EC6, policy initiatives arose more from the immediacy of agricultural and balance of trade problems than from a merging of idealized visions. The same goes for the reform of the CAP, which emanates more from cost and trade concerns than from a deep-seated expression of an environ-mental appreciation or a desire for a more 'wholesome' countryside. At the transnational level, then, rural policy has primarily been of the moment rather than bubbling-up from well-grounded historical traditions.

5

Economic globalization and rural transformation

In Chapter 2 we drew attention to contrasting images of rurality, some of which are increasingly 'consumption-driven', especially in an age of mass tourism, while others continue to retain production-related activities at their core. Both these conceptions have long histories, and with some adaptation in tone they have often survived considerable change in the demographic, economic, political and social character of the rural places they are attached to. Changes in patterns of production in rural Europe have been particularly dramatic in the postwar period. For example, the agricultural sector has undergone alterations encompassing mechanization, specialization, greater use of biochemical inputs, decreased labour inputs, heightened governmental intervention and increased direct corporate involvement (Munton, 1992a; Marsden *et al.*, 1993). In similar vein, for manufacturing, Aydalot and Keeble (1988, 1) talk of '... major changes ... at the spatial level', which set dynamic processes of rural industrialization in a dissimilar statistical frame from urban manufacturing decline. Meanwhile, changes that are evident almost globally have seen decline in the share of total production from agriculture and, more recently in much of Europe, also in manufacturing, as a 'post-industrial' service-centred employment structure has emerged. Whatever the employment sector, 'new economic landscapes' are being created in response to pressures for more flexible patterns of work organization (Scott, 1988; Storper and Walker, 1989). For some these new landscapes are argued to be essential if enterprises are to adapt successfully to new technologies, changing consumer preferences and improved input supplies. This is because transformations of productive spaces are said to be required if capital accumulation is to be sustained in the face of a global intensification of market competition (Chase-Dunn, 1989; Dicken, 1992).

It is easy to be persuaded by new tendencies that all is changing; all the more so when academic researchers seek to demonstrate their position at the research frontier by capturing the emergence of new currents in the sea of socioeconomic change. However, with particular regard to the economic transformation of rural space, this should prompt researchers to impose on their analysis two important cautions. The first is that we should not be carried away by the greater emphasis that post-industrial economies seem to place on the (non-agricultural) productive capacity of rural regions. Given decades of writing on the slight prospects of rural zones for attracting significant economic growth (e.g. Berry, 1972), we must resist the inclination to see the shift from an industrial to a post-industrial economy as heralding a clean break from the past which signifies a new economic centrality for rural locations. In reality, metropolitan centres are still advantaged by the workings of the capitalist system, not only by their concentrations of skilled people and their denser, superior telecommunications networks, but also because they enable commanders of economic and political affairs to engage easily in face-to-face contact, as well as accessing more specialized service skills (Dunford and Perrons, 1994). Hence, while economic enterprises might be less bound to primary urban centres than in the past, in significant measure mainstream economic change still binds rural places to proximate urban zones (Cheshire and Hay, 1989; Dunford and Perrons, 1994). A second note of caution concerns the temptation to see new(er) forms of economic organization, like flexible accumulation, lean production and just-in-time systems, as leading to new geographical surfaces. To date, researchers have not produced a bounty of locality-specific evidence that such post-Fordist approaches are widespread (e.g. Hadjimichalis and Papamichos, 1990; Vàzquez-Barquero, 1992b). Moreover, beyond abstract theorization and detailed case studies of noteworthy post-Fordist impulses, the realm of general trends in economic change reveals a continuing strength in Fordist, mass production that brings into question the fundamental attractions of new production spaces (Pecqueur and Silva, 1992; Rodríguez-Pose, 1994).

Central to our discussion of economic change in rural Europe in this chapter is the identification of trends affecting rural areas in various sectors of the European economy. We focus in turn on agricultural change, adjustments in manufacturing industries (including multinational corporations and the recent emphasis on small and medium sized enterprises, which are developments that extend beyond manufacturing), and the increasing economic significance of service industries (and, for rural areas especially, recreation and tourism). In so far as the appearance of new economic spaces can be identified, we question the extent to which this stems from the peculiarities of localities, set within the context of globalization processes, or if amongst their significant ingredients we find nation-specific elements. That general trends take divergent forms and have different impacts in specific nations and localities is clear, but whether they have occurred in a nation-specific manner is a moot point. Very evidently, new formations are unfolding at different paces across economic sectors and between nations. How far these draw on events and practices at a national level will be considered in the conclusion to this chapter. Here again, we will ask if the imprint of national traditions, ideologies and policies have shaped the rate and trajectory of rural change.

Transformation and retransformation in agriculture

Perhaps the most striking feature about agriculture in postwar Europe is its relative decline. The diminishing share of GDP and employment in most European countries since 1945 is readily apparent (e.g. Table 2.2, page 39; Table 2.7, page 59), and is well-documented in standard texts on the European economy (e.g. Williams, 1987; Tsoukalis, 1993). As a key recent EC survey on rural society noted, only 10 of the EC's 166 regions now has a farming sector that accounts for more than 30% of employment (CEC, 1988), with the average for the EC12 standing at just 5.4% in 1991. However, this process of decline has been accompanied by contradictory trends. As noted in Chapter 4, the years immediately after the Second World War were hardly auspicious for agriculture. Postwar food shortages were common and these helped persuade most European governments that strong intervention in the production and/or marketing of agricultural produce was needed, with policy priorities clearly emphasizing the adequacy of food supplies (Tracy, 1989; Milward, 1992). Within five years, production levels had risen above pre-war levels (Williams, 1987), and output in most countries subsequently rose rapidly. But fears about balance of payments problems remained, and the impetus favouring production increases was kept up, with particularly dramatic growth occurring in some nations. Indicative of the trend, cereal yields rose by an average of 30% between 1948/50 and 1958/60 in the 16 countries of western Europe, with increases nearer 50% in Austria, France and Greece; although Greece was starting from a rather low base (FAO, annual). Indeed, in comparison with growth in other economic sectors, the performance of agriculture in the early postwar years was outstanding, with all of the EC9 countries except the Republic of Ireland

Table 5.1 Milk yield per cow (Kilograms per cow) in selected European countries, 1948/50 – 1988/89

Country	1948/50	1988/89
Belgium-Luxembourg	3367	4192
Denmark	3117	6207
France	1730	2915
Germany	2173	4782
Ireland	1933	3960
Netherlands	3657	5876
Sweden	2703	6170
UK	2673	4801

Source: Grigg (1993, 122).

registering higher growth in output per worker in agriculture than in all other sectors combined (Tracy, 1989).

Over the postwar period as a whole, agricultural productivity, both of labour and of land, continued to grow largely without interruption in most of western Europe, and indeed accelerated in many cases in the 1960s and 1970s, although the pace and timing of change varied across states. In the case of cereals, yields more than doubled between 1951/53 and 1991/93 in every EC member state except Denmark, where yields were the highest in Europe at the start of the period. In France, cereal yields nearly quadrupled, from 1.7 tonnes/ha in 1951/53 to 6.5 tonnes/ha in 1991/93, whilst productivity improvements for certain cereals, notably wheat, were even higher. The dairy and drystock sectors saw similar dramatic improvements in both production and productivity. Milk yield per cow rose by between 50% and 100% from 1948/50 to 1988/89 (see Table 5.1, page 146) and production of beef and veal increased eightfold over the same period in Greece, with output doubling in all western European countries except France, Sweden and the UK (see Table 5.2).

The reasons for this robust performance run more deeply than state encouragement, although this was important. Crucial to the changes that affected the sector were massive technological shifts in farming practice. These stretched from the mechanization of farm tasks to the introduction of

Table 5.2 Beef and veal production (1000 MT per year) in selected European countries, 1948/50–1988/90

Country	1948/50	1988/90
Belgium-Luxembourg	124	322
Denmark	101	208
France	938	1864
Germany	483	1659
Greece	10	81
Ireland	63	458
Italy	302	1162
Netherlands	101	499
Portugal	33	120
Spain	99	455
UK	561	977
Austria	70	216
Finland	38	114
Sweden	108	137

Source: FAO (annual).

improved crops and livestock breeds, along with greater utilization of chemical fertilizers, pesticides, insecticides and herbicides (Goodman *et al.*, 1987). Illustrative of the magnitude of change in production practices, the number of tractors in the 16 countries of western Europe grew from 1.8 million in 1955 to 4.6 million in 1965 (Clout, 1971), and had reached more than 7.5 million by 1992 (FAO, 1994 volume). For nitrogenous fertilizers, the average EC consumption per hectare rose by more than two-thirds in little more than a decade from 1969/71 to 1983 (CEC, 1988). Perhaps a more important effect on output came from the development of new technologies that increased yields from animals and crops. Hybrid varieties of cereals illustrate the advantages these technological changes brought, for they are more responsive to fertilizer, have a stronger immunity to disease and, for short-stem crop varieties, carry a larger head of grain. Selective breeding similarly improved the efficiency with which livestock converts feed into meat (or milk), whilst advances in knowledge on disease, plus the development of vaccines and antibiotics, have reduced the morbidity and mortality rates of livestock and so raised productivity and profitability (Grigg, 1993).

These trends towards mechanization and 'chemical farming' reflect a global trend that has seen agriculture become tightly bound with manufacturing (e.g. Table 5.3), which increased its agricultural commitments not simply as a processor of farm products but also as a supplier of inputs for farm production (Goodman *et al.*, 1987; Munton, 1992a). This is worth

Table 5.3 Percentage of agricultural products sold under contracts concluded in advance, 1991

	Pig meat	Calves	Poultry meat	Eggs	Milk	Sugar beets	Potatoes	Peas	Canned tomatoes
Belgium	55	90	90	70	0	100	20–25	98	0
Denmark	0	0	0	0	0	100	4	100	0
France	30	35	50	20	1	100	10	90	0
Germany	0	0	0	0	99	100	0	90	0
Greece	5	3	15	10	30	100	3	85	100
Ireland	0	0	90	30	10	100	10	100	0
Italy	*	*	*	*	*	100	*	*	100
Luxembourg	15	0	0	0	0	0	0	0	0
Netherlands	35	85	90	50	90	100	50	85	0
Portugal	0	0	0	0	0	0	0	95	100
Spain	*	*	*	*	*	100	*	*	*
UK	70	1	95	70	98	100	20	60	0

Source: CEC (1994f, T132). Note: * no data.

emphasizing, given the popular tendency to heap blame on the CAP for (almost all) the ills that are said to beset farming (Body, 1991), ranging from the increasing trend toward monoculture on farms to the removal of hedgerows (Westmacott and Worthington, 1974; Bowers and Cheshire, 1983). The truth is that market expansion preceded the development of the CAP. Moreover, even though the CAP deviated little from the agrarian policies of the EC6 (Milward, 1992), this does not mean that governmental policies were the primary cause for output growth and change in production practices. This is made abundantly clear for those commodities, like horticultural products, that have been scarcely covered by CAP policies and yet show the highest levels of commercialization (Ilbery and Bowler, 1994). In reality, then, economic change in agriculture occurred outside the specific context of CAP policy. Pointedly, we can note poorer performances in states that did not have a strongly developed sense of agrarianism, with sluggish productivity improvements in Greece, Portugal and Spain, that were beyond the CAP for much of the postwar period, along with the laggardly performance of Italian farms compared with its original EC partners (Table 2.4, page 41). Rather than offering the CAP as the scapegoat for agricultural production increases (and ultimately over-supply), we see the imprint of dissimilar rural traditions (Chapter 3), in which the Mediterranean states demonstrate a less systematic commitment to agrarianism than their northern European counterparts, irrespective of whether they have been inside or outside the CAP.

One indication of this is found in the contribution of family farms to agricultural output. Here Britain is in the forefront of farm 'modernization', with just 16.3% of its 1989 production coming from family farms.[1] This is followed by Denmark (38.7%), Portugal (38.8%), Spain (48.0%) and the Netherlands (49.0%). Except for Germany at 54.0%, all other member states recorded percentages above the EC average of 54.4% (Table 2.3, page 40). Non-family farm output does not mirror this pattern, for while Britain (55.4%), Spain (30.9%), Denmark (25.5%) and both the Netherlands and Portugal (22.5% each) again take the extreme positions, their order is not the same as for family farm production (Table 2.3, page 40). Nations with a substantial family farming presence are distinguished by Germany (10.4%) falling behind both Ireland (13.5%) and France (11.0%) in the share of its output coming from non-family farm units. The incongruous position of both Portugal and Spain in these tables should be noted. These nations have agricultural sectors characterized by swathes of barely productive small farms, accompanied by some giant units, with a tradition of the latter being farmed in an unproductive manner (Brenan, 1950; Cutileiro, 1971; Sáenz Lorite, 1988; Black, 1992). This clearly separates these nations from Britain,

---

1 The EC defines a family farm as one in which 95% of the total labour input comes from unpaid family labour, an intermediate farm as one on which 50%-95% of labour is unpaid and a non-family farm has more than 50% of its labour input from paid workers (CEC, 1993a, 138).

Denmark and the Netherlands; whose farm economies have been driven by modernization imperatives. Of course, in the surreal world of British, and perhaps other, popular newspapers, it is the French who are most guilty of failing to pursue productivity improvements. In our view, this interpretation is simply incorrect. For most of the postwar period, the French government has actively promoted farm restructuring in order to improve the efficiency of its farm production (e.g. Clout, 1968; Perry, 1969). Indeed, Lambert (1963) sees even early efforts on this front as more directed and effective than many of its European competitors. A more accurate statement is that France was late to engage in this push for modernization; albeit, as is revealed in its present farm structure (Table 2.3, page 40), output performance (Table 2.4, page 41) and farm incomes (Table 2.5, page 42), irrespective of any backlog, France has reached a point where its farm sector matches if not outperforms that of Germany. The same cannot be said for Greece, Portugal and Spain. For these nations there was a relative absence of policies directed at improving farm structures and agricultural performance, such that these nations now seem to apply EC measures in a much more active manner than other nations; effectively employing the CAP to fill the near void that existed before (Cruz Villalón, 1987; Arkleton Trust (Research) Ltd., 1992; Maraveyas, 1994).

In this regard these nations were some way behind northern European states in seeing state involvement as an active ingredient in the farm economy. For most of northern Europe, state intervention on a major scale came from periods of 'crisis' for economies and polities as a whole rather than for agriculture in particular. The 1930s depression provides a key illustration (Tracy, 1989; Grant, 1993; Just, 1994). Most evidently, until the Second World War stimulated a broadly-based interest in state intervention in national economies, agriculture had seen few government initiatives in periods of economic tranquillity or growth. This does not mean that the path agricultural progress has taken has been uniform across space. Thus, with relatively abundant labour, and often having to rely on small landholdings, early farm productivity gains in Europe generally came through the introduction of a 'nitrogen economy' of chemical inputs, rather than through the widespread utilization of mechanization, as in the United States (Goodman and Redclift, 1991).[2] Effectively, farmers adopted a package of capital-intensive agricultural technologies, which, along with increased market commercialization, provided a rural counterpart to the largely urban-centred mass production practices of industrial Fordism (Goodman *et al.*, 1987). It would be inappropriate to describe this transition as having led to a 'Fordist' system of agricultural production (Goodman and Watts, 1994). The key difference between farming and manufacturing arose from the structure of farm enterprises. They operated within a more insecure and competitive market system than that which is commonly associated with Fordist mass production (Hoggart and Buller, 1987), which provided a key necessity for state intervention to provide stability to farm commodity markets (O'Connor,

1973; Tarrant, 1992). We repeat that this does not mean that state policies were the driving force behind the changing character of farm production. Indeed, instances where governments sought to contradict market forces, as under Spain's Francoist policy of autarky that nearly brought economic collapse until it was reversed (Harding, 1984; Harrison, 1985), have tended to lead to inadequate food supplies and greater agricultural problems in the future (Etxezarreta and Viladomiu, 1989).[3]

Of course, in the same manner that pressures for agrarian 'modernization' have pointed towards more specialized production on single farms, so too have natural resource endowments and commercial pressures interacted to provide a regional specialization in commodity production (Bowler, 1986). Thus, Terluin (1991) notes that all EC countries except Germany produce certain crops at levels well above the European average. To some extent this could be due to differences in consumer food preferences (Table 1.3, page 17), which build on cultural traditions and the ease with which natural resource endowments produce farm products. However, additionally, commercial pressures favour the identification and exploitation of a market niche that enables benefits from any regional comparative advantage to be maximized, either naturally or through deliberate manipulation. Certainly, specialization has not had to depend on natural resource endowments for a region to forge a production advantage. The painstaking improvements to land quality and the nineteenth century transformation of the Danish farm economy away from crop production toward animal products provides a stark example of this (Kamp, 1975; Jones, 1986). A more recent illustration is seen in converting parts of the Mediterranean coast of Spain into large-scale supply zones for the fruit and vegetables markets of northern Europe (Tout, 1990; Salmon, 1992). While climatic conditions have played some part in this transformation, the basis of specialization has come to rely increasingly on the artificial environment of 'glasshouse' cultivation. Terluin (1991) offers a further indication of such changes when noting how Britain shifted its specialization in livestock to a higher concentration on cereals over the years 1974–1986, at a time when Luxembourg was simultaneously moving in the opposite direction. Such specialisms are readily apparent when we examine national self-sufficiency in single farm commodities. Notable in this regard are the emphasis on pig production in Denmark (whose

---

2 As time progressed, farmers came to use both these mechanisms for increasing productivity (as well as seeking to bind their farm practices more strongly to manufactured innovations). But with land availability restricted, this transformation called for a key change in the European rural space that necessitated sharp reductions in farm numbers, more industrialized production practices and major surgery on the landscape (e.g. Westmacott and Worthington, 1974; Bowers, and Cheshire, 1983; Kostrowicki, 1991).

3 We do not mean to imply here that there is no alternative to a capitalist mode of agricultural production. Merely to point out that the experience of Europe has been that when attempts have been made to 'buck' market trends, improvements in agricultural production and productivity have tended to suffer; with the USSR providing a pointed illustration of this (e.g. Shaposhnikov, 1990).

self-sufficiency value for 1991/92 was 381%), dairy products in Denmark and Ireland (363% and 352% respectively for cheese, and 165% and 1225% for butter), beef in Ireland (992%), veal in Britain (406%) and the Netherlands (513%), citrus fruit in Spain (231%), grain in France (226%) and vegetables in the Netherlands (254%; all figures from CEC, 1994d, 264–265).

The wider farm economy

Based on trends in productivity and commercialization, the picture that can be painted is of a thriving agriculture. However, it has to be borne in mind that production and productivity increases were essential for farm viability. They did not generate substantial increments in income that enabled the farm sector to lead the postwar recovery of Europe's rural areas (e.g. Collard, 1981). Indeed, while the income of farmers did follow a rising trend up to the mid-1970s at least, the economic performance of the farm economy was more problematic in the 1980s (CEC, 1993a), and has been even more troubled in the 1990s. One illustration of this is provided by the net value added by an average farm work unit over time (see Table 5.4, page 153 and 154). Here we see that agricultural performance in a few nations (like Greece, Ireland and Spain) has continued to improve, despite quota restrictions, price reductions and rising input costs (Table 2.5, page 42). For countries like Britain, Germany, Italy and the Netherlands, on the other hand, the period since the 1980s has seen a flat or even declining trend, with other nations not producing more than a shallow upturn in their standing (Table 5.4). And these figures are recorded for those who actually remain in the farm sector. For what has to be recalled is that the ability of the farm sector to show impressive economic figures per worker stems in good part from the huge reductions that have occurred in both farmer and farm labourer numbers (OECD, 1994b). Of course, the number of people working in agriculture has been in decline in most European countries since the nineteenth century. Nevertheless the postwar years have seen a notable acceleration of this trend, with an estimated 26 million agricultural jobs lost in western Europe between 1950 and 1980 (Grigg, 1993). What lay behind these figures were a variety of trends. As far as employed agricultural labourers are concerned, retention has long been recognized as being sensitive to wage rates in other economic sectors (Drudy, 1978; OECD, 1994b), so the availability of alternative nearby job opportunities has commonly prompted occupational change. For this reason, those local governments which are controlled by large-scale farm operators might be expected to resist the introduction of new job opportunities in their patch (Saunders *et al.*, 1978; Broady, 1980; Wilson, 1992);[4] for such labour losses often necessitate expensive investments in machinery in order to enable production to be sustained with a smaller workforce

Table 5.4 (Part A) Net value-added at factor cost per annual farm work unit in EC nations, 1974–1992 (average 1984/86 = 100)

	1974	1975	1976	1977	1978	1979	1980	1981	1982	1983
Belgium	81.9	85.8	101.1	84.1	90.7	82.1	86.8	95.2	100.4	108.3
Denmark	65.3	52.9	54.7	63.9	69.5	60.4	65.3	74.8	90.6	77.5
France	102.0	94.8	93.8	91.6	95.0	97.2	88.0	91.2	107.6	100.3
Germany	104.6	118.8	123.4	118.2	113.3	101.3	90.3	91.1	111.0	89.4
Greece	71.0	72.4	78.5	75.4	85.3	80.8	92.0	97.4	101.1	90.9
Ireland	87.4	107.2	101.8	125.3	129.8	108.2	88.3	88.6	96.7	101.1
Italy	87.8	90.1	84.3	88.9	89.9	96.0	109.3	106.7	106.9	112.0
Luxem.	58.7	64.6	57.2	71.2	70.9	73.6	68.0	76.5	105.4	93.2
Nether.	77.0	82.9	90.3	86.0	84.7	78.1	75.2	92.3	96.9	93.4
Portugal	*	*	*	*	*	*	95.7	90.0	100.5	97.3
Spain	63.6	71.6	76.3	87.1	88.3	80.3	86.4	76.9	88.9	89.1
UK	109.8	104.8	114.3	106.6	101.2	98.0	92.0	97.2	104.3	93.0

Source: CEC (1994f, T41). Note: * no data.

(G. Clark, 1991). Meanwhile, many farm workers have not left the sector of their own free will but have been shed by their employers in a quest to maintain farm profitability (e.g. Drudy, 1978). Of course, where farms are small, it is often the farmer who has had to leave, owing to the inability of the farm unit (or the farmer) to provide a living wage (Errington and Tranter, 1991). In instances where alternative job openings are available locally, this has commonly resulted in farmers taking off-farm work, as in much of central and southern Germany (Arkleton Trust (Research) Ltd., 1992). An alternative, encouraged more recently by government policies, is for farmers to seek to extend the range of activities they undertake on their own farm, such as moving into farm tourism (Gannon, 1994).[5] In the absence of such options, or if the farm population needing extra income outstrips the possibilities of meeting demand, the solution has often been long distance migration. Flows from southern to northern Europe are particularly notable in this regard in the 1960s and 1970s (Douglas, 1976; Salt and Clout, 1976; Manganara, 1977). Such outflows could have particularly important rural development

4 Alternatively, as Groenendijk (1988) reports for the Netherlands, it might simply be that agricultural leaders wish to see local resources used to assist agricultural advancement, so that issues like service provision or alternative jobs receive little attention.

5 As work published for the CEC by the Arkleton Trust (Research) Ltd. (1992) showed, it is nonetheless the case that opportunities for off-farm work and for on-farm non-agricultural income generation are more highly constrained in southern Europe, where farm incomes are generally lower, than in the northern states of the EC.

Table 5.4 (Part B) Net value-added at factor cost per annual farm work unit in EC nations, 1974–1992 (average 1984/86=100)

	1984	1985	1986	1987	1988	1989	1990	1991	1992
Belgium	104.4	99.4	96.2	90.8	98.6	123.2	111.8	110.5	104.6
Denmark	103.3	96.2	100.5	81.6	83.3	99.5	93.5	85.6	76.6
France	99.4	100.0	100.5	101.5	99.1	115.1	120.1	115.6	114.6
Germany	102.4	92.6	105.0	87.8	109.1	129.7	115.4	108.1	110.8
Greece	98.8	101.3	99.9	102.2	112.4	124.4	104.5	131.4	118.1
Ireland	112.3	97.7	89.9	109.8	128.6	131.5	134.2	124.3	144.9
Italy	101.0	101.4	97.5	98.8	93.3	99.2	91.0	102.0	97.9
Luxem.	96.6	100.0	103.4	105.0	106.8	122.6	115.3	98.2	105.0
Nether.	100.9	95.6	103.5	85.5	88.3	103.1	98.8	98.6	86.6
Portugal	99.6	98.4	102.1	99.8	84.0	98.2	104.7	95.7	87.4
Spain	100.0	102.4	97.6	104.6	120.5	121.1	127.0	129.4	117.0
UK	112.9	89.9	97.2	95.7	86.0	97.9	98.9	95.8	97.9

Source: CEC (1994f, T41). Note: * no data.

consequences, for the emigrants often sent substantial remittances to their family (see Table 5.5, page 155), as well as returning at festival times to their home village to engage in rather lavish displays of conspicuous consumption that provided a needed economic lifeline for local bars and shops (Mansvelt Beck, 1988; Cavaco, 1993). In certain instances, the outflow of workers also released land that enabled those farmers that remained to rent or buy extensions to their productive capacity (Brandes, 1976; Reis and Nave, 1986).[6] None of this points toward emigration inevitably bringing positive benefits for economically troubled rural communities, for it has been long recognized that population losses can provoke a downward spiral in service provision and employment prospects (Parr, 1966). In addition, while some parts of Europe saw labour shortages from emigration encourage farm mechanization (e.g. Collier, 1987), the loss of a significant proportion of the agricultural workforce could undermine efforts to modernize the sector (King, 1984; Black, 1992), especially when this was associated with the loss of younger, more highly skilled or productive workers.

6 Potentially, there is a temporal specificity to these points, which is most evident for emigrant remittances, which declined significantly when 1970s economic recession led to many migrants returning home (e.g. Table 5.5). At the same time, farm improvement was not everywhere apparent, as many emigrants wished to retain their farm plots in order to have a base for their return to their local community (e.g. Manganara, 1977; Took, 1986).

Table 5.5 Percentage contribution of invisible earnings in the Greek economy, 1970–1990

	1970	1975	1980	1985	1990
Tourism	20.4	24.5	32.1	37.6	36.6
Nautical	29.1	29.6	29.5	23.6	17.4
Emigrant remittances	36.2	27.4	17.6	18.2	18.0
Other	14.3	18.5	20.8	20.6	28.0

Source: Briassoulis (1993, 294).

In reality, differing agricultural performances over both space and time relate in part to the dissimilar economic circumstances and farmer strategies in various European countries, for in effect '… the space/time disparities currently observed in Community agriculture reveal two contrasting models, one based on growth and investment, the other on an exodus of labour and input reduction in a context of stagnating production' (Bureau et *al.*, 1991, 3). Examples of the former include the Netherlands and Ireland, which over the period from 1967–1987 saw higher than average growth in production and productivity based on high levels of capital investment and, for Ireland, particularly favourable circumstances for the dairy industry immediately after joining the EC. In contrast, countries as diverse in their agricultural base as Belgium, Luxembourg, Italy and Spain have seen slow production growth, with increases in productivity based on the removal of labour from the sector. However, care must be taken in allocating nations to one or other of these categories too rigidly. For example, in an analysis of growth in the use of three key agricultural inputs, feedstuffs, fertilizers, and energy, from 1974–1987, Terluin (1991) notes that Spain shows the highest rate of increase in the first two, and the second highest rate of increase for the third after Greece, whilst Italy too has seen a significant increase in energy inputs into farming. Meanwhile, amongst the group of countries assessed by Bureau and colleagues (1991) to have low levels of investment, impacts on farm incomes have varied (see Table 5.6, page 156). Thus Luxembourg and Spain have both seen losses of labour from the agricultural sector combined with rising incomes for those remaining in the sector, whereas at the other extreme, Italy's remaining farmers suffered a fall in net income, at least during the 1980s.

The 'industrialization' of agriculture has certainly been beneficial for some farm enterprises, but it has also had destructive effects. Perhaps this is the nature of any economic change, yet it has produced a peculiar diversity of responses within the farm sector. Marsden and colleagues (1992b) suggest that these can be characterized as three broad types of agricultural endeavour; namely, 'agricultural reproduction', 'professionalization' and 'disengagement'. The first represents a 'retreat' into a more subsistence-based

Table 5.6 Farm incomes in the European Community, 1982–1989

Agricultural incomes in 1982	Trends in agricultural incomes, 1982–1989		
	Increase above EC average	Increase below EC average	Incomes decline
More than 150% of EC average	Luxembourg Netherlands	Belgium	
Between 75% and 150% of EC average	Denmark Germany Ireland	France	Italy UK
Less than 75% of EC average	Greece Spain		Portugal

Source: CEC (1993a).

production, with non-agricultural activities providing additional income. Professionalization is associated with the full adoption of business methods, although again perhaps accompanied by non-agricultural activities as part of a business plan for the farm enterprise. Disengagement represents a complete move away from agriculture as a serious economic concern, although land may be retained as a hobby farm or to supplement income earned elsewhere.

Certainly, those farmers who adopt a professionalized role do not have only one option to follow. Most commonly, however, the path that has been trodden has gone beyond the introduction of mass production techniques based on packages of capital intensive technologies. Often a level of industrial control and organization has been introduced that would make the average citizen feel that the label 'agricultural' was hardly appropriate at all (Bowler, 1992). Goodman and Redclift (1991) divided this process into two distinct elements; namely, 'appropriation', or the transformation of agricultural activities into industrial processes based on purchased inputs, and 'substitution', which principally involves replacing 'natural' farm products with chemical and synthetic raw materials. An example of the former is provided by horticultural production in parts of the Netherlands, such as around the Hook of Holland or in the Leiden-Haarlem-Amsterdam region, where the cultivation of flowers and vegetables is undertaken in artificially controlled environments that are divorced from seasonal rhythms or local natural conditions such as soil, temperature and rainfall (Bowler, 1992). A comparable situation exists for livestock production, especially for pigs and poultry, where industrial style buildings house the animals and manufactured feedstock has replaced the use of 'natural' pasture (e.g. White and Watts, 1977). Meanwhile, the development of biotechnology promises an almost complete separation of the production of food from physical resources such as land, soil and climate (Goodman and Wilkinson, 1990).

The question that arises for the future is whether such practices can continue. As discussed in Chapter 4, budget costs and pressures from trading partners, amongst other factors, have already combined to provoke the EC to lessen its agricultural import restrictions. Further strengthening of the global nature of agricultural markets will reveal the uneven ability of European farm sectors to compete on world markets, with the result that a shift in commodity production should occur. Where European costs of production are comparatively high, farmers might still be able to retain their interest in a production sphere if they change their farming practices. This does not have to mean an intensification of industrialized farming mechanisms, which is likely to be expensive if farmers manage a significant reduction in their costs such that they can compete effectively against producers with larger farms and more favourable environmental conditions for crop growth (as in the case of European cereal growers competing with their counterparts in the USA). Rather the option might be to develop a market niche, such as through focusing on high quality, prestige products (Moran, 1993), or else change the central principles of their existing farming methods to tap into the as yet small but growing market for organic food (Clunies-Ross and Cox, 1994). The extent to which this will occur is questionable, and almost certainly will rely in good measure on state encouragement (Peterson, 1990). For individual farmers, evidence is already confused over the differential willingness of groups to adopt organic or biodynamic farming practices. In Britain, for instance, support seems to favour the idea that it will be older farmers that will more readily adopt environmentally sound farming practices, especially if they do not have a successor (Potter and Gasson, 1988; Potter and Lobley, 1992). In contrast, German evidence suggests that younger farmers are more prone to accept such practices; although the real distinguishing feature in attitudes towards environmentally sensitive farming is reported to be their greater acceptance by women (Schmitt, 1994). But neither of these groups point toward widespread acceptance of such practices. In Britain this is because adoption of organic farming appears to involve those who are leaving farming and predominantly occupy less productive farms anyway. In Germany, meanwhile, this is because few single-operator farmers are run by women and few partnerships are comprised solely of women.[7] In turn, the prospects of women bringing influence to bear on the production decisions of farms governed by male-female partnerships seems unlikely, for although some foresee a greater future role for women in farm businesses (Gasson, 1992), others identify a withdrawal of women's labour from farm operations (Siiskonen, 1988; Repassy, 1991; Hillebrand and Blom, 1993); with little evidence as yet of gender equality in decision-making on farm production issues (Whatmore, 1991; Haugen, 1994).

---

7 Even in socialist Yugoslavia, First-Dilic (1978) found that only one-eighth of farms were headed by females, and then they tended to be landholdings on average smaller than those controlled by males (and this was despite the fact that women made up 54.6% of the agricultural workforce in 1971). Providing some comparable figures, Gasson (1981) reports that 11% of farmers in England and Wales in 1971 were female, a rise from the 1951 figure of 7%.

But it is not perhaps the actions of those farmers who adopt a professionalized role that points most forcefully toward the future for agriculture in the rural economy. With falling incomes in a number of farm sectors (CEC, 1992a), it is no surprise that a major trend is now for many farmers to seek disengagement from agricultural production (Errington and Tranter, 1991; Marsden *et al.*, 1992b). This does not have to result in them leaving agriculture, for as Marsden and associates (1992b) note, farmers can adjust to changes in the farm economy through pluriactivity. A classic exposition of this is Franklin's (1969) work in Baden-Württemberg, which was one of the first to identify a group of 'worker-peasants' who maintained a foothold in agriculture, while taking jobs in the rapidly expanding manufacturing sector (see also Holmes, 1989). For Saraceno (1994), commenting on the Udine region of Italy, such off-farm employment has helped fossilize traditional labour patterns within agriculture. But as a general observation we need to treat this point with care, for patterns of off-farm (and non-agricultural on-farm) work are quite diverse, as well as being regionally specific. Thus, in Greece, Efstratoglou-Todoulou (1990) found evidence that pluriactivity amongst farmers was promoted by necessity in areas of small-holdings which provided an insufficient income on their own, but was also associated more with choice in favoured agricultural zones. Providing a French illustration of similar regional differences, Campagne and colleagues (1990) record how the larger scale farm operations of Picardie are associated with the utilization of farm resources to improve income generation from non-agricultural work, whereas non-agricultural resources are essential to provide the funds to improve farm capacity in Languedoc, and in Savoie pluriactivity is needed for economic survival. At an individual farm level, evidence of similar considerations has been provided in a three county English study by Evans and Ilbery (1992). They found that large-scale farm operators offered farm-based accommodation as part of a strategy for capital accumulation, whereas more marginal farm operations were likely to engage in this activity because their agriculturally-generated income was low. Viewed across the continent, a baseline survey of some 6000 farm households in 24 study areas around Europe found that on more than half of all farms surveyed, farming provided less than half of household income, with small farms generally showing the lowest proportional contribution from agriculture (Mackinnon *et al.*, 1991). Hence, while the cost-price squeeze on farm household economies points toward the necessity for smaller producers to leave agriculture, across Europe we find evidence that some farmers have adjusted to these pressures by engaging in pluriactivity. While in some areas this involved the utilization of potential agricultural resources for broader purposes, such as farm accommodation, in many instances the ability to stay in farming was conditioned by what was happening in other sectors. Here, farmers undoubtedly benefited from decentralization trends within the manufacturing sector.

Established manufacturers and diffuse industrialization

As agriculture saw its employment and economic significance decline (Table 2.2, page 39; Table 2.7, page 59), rural areas were faced with mass depopulation (as in much of the twentieth century; e.g. Saville, 1957; Bontron and Mathieu, 1977), unless alternative means were found for sustaining local populations. In the early post-war decades, both researchers and government officials seemed to hold out little hope that manufacturing would be instrumental in meeting this need. For one thing, local economic leaders were held to be disinclined to promote manufacturing growth, either because their own economic interests could be threatened (Graziano, 1978; Saunders *et al.*, 1978; Broady, 1980) or else because they were unwilling to take the risks to establish their own enterprises or were insufficiently persuasive to attract outside investors (Paine, 1963; Chadwick *et al.*, 1972). These restraints were matched by a general sense that rural centres provided an insufficient infrastructural base to support manufacturing. This was said to be due to the workforce not being trained for factory work, to inadequate transport, communication and social facilities, and an inability to provide a network of needed business contacts, a sufficiently innovative business climate and the complex of business services that larger urban centres offered (Berry, 1972). Such was the tone of this work that for many rural areas the arrival of a new manufacturing plant was something to be lauded over, with considerable analysis of their impact over extensive geographical areas and various economic sectors (e.g. Lucy and Kaldor, 1969; Eccles and Fuller, 1970). Yet by the 1980s the weight of evidence was that rural manufacturing growth was commonplace, or at least that rural areas were holding their own in the face of massive job and output losses in cities (Keeble *et al.*, 1983). This turnaround was identified as a major factor behind population growth in areas that had long experienced depopulation (Fielding, 1982; Cross, 1990).

One feature that was said to lie behind this shift was dramatic change in the organization and operation of manufacturing activities. These innovations have had profound effects that extend into every sphere of economic, and indeed political and social, life, such that simple labels like the industrialization of the countryside or the emergence of a post-industrial economy, do not capture the magnitude of economic change which has occurred and is occurring. Since 1945 an array of interconnected processes have brought about these profound changes. These include: the growth and concentration of economic power in the hands of multinational corporations (Chase-Dunn, 1989; Dicken, 1992); the changing balance of national and corporate economic power (Taylor, 1989; Tussie, 1991; O Tuathail, 1993), including strong challenges to the industrial base of Europe from newly industrializing

countries (Auty, 1993; Hobday, 1994); the increasing integration of manufacturing units across national frontiers (Clark, 1993; Dicken, 1994) and the
associated importance attached to 'inward investment' by governments
(Netherlands Economic Institute, 1993); productive decentralization; the
shift from 'Fordism' to 'flexible accumulation' (Scott, 1988); the growing
centrality of financial markets in the world economy (Corbridge *et al.*,
1994); and changes in the role of small- and medium-scale enterprises (SMEs;
Bagnasco, 1977; Andrikopoulou, 1987; Vàzquez-Barquero, 1992b). While
none of these have any special rural orientation, they have made their
presence felt in major ways in many rural areas, leading in some cases to
urbanization of the countryside and a 'loss' of rural space (Paloscia, 1991),
but in others to a changed rural character which nonetheless remains
distinctively rural (Lewis and Williams, 1988).

The decentralization of manufacturing from physically urbanized zones,
in which rural regions 'gain' most in terms of output and employment, is seen
by Keeble and associates (1983) as one of the key economic changes in
Europe in recent decades. Taking-off in the 1960s and 1970s, although with
earlier roots in certain countries in northern Europe, production decentralization has been characterized by Williams (1987) as having four principal
elements: direct decentralization of 'branch plant' activities to areas that can
provide cheaper labour (Peck and Stone, 1994); direct decentralization to
greenfield sites to expand from a cramped urban factory or introduce new
technologies requiring different plant building configurations (Fothergill
et al., 1987); indirect decentralization of productive activities to the informal economy in metropolitan areas; and indirect decentralization to SMEs in
nonmetropolitan regions (Andrikopoulou, 1987). Examples of productive
decentralization are cited by Williams mainly in a southern European context, where the rural industrialization of the so-called Third Italy (Bagnasco,
1977) provides the classic example, but which can be illustrated as well by
events in Spain (Precedo Ledo and Grimes, 1991; Granados Cabezas, 1992;
Vàzquez-Barquero, 1992b), Greece (Andrikopoulou, 1987; Hadjimichalis
and Papamichos, 1990) and Portugal (Lewis and Williams, 1987). But
productive decentralization is by no means confined to southern Europe,
with perhaps the most comprehensive evidence of specifically rural manufacturing development being available for the UK (Keeble *et al.*, 1992; Townsend,
1986, 1993).

In a British context, David Keeble and his colleagues identified that the
birth and growth of small, new manufacturing enterprises held the key to
rural industrialization in the 1970s and 1980s, rather than relocation or
inward investment by externally-owned firms. Indeed, there has even been a
suggestion that rural areas have higher rates of new firm formation than
urban centres (Gould and Keeble, 1984); although more recent research
provides less convincing support (Keeble and Walker, 1994). Certainly,
across Europe as a whole, evidence on spatial differences in new firm
formation between rural and urban areas is not clear cut, albeit part of the

problem here is the difficulty of distinguishing 'rural' areas in the statistics used. For example, in Ireland, O'Farrell and Croutchley (1984) identified a strong negative correlation between the rate of new firm formation for 1973–1981 and the proportion of the population living in towns of more than 5000 inhabitants. This association between rurality and high levels of new firm formation is confirmed in a more recent study by Hart and Gudgin (1994) covering 1980–1990, with the presence of professional and managerial workers in a region also proving influential. In contrast, three measures of 'urbanity' used by Garofoli (1994) in a study of Italy, namely population density, net in-migration and the proportion of young people in the population, show no significant correlation with new firm formation, whilst there is a significant positive correlation with population density in Germany (Andretsch and Fritsch, 1994).

Adding further confusion to any attempt to generalize across nations, a complicated picture emerges in the literature on France. Guesnier (1994) confirms the importance of 'human capital' in new firm formation, with the proportion of the population with university degrees and the proportion of middle managers being especially important. However, also of note is the rate of population growth. Indeed, this factor shows the highest correlation with rates of new firm formation, and reflects the emergence of new SMEs in expanding urban-centred regions. In fact, the position of rurality and urbanity is unclear, with rural Corse topping the list of *départements* with high rates of new firm formation per active worker, whereas the Paris-centred Ile de France heads the list when new firm formation is standardized by the number of existing firms (Guesnier, 1994). With the three southern regions of Provence-Alpes-Côtes d'Azur, Languedoc-Rousillon and Aquitaine also having high rates of new firm creation, we see a similar rural and urban contrast to the Corse – Ile de France comparison. Part of the problem, as many analysts of new firm formation indicate, lies in the geographical units for which data are available. In Germany, for instance, the rural Sauerland provides an example of rural industrialization – in this case based on specialized light engineering – but as this area lies in the highly urbanized state of Nordrhein-Westfalen a 'rural effect' is unlikely to show-up given that statistics are only available for administrative regions (Andretsch and Fritsch, 1994). But even were data available, simplistic ascriptions of firm formation rates to geographical milieu are questionable, as the legislative, financial and sectoral context of manufacturing companies are likely to be central to the patterns found. This is readily apparent in the contrasting findings that turbulence in firm numbers (high births and deaths) is associated with later economic decline in Germany (Audretsch and Fritsch, 1994) but is positively allied to economic growth in Sweden (Davidsson et al., 1994), with both firm creation and product innovations being identified as being dependent upon the economic sector in which companies operate (Davidsson *et al.*, 1994; Fischer *et al.*, 1994). What we also see in the pattern of firm creation rates is a clear nation-specific imprint. In France this is said to result from the long-enduring attachment to Fordist production principles, which restricted the innovative potential of SMEs (Guesnier, 1994). In Italy precisely the opposite applies. Here, building on governmental mistrust and even antagonism toward large business organizations, policies have long been highly advantageous for SMEs (Furlong, 1994).[8]

A further example of state influence on manufacturing policy is provided by Ireland. Here, despite a series of policy shifts, significant periods of state initiative can be identified in which rural areas in the west of the country were targeted for industrial development, and where such development did filter through to even the most remote rural regions (Gillmor, 1986). These policies can be traced back to a 'ruralist' ideology that has characterized most Irish governments since the nation's independence from Britain (Leeuwis, 1989; Breen *et al.*, 1990). They took their first concrete form with the election in 1932 of a Fianna Fail government committed to rural industrialization as part of an import-substitution policy for the economy as a whole (Curtin and Varley, 1991). This policy however proceeded in fits and starts. In 1952, for example, the Undeveloped Areas Act provided preferential government grants to rural areas in the west of Ireland, but the rural and regional element of this policy was watered down in 1956 as grants were extended to all areas of the country, and a 'growth pole' approach was adopted to development. The most important period of pro-rural industrial policy then followed from 1964 to the mid-1980s, as the Second Programme of Economic Expansion provided special grant aid for the dispersion of industrial sites, overseen by the Industrial Development Authority (IDA). Since 1984, policy has switched away from rural areas to concentrate on attracting inward investment in high technology industries to serve an international market, resulting in a re-favouring of urban centres in the east and south of the country.

The lesson is that the 'policy-on' period favouring rural development did see expansion in rural manufacturing at its strongest (Curtin and Varley, 1991), with nearly a doubling of employment between 1961 and 1983 in 'designated areas', principally in the rural west of the country. This accounted for two-thirds of the growth in manufacturing jobs in Ireland during this period. Moreover, much of this development occurred in relatively small population centres, in accordance with the aims of the Third Programme for Economic and Social Development (1969), which noted that most small industries '... can operate effectively in relatively small centres thus providing a valuable outlet for rural manpower' (Curtin and Varley, 1991, 106),

8 As Furlong (1994) makes clear this should in no sense be read as meaning that governmental policies were behind the emergence of flexible production units in the so-called Third Italy (Amin, 1989). What government policies did do was set up an environment that was more conducive to such enterprises than that of, say, France. Yet this was not a deliberate economic policy but owed much to the ability to use small companies for the political purpose of patronage distribution (Silverman, 1965; Littlewood, 1981; Furlong, 1994).

although the role of factors such as incentives provided by the Shannon Free Airport Development Company (SFADCo) and Ireland's entry into the EC in 1973 must also be taken into account (Lucy and Kaldor, 1969). A similar conclusion about the potentially beneficial role of government rural policy on new firm formation is reached by Keeble and Walker (1994), who argue that there is a 'Wales effect' in the spatial distribution of new firms in the UK, possibly resulting largely from the activities of government quangos like Mid Wales Development and the Welsh Development Agency. By contrast, it is claimed that in France and the Netherlands processes of endogenous economic growth and local policies to encourage them are proving more effective mechanisms of economic decentralization than 30 years of state-led regional policy (Cappelin, 1992; Stöhr, 1992).

Apart from possible governmental influences, another reason for caution in linking geographical environments to economic initiatives comes from a lack of uniformity within rural areas (Hoggart, 1988; Hodge and Monk, 1991). In the European Commission's assessment of the socioeconomic challenges facing rural areas (CEC, 1988), specific comparison was drawn between:

1 rural areas under 'pressure of modern life', which commonly meant regions '... within easy access of large urban areas' have enjoy a relatively favourable economic performance (CEC, 1988, 28)
2 rural regions in decline, which tend to experience high rates of unemployment and outmigration, and
3 'very marginal areas', where rural decline is even more marked and the potential for economic diversification is highly limited (see also OECD, 1994d).

This last group is characterized most notably by mountainous and island regions on Europe's periphery. It is the first of these rural classes that corresponds most closely with the areas that are discussed in the literature on diffuse industrialization, with many of these places effectively gaining from spill-over pressures from highly urbanized manufacturing regions (Holmes, 1989), benefiting from growth in other economic sectors, such as tourism (Vàzquez-Barquero, 1992a), or existing in a highly urbanized countryside, where access to urban business and research facilities is easy (Gould and Keeble, 1984). Certainly, it would be difficult to contradict Reynolds and associates' (1994, 451) conclusion that '... the highest birth rates in manufacturing in Germany and the UK occur in rural regions adjacent to the major urban regions', for key factors like lower costs for land and labour (compared to cities), efficient transport systems and a high quality of life, are often seen as a potent combination in such locations. In contrast, more distant places are relatively handicapped by poorer access, for example, to cultural facilities (Perry et *al.*, 1986), and evidence points towards lesser efforts on behalf of local governments in rural areas to address this lacuna (e.g. Williams et *al.*, 1995).

Despite the disadvantages of greater distances to large urban centres, the rural character of peripheral regions often provides a key to their success. This is not only because lifestyle factors attract people to such places, some of whom establish new companies (Thrift, 1987), albeit not necessarily in manufacturing (e.g. Williams *et al.*, 1989), but also because weaker Fordist principles in labour market organization tends to bestow on remoter places a presumption that their labour force is placid and malleable (i.e. flexible). Indeed, the attraction of rural areas in terms of the opportunity for firms to shape local labour market structures can be seen in much earlier times than the recent move towards productive decentralization. Certain places that have experienced manufacturing expansion have also benefited from an ideologically supportive social environment, in which small, family-run businesses predominate (Pecqueur and Silva, 1992). It is hardly surprising then, that Keeble's (1989) review of the 'new regional dynamisms' in Europe lists such seemingly peripheral locations as the Midi of southern France, south west England, rural Wales and even rural Portugal as places that have gained from diffuse industrialization, with proximity to major urban centres being a minor consideration for all of these regions.

With particular reference to inward investment in manufacturing or other industrial activity in rural areas, the ability of incoming firms to dominate geographically isolated labour markets, because they are characterized by a lack of militancy, has been seen as one key factor in productive decentralization (Keeble, 1989; Pecqueur and Silva, 1992). However, a note of caution is required, both because this is not necessarily a new phenomenon, and because evidence on productive decentralization appears equally likely to report that some companies prefer to work with established trade unions and to pay high wages, provided they are able to find productive strength through high productivity (Jaeger and Dürrenberger, 1991). Additionally, the ultimate character of emerging local labour markets depends on a number of intervening factors. Thus, Johnston (1991) notes that the development of British coalfields occurred away from established population centres in predominantly rural areas, with pit villages built to house workers in the new mines. On this new economic surface, distinct sets of labour relations emerged in comparable regions, like the Yorkshire and Nottinghamshire coalfields. Johnston suggests that this was a result of different systems of work organization, which ultimately fed into distinct local political cultures. In the case of the Dukeries coalfield in Nottinghamshire, a relative lack of militancy owed much to a form of work organization and dominance by mine owners within geographically isolated communities that would seem to have much in common with more recent trends towards productive decentralization. Yet the attractions of this explanation need to be handled with care, with Rees (1986) providing an alternative analysis which suggests that the link between labour organization and resulting sociopolitical cultures is more complex in the coalfields. Pointing strongly in this direction, Andrikopoulou (1987) notes how decentralized manufacturing enterprises in Thraki, Greece, reveal divergent employer work systems with the same region. For incoming enterprises, the hallmark of Fordism is still strong, with a clear preference to employ return migrants who have learnt the discipline of industrial work practices in long-established manufacturing areas. By contrast, locally-controlled SMEs are drawn to those with little industrial experience, often seeking to avoid proper contracts with workers to gain maximum flexibility, while investing the minimum possible in wages and other labour costs. As these examples show, the manner in which styles of production and capitalist organization are tied to the specifics of localities is much more complex than our current theorizations appear capable of coping with.

The emergence of rural industrial districts

Even so, there is no doubt that the prospect of close-knit industrial districts drawing sustenance from traditions of local entrepreneurship has generated a great deal of theoretical interest in recent years (e.g. Amin, 1989; Hadjimichalis and Papamichos, 1990; van de Ploeg and Long, 1994). But, for our purposes, a number of questions need to be asked on the character of these new industrial spaces. The first of these is whether such areas can be described as rural. At least for Paloscia (1991), the industrial districts of Toscana (Tuscany) are more an urbanized countryside, with Third Italy centres like Bologna hardly fitting a rural billing. Similarly, Lazerson's (1993) study of knitwear industries in the agro-industrial district of Modena notes that 90% of the firms are located in the immediate vicinity of Carpi and two other townships. However, frequently the agricultural traditions of such regions are based on farm landlords living in urban centres, so these urban-centred networks having developed from earlier economic and social arrangements. Paloscia (1991) makes this point when arguing that the manufacturing development in Toscana has been sustained by the prior existence of the rural *mezzadria* social system. This was a sharecropping system wherein farmers had day-to-day control over production decisions, engaged in a range of economic activities, including artisan production, but where overall control was maintained by an agricultural bourgeoisie based in urban centres (Snowden, 1979). Similar patterns are characteristic of the Third Italy as a whole, for a tradition of petit bourgeois and peasant economic units provided a social base of often underemployed workers, who had a tradition of needing to adapt to new income earning prospects in order to sustain their higher goal of retaining an agricultural occupation (Holmes, 1989). In addition, as King (1992) notes, a rural basis for manufacturing expansion in Emilia-Romagna can be traced to its fertile soils and efficient agricultural systems that provided a base for food industries, such as the production of Lambrusco wine, and Parma ham and cheese.

If we accept that certain aspects at least of the Third Italy have inherent features that enable it to be taken as a process of rural development, then a principal question is how far this development pattern has been extended to other districts with similar rural traditions. According to Marsden and colleagues (1993, 8), the decentralized industrialization of the Third Italy shares features with other 'economically buoyant' rural areas like East Anglia and Bavaria, where '... the spatial and social structures established around agriculture and other forms of land-based production can offer advantages to both consumers and producers in the shift towards more flexible systems of production and service provision'. Emphasizing the attractions of a rural base for such regions, Keeble (1989, 5) argued that '... in various parts of southern Europe, a tradition of vigorous agriculturally-based entrepreneurship, often historically associated with sharecropping tenurial systems, has generated growing numbers of indigenous entrepreneurs and small firms'. And while Keeble only cites Italian examples to back his point, other researchers have been able to pinpoint similar patterns in other Mediterranean countries (e.g. Andrikopoulou, 1987; Hadjimichalis and Papamichos, 1990; Pecqueur and Silva, 1992; Vàzquez-Barquero, 1992b). But as both Amin (1989) and Hadjimichalis and Papamichos (1990) make clear, we should not interpret this as indicating that specific socioeconomic features or political-cultural traditions unequivocally either feed indigenous manufacturing expansion or attract urban decentralization.

Even within southern Europe it seems that the attributes of those places that have developed into successful industrial districts are highly place-specific (Cooke and Rosa Pires, 1985; Amin, 1989), with areas of seemingly similar type having very uneven success in promoting (or attracting) manufacturing growth (Vàzquez-Barquero, 1992b; Rodríguez-Pose, 1994). Another side to urban-rural manufacturing shifts is provided by the 'restructuring thesis', that sees firms attempting to lower operating costs through the engagement of less skilled, less unionized and cheaper rural labour forces (e.g. Kiljunen, 1992; Pecqueur and Silva, 1992). Such restructuring might involve the relocation of a production plant but, in an age of increasing use of contract suppliers, it can also involve letting contracts to companies who provide the cheapest quality product, which today could equally benefit firms in rural locations. Certainly, within southern Europe there does appear to be some support for the view that cost considerations, including illegal work practices (Hadjimichalis and Papamichos, 1990), are instrumental in some instances of manufacturing expansion (e.g. Cooke and Rosa Pires, 1985; Holmes, 1989). However, when we extend our geographical frame to northern Europe we should be cautioned by Keeble's (1989) reference to the 'rather different' processes driving new firm formation in north European rural regions. For one, while evidence is not plentiful, there is a suggestion that rates of union membership are not much different across the rural-urban divide in nations with longer industrial traditions, so cost disparities might be much less than imagined (Church and Stevens, 1994). Moreover, in the northern EC states, many of the industrial sectors that are relocating in rural areas commonly seem to need small and skilled workforces (e.g. Fothergill *et al.*, 1987). This has meant that a new appreciation of the environmental and amenity value of rural areas has emerged as a deciding factor in the location of some companies (Thrift, 1987). However, as Lenormand (1994) demonstrates, this is far from the case in France. Here, the tendency towards metropolization in the nation has led commentators to suggest that the 'centre no longer needs the periphery' (Kayser *et al.*, 1994, 26).

It follows that we might well be seeing distinctive paths to regional manufacturing performance across Europe. Cooke (1992) offers some pointers on this in presenting three models on technology transfer in regional growth. In the *grassroots approach* the need for technological change is identified at a local level, in a situation that approximates the endogenous growth identified for the Third Italy (Brusco, 1986), but is seen by Cooke as being more characteristic of parts of Japan. This is the kind of local dynamic that researchers find so difficult to see as a general model of areal advancement (Amin, 1989; Hadjimichalis and Papamichos, 1990). By contrast, the *network approach*, which is seen as characteristic of growth in Baden-Württemberg, finds local initiatives crucially supported by regional and national governments (cf. Herrigel, 1993). Once again, no easy mechanism has been identified for capturing or reproducing the successes of other regions (e.g. Vàzquez-Barquero, 1992a; Hadjimichalis, 1994). At another extreme, the *dirigiste approach* is said to characterize France, where top-down initiatives only seem to have been successful in the Rhône-Alpes region and might even have restricted growth in other regions (Guesnier, 1994). The reality of high growth gaining succour from flexible workforces and SMEs is a reality in European rural areas, but the truth is that this reality occurs in sporadic, highly individualized circumstances, and has proved difficult to encourage in targeted areas.

Beyond this, while rural economic development has been far from uniform across rural Europe, notions of core and periphery do not fit easily into present development patterns (Hadjimichalis, 1994). Thus, while large multinational corporations reportedly express reservations over locating new manufacturing investments in the peripheral EC countries of Greece, Ireland and Portugal (CEC, 1994c), manufacturing expansion in the west of Ireland from the mid-1960s to the mid-1980s was dominated by the arrival of mainly US-based multinationals (Curtin and Varley, 1991), with small indigenous firms often finding a role as suppliers to this external sector (Gillmor, 1986; O'Cinnéide and Keane, 1987). Such inward investments have proved particularly significant for some rural regions, like the west of Ireland or Shetland and Orkney in Scotland, where major economic commitments have been based around single, large-scale projects such as Shannon Airport and the Sutton Voe oil terminal, respectively (e.g. Button, 1976). But the geography of inward investment does not mirror that of new firm formation or *in situ* growth in SMEs. One reason for this is that such investments are often associated with national government policies; as with the large-scale capital-intensive investments that were made by the Italian government as part of its development strategy for the Mezzogiorno and Sardegna (Wade, 1979; Schweizer, 1988) or the location of controversial public facilities away from high density populations (Davies, 1978). Added to this, externally-originating investments are commonly made as part of a global corporate strategy. Thus, major investments by the automobile industry in southern Italy cannot be divorced from labour unrest in the factories of Fiat and the need for a low cost production site by Alfa Romeo after its decision to compete with Fiat in the medium-size car market (Dunford and Perrons, 1984). Such large-scale investments are often surrounded with controversy, and have been criticized for their meagre interaction with existing local enterprises, such that they generate few spin-offs for local companies and even comparatively little employment (Davis, 1973). It would be wrong to exaggerate this point, although the absence of local input linkages is not uncommon amongst small firms that move into rural areas (Moseley, 1973). One problem is that the character of low density areas means that there is unlikely to be an existing network of suitable suppliers for a new company. And if the company that is moving in is well-established elsewhere, or is part of a large corporation, it is likely to have a network of suppliers in which it has confidence already (Moseley and Townroe, 1973), so there is little need to look locally.

Agro-industries and rural development

An obvious exception to situations in which there are few local linkages arises when the new manufacturing plant is attracted by indigenous resources, for here local agents are often in demand to supply inputs for the new manufacturing plant (e.g. Eccles and Fuller, 1970). Most evident in this regard is where industrial development is linked to the farming sector, for processors of many agricultural products (such as fruit and vegetables) require their farm suppliers to be close to their plant in order to guarantee a high quality finished product (White and Watts, 1977; Hart, 1978). As a result, it is noticeable that while analysts generally find weaker local linkages for externally-controlled manufacturing plants than for locally-owned ones, in the case of enterprises engaged in food processing even externally-controlled companies have notable local linkages (e.g. Stewart, 1976). It should not be inferred from this that agro-processing necessarily provides a solid base for rural economic expansion. As with other sectors of the European economy, agro-processing operates in an environment in which corporate restructuring and market (and ownership) globalization are increasingly important (Whatmore, 1994). Thus economic uncertainty is likely to be an increasingly

Table 5.7 Increases in production output in the food and drink industry, 1985–1993 (1985 = 100)

	1986	1987	1988	1989	1990	1991	1992	1993	All in 1993[1]
Belgium	102.7	108.1	110.7	115.9	121.6	123.5	125.5	124.3	109.6
Denmark	104.2	102.2	105.2	104.4	107.2	109.8	112.6	114.4	111.1
France	100.2	100.9	104.2	107.4	110.3	112.6	112.7	113.7	107.7
Germany	103.2	109.1	112.8	116.6	127.9	127.5	127.0	126.6	103.1
Greece	90.8	84.0	92.4	100.8	93.8	99.5	106.5	106.4	99.1
Ireland	104.1	108.4	115.3	125.3	130.9	138.8	144.8	154.5	156.9
Italy	103.0	106.4	111.2	123.6	113.5	116.6	115.9	117.4	111.7
Luxembourg	100.5	98.0	97.4	104.3	108.2	111.8	107.4	110.7	113.8
Netherlands	106.1	106.5	107.9	112.3	116.9	118.8	122.5	123.7	112.1
Portugal	108.2	114.3	121.0	125.4	134.0	137.5	129.6	132.8	119.9
Spain	98.4	109.1	112.2	110.5	115.6	117.8	113.5	114.0	100.7
UK	100.1	104.3	106.2	106.4	105.6	103.8	103.4	104.3	101.0
EC	101.9	106.3	109.8	113.7	115.3	116.6	116.4	117.4	106.5

Source: calculated from Eurostat (monthly). Note: [1] These figures are for the 1993 production levels of all industries exclusive of construction, with 1985 = 100.

apparent feature of such plant operations in the future. Already we see elements of this in the rationalization of production units. In Belgium, for instance, the number of dairy processing plants fell from 256 in 1950 to just 58 in 1986 (van Doninck, 1992). And while agro-processing has recently performed better than manufacturing as a whole, fluctuations in employment and output are notable (Table 5.7). Thus Errington and Harrison (1990) report that UK employment in both flour production and fruit and vegetable processing increased over the 1970–1974 period, then fell until the mid-1980s, when it began to rise again. The rapidity of production change is brought home with force by the drop in flour output from £1500 million in 1973 to £1040 million in 1974, with further fluctuations between 1975 and 1978 before it stabilized at around the £1200 million mark up to 1986. This is not the whole story, for employment in the milk processing industry showed a steady downward trend over the 1970–1986 period, when the workforce in full-time equivalents fell from 72 000 to 38 000, even though milk output was fairly stable at around £700 million a year. This job decline in some measure reflects the pattern for the manufacturing sector as a whole, and certainly represents the general picture for British manufacturers of farm inputs, amongst which fertilizer production, agricultural machinery and animal feed output all recorded significant declines from 1970–1986.[9]

Of course, while it is commonly assumed that agricultural processors are found in rural locations, if only to ensure that produce is fresh on processing (Hart, 1978), the same cannot be said for farm input suppliers, which are less obviously bound to country locations. It could also be added that the whole question of the spatial distribution of agro-industries remains to be assessed. As Errington and Harrison (1990) point out, little is known about the geography of production in either agro-processing or farm input manufacture. We would not disagree with this sentiment, but as Hart (1978) illustrates, there is some information that points to a rural bias in processing activity. One indication of this tendency is given by the location of olive oil processing in Andalucía, where the most rural provinces of Córdoba and Jaén accounted for 402 000 of the 501 040 tonnes that were produced in Andalucía in 1990. This represented some 68.4% of total Spanish output in that year (Caldentey Albert, 1993, 380).

What is significant is not so much the potential for an urban-rural distinction, although this needs to be noted, but the manner in which substantial numbers of workers who seem far removed from agriculture and the countryside are nonetheless integral to agricultural production (Munton, 1992b). This binding of agriculture-related manufacturers into a general industrial nexus also draws attention to the links that should exist between national manufacturing performance and, for example, agro-processing activity. Put simply, if a nation is characterized by a particular manufacturing ethos, then we should ask whether this penetrates the *modus operandi* of rural-based manufacturers. For Errington and Harrison (1990) traces of this interchange can certainly be posited for the UK tractor industry. Here, a neglect of R&D and low productivity (and profitability) is seen as offering a cameo of slow adoption of technological advance and the higher priority given to short-term profit over long-term gain that sets Britain's manufacturing performance apart from its main competitors and helps explain its relative economic decline in the twentieth century (Wiener, 1981; Scott, 1985; Taylor, 1989). How strong such nation-specific tendencies are is difficult to assess, for their impact is unlikely to be uniform across production sectors. Yet it is instructive to note that, while Greek membership of the EC has revealed structural weaknesses and a lack of innovation in the nation's manufacturing industries (Tsoukalis, 1993; Giannitsis, 1994), the agro-processing sector proved to be one of the few dynamic sectors after accession (Konsalas, 1992). In like manner, the entry of Spain into the EC has seen some significant improvements in agro-processing activity, with output for olive oil increasing from 399 381 tonnes in 1988 to 586 500 tonnes in 1990,

---

9 For agricultural machinery there was also a substantial fall in output, although the picture for fertilizer reveals a pattern of annual fluctuations, while animal feedstuff production actually increased (Errington and Harrison, 1990).

while wine production rose from 22 128 000 to 41 158 000 hectolitres over the same period (Caldentey Albert, 1993). Whether this pattern applies generally to peripheral regions in Europe is a research question that has to be answered. With competition amongst agro-processors and farm input suppliers increasing significantly on a global scale (Le Heron, 1988; Errington and Harrison, 1990; Whatmore, 1994), there is every reason to believe that the availability of a cheap production option, as both Greece and Spain represent compared with northern Europe, could shift output towards the periphery. There is already evidence of the importance of this search for cheaper production options within the food and drink industry, with nations on the EC periphery in general recording more significant production increases in this sector over time, especially compared with their overall manufacturing performances (Table 5.7). However, there is also evidence that some of the expansion that is occurring in agro-processing is associated with the new practices of flexible production and manufacturing decentralization (e.g. Precedo Ledo and Grimes, 1991). So not only has European rural industrialization involved innovative, high-tech enterprises of a capital intensive kind (Keeble *et al.*, 1983), but it has also seen employment gains coming from an ability to produce standard articles, including agriculturally-based items, more cheaply than elsewhere (Andrikopoulou, 1987; Keeble, 1989; Kiljunen, 1992).

New industrial landscapes

Set amongst this variety of trends, there is little doubt that locality-specific forces are at work. This is readily seen in the very uneven incidence of new industrial districts, as well as in research assessments that point to unique socioeconomic and political-cultural traditions in the localities concerned (Amin, 1989; Hadjimichalis and Papamichos, 1990). Perhaps it is true that no matter whether we look at multinational corporate investment (Netherlands Economic Institute, 1993) or high-tech industries in general (Hall *et al.*, 1987), no single, and certainly no simple, theoretical model offers a convincing explanation for changes in the location of manufacturing production. Yet it is apparent that certain localities are likely to have considerable importance in national development strategies in the future. For one, if the strategies of large European companies continue to be informed by lessons derived from their Japanese competitors, then a new form of industrial organization that could unfold, and has already begun to in the motor vehicle industry (Mair, 1993), is the creation of a 'company region'. Here, just-in-time production practices are eased by the spatial proximity of assembly plants and associated supplier networks. Mair (1993) identified the beginnings of such regions in Niedersachsen (VW) and Bavaria (BMW) in Germany, in north west France (Renault), northern Italy (Fiat) and perhaps eventually north east England

(Nissan). But while such local complexes might emphasize the significance of locality, as with the Third Italy (Brusco, 1986), we should recognize the sectoral specificity of such interconnections. As the semi-conductor industry shows (Angel, 1994), in other manufacturing industries spatial proximity might be a much less acute issue.

This is a critical point to make, for there is perhaps an inevitable tendency in exploring 'new' developments to provide a somewhat unbalanced account of general trends. Indeed, work on new industrial spaces is as prone to this as other research ventures (Tickell and Peck, 1992). In this respect, the work of Rodríguez-Pose (1994) is particularly important. Addressing the question of how far regional change in Europe has been informed by the new territorial model of flexible accumulation and locality-specific production complexes, Rodríguez-Pose concludes that economic change in the EC is still deeply penetrated by Fordist production practices. Reminiscent of Pompili's (1994) observations, Rodríguez-Pose goes on to highlight how there is a nation-specific impetus to development patterns. Comparison of Figure 2.5, page 44 and Figure 2.6, page 46 provides one example that illustrates this conclusion, for Rodríguez-Pose notes how nations on the European periphery had distinctive growth patterns over the 1980s. Compared with the European core, which the 'vital axis' captures (Figure 2.9, page 50), with the exception of Greece the peripheral nations of southern Europe grew more rapidly than countries like Belgium, France, Germany and the Netherlands. Yet, whereas regions within these core nations had comparatively even growth rates, amongst peripheral nations the areas that grew most rapidly did so in the opposite manner to their national economies. So, while the growth rates of Italy and Spain could be said to be catching-up with the core nations of the EC, within those nations it was not the periphery that grew most but core regions (Emilia-Romagna, Lombardia and Valle d'Aosta in Italy and the Islas Baleares, Cataluña and Madrid in Spain). Examples of such national effects are even seen in local development endeavours. Contrast the situation in Portugal and Britain, for instance. For the former, as Syrett (1994) shows, local government development initiatives are more likely to occur in the industrial field in the rural interior than in more urbanized littoral region that has the added advantage of tourist activity (see Table 5.8, page 173). By contrast, in line with case studies that show a continuing reluctance on the part of some local governments to promote local manufacturing activity (e.g. Boucher *et al.*, 1991), it seems that the more rural districts in Britain engage less heavily in economic development initiatives (Hoggart, 1994). Significantly, the divergences that can be identified in two nation comparisons are equally apparent in Europe-wide surveys, where local development initiatives are found to be informed in crucial ways by national government inputs (Bennett and Krebs, 1994). Indeed, local initiatives in some nations appear to be subsumed by national government influences (e.g. Georgiou, 1994; Wenturis, 1994).

As happens so often in research on rural Europe, as soon as a researcher begins to pose questions a shortage of material restricts answers to the issues

Table 5.8 Local authorities undertaking economic development initiatives in central Portugal

	Percent local authorities undertaking initiatives in the	
	littoral	interior
Manufacturing	63.2	75.8
Tourism	36.8	42.4
Training	57.9	33.3
Agriculture	21.0	9.1
Promotion	31.6	18.3
Number in study	19	33

Source: Syrett (1994).

at hand. In part this might arise from language difficulties and the problem of establishing what research studies have been undertaken in other nations. Perhaps more important, at least at the level of moving beyond case studies to an assessment of national trends, the issue of data availability introduces further uncertainties. Errington and Harrison's point that we do not really know whether agriculturally-related manufacturers are located in rural areas or not provides an appropriate example. Significantly for manufacturing, our reading of the literature is that research on general trends in rural areas is far from plentiful. So while researchers have been able to identify a growing manufacturing interest in rural locations (e.g. Keeble *et al.*, 1983), they have also noted that rural complexes of flexible production practices emerge from local traditions of artisanship in a restricted number of places, so they do not offer a valid general development model for the future (Hadjimichalis, 1994). And if the decentralization of manufacturing into rural centres carries the strong imprint of large corporate involvement (Curtin and Varley, 1991; Peck and Stone, 1994), does this really represent much that is different from the past? In our view it does only in limited ways, for while flexible accumulation and the utilization of small companies as suppliers for large firms perhaps provides new locational opportunities, its main characteristic lies in passing on the uncertainties of market fluctuations from large corporations to small firms. This is demonstrated by Japanese corporations who early recognized the advantages of contracted production units, and found that their utilization of networks of small supply firms in rural areas was made easy by the availability of large numbers of underemployed Japanese farmers (Wiltshire, 1992). Hence, while new openings are being created in some rural areas, in an economic climate in which manufacturing as a whole is in decline, the attractions of close urban proximity have not disappeared for many industries (Dunford and Perrons, 1994). Moreover, in industries like food processing, the evidence points most strongly towards market dominance by large (often transnational) corporate organizations (Le Heron, 1988). It is no surprise therefore that investigations of general trends in European economic change find Fordist principles still carry substantial weight in locational patterns (Keeble, 1989; Rodríguez-Pose, 1994). In research terms, this pinpoints holes in our knowledge, for as rural researchers we appear to know more about the new territorial spaces associated with a few geographical areas of productive innovation, new firm formation and high-tech output, than we do about changes in manufacturing that has characteristically been located in rural areas. Significantly, a key underlying theme of these new territorial spaces is the forging of links between economic globalization and specific localities. As we have sought to show in this section, while such links do exist, they are spatially restricted and, when compared with general trends in economic activity, they carry insufficient weight to override national imprints on manufacturing activity. The imprint of German unification provides a sharp reminder of how overtly this national power can be expressed (e.g. Wild and Jones, 1994).

Rurality and the service economy

The dangers of over-interpreting the impact new trends in advanced economies have on rural areas is as relevant for service industries as for manufacturing. It might be justified to talk of post-industrial service-based economies in employment terms, but the justification for doing so for rural growth promotion is much less certain. Rodríguez-Pose (1994) provides one indication of this, when noting that service-based employment has a negative association with general patterns of economic growth in the European regions, with manufacturing expansion still expressing its impact most strongly on geographical surfaces of economic wellbeing. The pertinence of this for rural areas is strong, for lower density settlements experience deficiencies in the capacity of the service sector to promote economic growth on account of the volume and character of the service industries they house. In addition, as Rodríguez-Pose (1994) notes, there is a difference in service growth between high growth areas and economically lagging regions. In particular, the 'tertiarization' of lagging regions is quite different to the processes occurring closer to economic core regions, as service growth here can be a substitute for unemployment, that absorbs workers from the crisis of agriculture or manufacturing decline. In effect, even if service industries are instrumental in the generation of sustained economic growth, rural recipients of its bounties are highly reliant on non-local and predominantly urbanized zones; whether from service decentralization or tourist inflows.

Illustrating the limited growth generating capacity of rural-based service sectors, Persson (1992) notes the uneven geography of Sweden's fastest growing industrial sectors – the knowledge-based and R&D industries. Here, compared with a national figure of 5.5% of the workforce in knowledge-based industries and 2.6% in R&D industry, figures for four peripheral counties are 3.8% and 2.0%, respectively. Such figures are instructive because they refer to newer service industries, but even when more traditional services, like public services, retailing and tourism are added, the service sector constitutes a smaller proportion of employment than in urban centres. And this is in a nation in which principles of egalitarianism and compensatory action are deeply embedded, so that public services in rural areas are highly developed compared with rural areas elsewhere in Europe (Persson, 1992). It also refers to a time prior to recent cut-backs in public sector provision (Persson and Westholm, 1993). More generally, the literature on both private and public provision of consumer oriented services in rural areas has been dominated by evidence of under-provision that potentially affects future economic advancement (Stone, 1990; D.M. Clark, 1991) and, with the exception of public utilities (e.g. Shiel, 1988), the message is one of facility decline (e.g. Mackay and Laing, 1982). Indeed, rather than promoting service growth, rural communities appear to be searching for more innovative ways to preserve the facilities they already have (e.g. Woollett, 1982; Nutley, 1988). Within the public sector, contemporary emphases on clawing-back state provision and privatizing public services holds out little prospect of any improvement in this situation (Bell and Cloke, 1989; Persson and Westholm, 1993), with population growth also offering little encouragement (e.g. Stockdale, 1993), and while the geographical centralization of public services might provide a few centres with a service package sufficient to attract some outside investment, it creates significant social problems for those living away from these favoured centres (Garner, 1975; Sjøholt, 1988; Persson, 1992). In sectors that generally have never been in public ownership, such as retailing, the picture is much the same. Here, the growing importance of large corporate ownership in much of Europe,[10] and the relative inability of low density areas to satisfy the corporate thirst for profit, has meant that the structure of available services leads to rural options that are both more expensive and give less choice. The only exception is in places that are proximate to urban centres that are more likely to house corporate household services (Guy, 1991). Beyond such urban fields, the economics of private sector services tends to be precarious, with limited opportunities to establish prosperous businesses. In such circum-

---

10 Different patterns can certainly be observed across Europe, as they can within single nations. Reflecting something of the continental north-south divide in patterns of retail provision, in Italy there are notable differences in the kind of retail outlets that are found across the nation. Thus, while the North housed 56.6% of supermarkets in 1990, with the South accounting for 23.9%, when it comes to small retail outlets and mobile shops their respective percentages are 42.4 and 44.2, for the North, and 37.6 and 34.4, for the South (Ministero Dell'Agricultura e Delle Foreste, 1991, 37). Indicative of the 'old fashioned' structure of retailing in the South, figures for change in the number of outlets for these types over the 1981–1990 period were: +108.8% for supermarkets; –13.4% for retail outlets and –38.2% for mobile shops.

stances, service provider initiative or their willingness to operate on slight economic returns plays a key part in facility preservation (e.g. Jussila *et al.*, 1992).

However, mainstream consumer industries do not capture the whole picture for rural Europe, as there has been service decentralization of producer services as well as newer service options arising from changing patterns of consumer behaviour. It is true that service industries with the greatest likelihood of promoting economic advancement still cling to large city locations (Buck *et al.*, 1992; Thrift and Leyshon, 1992). However, for Camagni (1988) a movement of these producer services away from cities has begun to occur, with strong spatial synergies in diffused manufacturing growth between production, organizational, finance and sales-based companies. The prospect that this trend will breath new life into more peripheral rural regions should not be exaggerated. Marshall and Raybould (1993) illustrate this, when examining the manner in which pressures on cost-cutting are encouraging corporations to decentralize their white collar workers from London, but significantly barely at all beyond the South East of England.[11] Marshall (1988, 84–88) offers a more general view on trends in the productive service sector that confirms this picture, but the observation that nations like Germany are seeing more rapid growth in producer services in peripheral areas than in intermediate or metropolitan ones, prompts us to hold back on any inclination to generalize across states (Jaeger and Dürrenberger, 1991). Even so, as Reynolds and associates (1994) note, it also has to be remembered that service activities are less likely to decentralize from cities than manufacturing plants. This arises not only because they are less pressured by space considerations (Fothergill *et al.*, 1987), but also because they are commonly attracted to servicing local populations or require access to concentrations of economic decision-makers in order to be more effective in providing specialized support for other businesses. This localization is one reason why the EC has seen legislation covering service industries as a matter largely for national governments (Molle, 1988). Not surprisingly, therefore, evidence of the spread of intermediate service activities across Europe, notes modest growth in peripheral areas, with those gains that are won being said mainly to concern '... relatively routine functions, often with a programmed character' (van Dinteren *et al.*, 1994, 292).

The general issue here is little different from that for manufacturing; namely, that individual industries, or sections of the service industry if this is seen as one unit, have different propensities both to generate economic growth and to move away from larger urban centres. This is made clear in the

11 The pattern described by Marshall and Raybould (1993) raises the question of what service employment actually is, given that corporate decentralization often involves office workers in manufacturing companies. We tend to see these employees as belonging to the service sector largely because large corporations are now so diversified across economic sectors that it can be difficult to categorize companies as a whole either as service or manufacturing oriented, let alone the workers they employ.

distinction drawn by Todoir (1994) between different types of business service in the information-intensive service sector, each of which have a different propensity for dispersal to rural areas. Thus, *sparring services*, such as public relations, or organizational and financial consulting, require inter- action with business clients during the course of day-to-day operations. As a result, these high-income earners tend to remain in large conurbations, given that proximity to clients is necessary to generate business and fulfil contracts. In contrast, for *jobbing services*, where a job is allocated to a service company which carries out the work relatively independently, and usually for less cost than sparring services (e.g. proof-reading, economic analysis and engineering consultancies), have less need for geographical proximity. In this case, a dispersed rural distribution that enables workers to benefit from cheaper rents and quality of life considerations is more feasible. We can recognize a third category in *frequent-use services*, that are more obviously oriented towards consumers, although they also have business customers. Here, even higher-waged occupations have seen some dispersal from cities. Blacksell and Fussell (1994) offer one illustration of this when noting that the highest rate of 1970–1985 growth in the number of barristers and solicitors in the UK occurred in two predominantly rural regions, East Anglia and the South West, with London having one of the lowest rates of growth in the country. But it is characteristic of these frequent-use services that they tend to follow population and employment moves, rather than leading them (Molle, 1988). As such we need look little farther than the decentralization of population from London (Warnes, 1991) and so-called counterurbanization migrant flows to appreciate the movement of locational change in the British legal profession (e.g. Williams *et al.*, 1989; Bolton and Chalkley, 1990).

Tourism and rural economies

Although growth in business services has been of some importance to Europe's rural areas in recent decades, most service growth has generally not been in this sector. Commonly, leisure-oriented activities have dominated employment increases. In some cases, this growth has been based on the attractions of the 'countryside' as a basic resource for tourism. In other cases, what started as a rural area has been transformed by major inflows of capital and people, which has revolutionized both the settlement structure and the economic base of an area. Most evidently we see the latter of these changes occurring in the emergence of centres of mass tourism in southern Europe. Offering one illustration of this, Benidorm has grown from a settlement of just 2726 inhabitants in 1950 to a centre of more than 60 000 year-round residents, with high-rise blocks now dominating the skyline. In the process its economic base shifted from that of fishing village with small-scale, individually arranged tourist visits to a town that sees many of its tourists arriving on package holidays organized by large companies outside Spain. Such large-scale expansions are not only important on account of the way in which they transform the immediate landscape but also due to their impact on other economic sectors, both within and beyond the tourist settlement itself. For one, the jobs that tourism creates provide opportunities for retaining populations in areas that might otherwise experience depopulation (Grafton, 1984; White, 1985), and even enable areas to be repopulated, as Ciaccio (1990) found in Sicilia. These jobs come not simply from accommodation, food and leisure outlets but also from local craft industries that receive a sales boost from tourist inflows. However, the impact on other economic sectors is not always easy to predict.

It is not necessary for significant tourist stays to result in the urbanization of a rural landscape, but the greater the inward movement the more likely that the settlement structure will be transformed in order to house a greater population. In Austria and Switzerland major tourist inflows have largely been sustained within a rural landscape, although this has been so altered by a sprawl of new constructions as to have changed its character significantly (e.g. Barker, 1982). Also important in this regard is the nature of the tourist activities that have been generated in these areas, for the significance of skiing in their tourist calendars has favoured a rural landscape, which provides easier access to ski slopes and lessens the number of people at a single site. In most of Europe such ski centres have been grafted onto existing settlements, but in others they have been built from scratch, as with Val d'Isère in France (Pearce, 1995). Self-contained complexes of this character are not restricted to skiing resorts, for they have also been created in coastal locations, as in Corse (Kofman, 1985) and Sardegna (Schweizer, 1988). For King (1991) such government-inspired tourist developments in southern Italy often have tenuous links with local economies, appearing, as for much of the Italian government's investment in manufacturing in the South, as 'cathedrals in a desert'. Despite the obvious gains that have accrued to areas that have gained a reputation as skiing centres, this image of relatively slight gains from tourist activity is one that has to be taken seriously. By the very nature of their low population densities, accompanied as this is by a relatively narrow range of economic activities, rural areas have significantly lower multiplier effects from tourist expenditure (PA Cambridge Economic Consultants Ltd., 1987). Most commonly they also suffer from a highly seasonal pattern of tourist inflow, which further limits the economic gains that accrue (Getz, 1981; Harrigan, 1994). However, as with the debate over the impact of second homes (e.g. Coppock, 1977), whether or not the outcome is regarded as positive or negative for communities can depend heavily upon local circumstances. It can also depend on who is asked, with general tourist developments being restricted in some regions in order to preserve a more specialist recreational resource, such as hunting, for local socioeconomic elites and those of similar class standing from outside the region (e.g. Wilson, 1992).

In general, however, analysts and authorities see tourism as a major development option, even for remoter rural areas (e.g. Moore, 1976; Ciacco, 1990). Significantly, for instance, some 69% of Spanish project proposals under the EC LEADER programme, which seeks to promote rural regeneration, have been for tourism (Barke and Newton, 1994). This perhaps is not surprising, for in areas like the Alpujarras the decade before the LEADER programme began saw employment in service industries rise from 25% to 66% of the workforce, with increases in available accommodation of 150% and a twelvefold rise in tourism turnover (Barke and Newton, 1994). In Greece also, areas that previously recorded an 'intermediate' level of growth found that in the 1980s the most significant factor that advanced their economy was the '... development of private property (land and housing), especially in areas of second home and tourist demand' (Getimis and Kafkalas, 1992, 80; for an earlier conclusion along these lines see Collard, 1981). Moreover, the failure of past manufacturing-led policies aimed at rural and national development has promoted a switch in Irish government emphasis in favour of tourism. The potential for this sector can be inferred from the fact that it provided 75% of new jobs in the Irish economy over the 1989–1992 period. Part of the attraction of this sector is that it is capable of being organized and sustained using small enterprises that are available for local ownership (Pearce, 1990). Of course this characteristic also offered former city or urbanized region dwellers an opening to life in the countryside, for the cost of establishing a tourist enterprise is comparatively low, especially if the proceeds from the sale of a home in a more costly urban region can be employed (Williams *et al* ., 1989; Buller and Hoggart, 1994a).[12] Again we are faced with a shortage of available empirical evidence to assess the long-term durability of such enterprises or indeed to assess with confidence their local impact. At a social level, Getz (1981) found that in the Aviemore area of Scotland the ownership of tourist enterprises by 'outsiders' who had migrated into the area was resented. But even where the 'intruders' were genuine 'foreigners' Buller and Hoggart (1994b) found no evidence of this for British migrants in rural France. Indeed, as has been long recognized in Ireland, tourists bring numbers, money and reassurance to those who live in isolated rural communities, with an enhancement to self-esteem perhaps having the most profound effect (Messenger, 1969; Brody, 1973).

Significant points that need to be grasped about the links between tourism and rurality include recognizing that although large-scale package tourism is directed at quite restricted geographical areas (Pearce, 1987; Valenzuela, 1991), beyond these areas it is often the case that relatively small numbers of tourists can have a significant impact on local economies (e.g. Moore, 1976).

12 It merits noting that Keeble and Walker (1994) have recorded that the geography of new firm formation across all economic sectors in the UK is linked to housing wealth, which provides a capital stock that can be used as collateral or as a saleable asset to fund a new economic endeavour. Further research will be needed to see if this effect is limited to the UK or is found more widely across Europe.

Moreover, even the seasonality of tourist movement can be beneficial, if this meshes conveniently with other economic activities, so that tourist earnings can be gathered at otherwise slack times (Neate, 1987). For the future it also has to be recognized that the character of tourist behaviour is itself changing, with a growing emphasis on alternative forms of tourism, in which ecotourism and other country-related activities are finding an increasing number of adherents (Eadington and Smith, 1992). In addition, in much of northern Europe at least there seems to be an association in many city dwellers' minds of rural areas and a serene, idyllic existence that is synonymous with the good life, which is a powerful draw for tourists (Pigram, 1993). This association has the potential to be particularly important in southern Europe, as well as in the northern parts of the continent, for while states in the south see a larger part of their national income coming from tourism (Table 2.9, page 71), the generation of international tourist flows that bring these benefits owe their origins primarily to northern states (see Table 5.9). For the future we should also watch the manner in which tourist flows are converted into permanent residence. The last decade has already seen an increasing flow of northern

Table 5.9 Levels of trip-making in Europe by age group (percentage population travelling in each age bracket)

	All travel			Foreign travel		
	15–34 years	35–54 years	55 plus years	15–34 years	35–54 years	55 plus years
Belgium	76	67	53	63	54	40
Denmark	68	69	50	47	48	32
France	67	64	45	24	21	15
Germany (West)	67	71	53	47	44	24
Greece	63	57	38	13	16	9
Ireland	69	63	52	33	33	27
Italy	70	63	47	22	16	8
Luxembourg	68	69	58	65	67	55
Netherlands	80	75	62	60	55	40
Portugal	58	53	39	16	14	7
Spain	85	75	65	19	15	6
UK	68	72	63	31	28	24
Austria	67	65	44	45	45	30
Finland	86	89	70	54	60	36
Sweden	81	88	67	50	53	36

Source: CEC DG-XXIII Tourism Unit (1994, 106)

European residents taking up homes in warmer southern European climates (King and Rybaczuk, 1993). Indeed, it is clear that southern European destinations have become quite significant residential locations for northern Europeans (see Table 5.10, page 182). While some of these inflows will result from company-related job moves, investigations already show that a significant number of these international migrants are driven more by consumption considerations. In this, inflows are targeted toward tourist zones, whether for retirement migration (Valero Escandell, 1992) or across a broader age-range (King, 1991; Buller and Hoggart, 1994a). But the character of tourist destinations does vary. While coastal zones are still a high priority there is a notable trend toward more upland, scenic areas, as well as to distant places from larger urban centres, with both types of area perhaps having experienced extensive depopulation in the past (King, 1991). In some locations prior population losses are one factor that helps attract outsiders, for they can acquire traditional homes that match their images of a serene rurality for relatively cheap prices. With frequently abandoned properties being taken up, one consequence of such inward movement is a marked improvement in local housing quality, along with increased job opportunities in construction and home maintenance (Hoggart and Buller, 1995b). Yet these gains do not have to rely on a stock of cheap, traditional housing, for the tendency in certain regions is for second home and new in-migrant populations to construct new properties in their chosen destinations (Weatherley, 1982; Bontron, 1989). Possibly we could draw distinctions between trends across nations, as well as for home and foreign populations, but available evidence is currently insufficient to do so with conviction.

Caution needs to be exercised against over-interpreting the apparently beneficial effects on rural economic and social viability of the growth of tourism. For example, it is noteworthy that in siting the largest single tourist development in recent decades in Europe, the planners of EuroDisney chose a rain-swept plain north of Paris over the attraction of sunnier rural areas to the south. Despite initial difficulties in attracting custom, the reason for this choice was clear, for ease of access meant that a location in the northern European 'industrial triangle' was the key to encouraging short-stay, high-spend mass tourism (Figure 2.2, page 32). Thus although there is evidence that points towards a growing tendency for increased internationalization of tourist impacts, with visits and even permanent residence favouring rural areas, whether or not the apparent governmental emphasis on tourism runs the risk of stretching the market for specifically rural tourism beyond its capacity has yet to be seen. Even so, for many remoter areas tourism still seems to be the main option for increasing local incomes.

Table 5.10 Four main emigrant destinations from source European nations

Source	Destination	Emigrants (000s)	Destination	Emigrants (000s)	Destination	Emigrants (000s)	Destination	Emigrants (000s)
Belgium	Netherlands	23.9	Luxembourg	9.4	Spain	6.7	Belgium	2.6
Denmark	Sweden	27.9	Norway	17.4	Spain	3.5	Italy	25.1
France	Belgium	94.9	Germany	88.9	UK	38.0	Spain	28.7
Germany	Netherlands	46.9	UK	41.0	Italy	40.3	Sweden	6.0
Greece	Germany	336.9	Belgium	20.6	Italy	17.5		
Ireland	UK	469.0	Belgium	2.5				
Italy	Germany	560.1	Belgium	240.0	France	87.5	UK	86.0
Luxembourg	Belgium	4.7						
Netherlands	Belgium	67.7	Spain	9.7	Norway	2.6	Sweden	2.6
Portugal	France	426.5	Germany	93.0	Luxembourg	34.0	Spain	25.4
Spain	Germany	135.2	France	81.6	Belgium	51.1	UK	30.0
UK	Germany	113.3	Spain	50.1	Netherlands	46.9	Italy	27.9
Austria	Germany	186.9	Sweden	2.8				
Finland	Sweden	115.0	Germany	11.2	Denmark	1.9		
Sweden	Norway	12.0	Denmark	8.3	Finland	6.4	Spain	5.0

Source: OECD (1994c). Note: Data by destination countries are for 1991, except for Finland (1992) and Luxembourg (1989). Data for France (and Austria, which does not appear here as a top four destination for any country) are for labour force members only. Not all source nations have four destinations listed, as data were only available for the largest inflows to each destination country. These data only include those who retain their home nation citizenship and were living in the destination country at the time the data were collected.

Divergent national impacts?

From this examination of three major areas of economic activity, the reader will have noted that our comments on national traditions and practices has not been plentiful. Of course, it would be easy to provide a long list of distinctive national government policies, with an analysis of their underlying causal impetus. As just one example, it is instructive to note the quite different approaches of the Portuguese and Spanish governments toward an ideal of full-employment and very different employment strategies they have adopted despite many similarities in underlying economic conditions (Brassloff, 1993). Similarly, we could devote more lines to the way in which rural traditions make themselves manifest in cross-national differences in economic policy. Here, agriculture is the most obvious example to use, given its overt rural orientation, but dissimilar attitudes and responses to manufacturing and commerce also find their way into distinct national cultures, with obvious spin-off effects for rural activities. However, with the exception of agriculture, where national traditions make their presence felt most obviously in transnational policies that seek to fine-tune global trends (Chapter 4), it is evident that it is extremely difficult to distinguish events in rural areas from their urban counterparts, so that a distinctive rural view is hard to grasp. Of course, we would be aided in this endeavour were there more studies of explicitly rural venues, but the literature is distinguished by the sparsity of research on a number of significant rural themes (e.g. service provision). This problem is not eased by the rapid changes that have occurred in the economic sphere as a result of globalization trends, with the formation of new configurations on the economic landscape having been forged in each of agriculture, manufacturing and service industries in recent years. As such, there are still many empirical questions that have to be addressed before we can adequately answer broad questions of causal priorities.

What can be said is that economic change has generated transnational forces that are provoking labour market adjustments in all nations. How states respond to these will partly depend on the base from which they start to make adjustments, with low wages offering attractions for certain employers in manufacturing (Hadjimichalis and Papamichos, 1990; Kiljunen, 1992), whereas skilled and innovative workers are in higher demand for other economic sectors (Fothergill *et al.*, 1987). Of course, as we see readily for producer services, the divergent demands of these employment requirements can often be met by locations within the same nation (van Dinteren *et al.*, 1994). As such, we do see a pattern of locality-specific responses (Amin, 1989; Hadjimichalis and Papamichos, 1990; van de Ploeg and Long, 1994). However, general surveys also suggest that they are set within a nationally-specific framework (e.g. Pompili, 1994; Rodríguez-Pose, 1994), with nationally distinct patterns recorded for support for SMEs over large corporations (compare France and Italy), in the character of local efforts to promote growth (Bennett and Krebs, 1994; Georgiou, 1994), and very obviously in the distinctive agricultural trajectories of the southern and northern European states. Indeed, given that we can recognize dissimilar national paths of economic development, with the decline of Britain an obvious example (Taylor, 1989), future research might be able to link dissimilar geographical patterns of economic change to distinctive traditions in national political economies. Is it an accident, for instance, that the British history of financial sector domination in its economic affairs (Ingham, 1988), coupled with the increased centralization of state power at the national level (Gamble, 1988), is associated with a concentration of producer service decentralization in the London urban field (Marshall, 1988), whereas a political economy with a history of more dispersed power holding in Germany is allied to the movement of producer services into more peripheral locations (Jaeger and Dürrenberger, 1991)? At this point we speculate. While we can trace elements of national political-cultural traditions in economic performance, at least as regards rural areas, we require much more research that contextualizes rural change processes in the context of their national power structures before we can trace out whether a distinctive national pattern makes a strong imprint on the economic landscapes of rural Europe.

6

Uneven development and social change

In the discussion of rural transformations so far in this book, a clear tension has emerged that is not unusual in either academic or popular discourse. In examining the character of rural Europe we have drawn attention to national rural traditions as a basis for a particular rural identity. Yet these traditions, embedded as they are in the peculiarities of a long historical process of socioeconomic, political-cultural change, are today confronted by a rapidity of change that some characterize as modernism, others as postmodernism, but which, for us, is more appropriately seen in terms of a dialectic of integration and differentiation within rural places. Significant elements of the socioeconomic adjustments that are being experienced in rural areas come from processes of 'integration'; whether into national polities and economies, as in earlier centuries, through forces of social homogenization such as the emergence of the mass media; or into the international arena through more recent transmutations of political economy, such as economic globalization and the assertion of transnational governance within the European Community. But set within this pattern of greater integration is a parallel differentiation of people and places. For some this is visible in a new localism (Urry, 1981), with the manner in which processes of 'productive decentralization' are grounded in local social practices offering one example (Chapter 5). However, while certain new socioeconomic spaces are a product of peculiarities of local social forms, others can be completely transformed with almost no account being taken of local character or tradition. It is for this reason that we have serious doubts about some claims that an understanding of social structure has to be embedded in localities. Under circumstances of socioeconomic change, even implicitly to provide the locality with an equal (or greater) causal importance than the extralocal forces that act upon it seems to us to deny the existence of inequities in power between human agents. At a very simple level we see this when areas receiving large in-migrant flows find their 'traditional' sociocultural existence swamped by the more assertive actions of new arrivals (Connell, 1978; Forsythe, 1980). At the same time, major transformations in the economic base of areas can occur without provoking substantial change in a locality; as with government-inspired major investments in economically backward parts of Italy, which are commonly grafted alongside a local socioeconomic system with little interaction between the new and the old (Davis, 1973; Schweizer, 1988), let alone any peculiar local socioeconomic features that encourage capitalists to commit themselves to that place (Wade, 1979). While we would wholeheartedly support the argument that local social organization is not the same as its national counterpart writ small (Pitt-Rivers, 1960; Urry, 1981), our note of caution is that much of what happens in localities is not decided by the specifics of local socioeconomic structures or traditions. This is perhaps especially true for rural locations, since all other things being equal, a small population size increases the chances that an outside force can have a major transformative effect.

None of this means that an interaction effect cannot be discerned for local-extralocal interchanges. There is no doubt that the spatio-temporal context of change processes provides one of the keys to understanding the social geography of Europe's rural areas. However, the place dimension of any context is not simply that of locality but also of nation (amongst others). Moreover, we know that places are not equivalent even within the same country. We need only examine the work of single researchers who have investigated different villages to readily appreciate this point. Williams provides a useful example, for despite seeming similarities with the economic base of the Cumbrian village of Gosforth (Williams, 1956), he found that the west country village of Ashworthy had a much more poorly developed social class structure (Williams, 1963). It follows that if we are to understand social change in rural Europe, we should start from recognition of the diversity that exists in the social character of rural communities, and then ask how that character is being transformed by broader processes of socioeconomic change. Of course, as with national rural traditions, at any one point in time we can identify different rural social formations, few of which present themselves as 'pure' essences of their character-type. Nevertheless, we believe that it is heuristically appropriate to envisage much of rural Europe as revealing one of five social tendencies. In adjusted form, three of these (contestation, paternalism, clientelism) are taken from the ideal-types that Murdoch and Marsden (1994, xi–xii) present to describe processes of change in rural Britain. The fourth (competitive) has been added to provide a clear baseline against which social change in 'traditional' communities based on commercial agriculture can be assessed, while the fifth (peasantism) acknowledges differences in rural traditions between Britain and continental Europe (Chapter 3). Our intention in this chapter is to examine each of these in turn, questioning their present-day occurrence and historical roots, before refocusing the analysis on key trends in rural social transformation in recent decades. Here we will draw on a fourth category of Murdoch and Marsden (1994), namely preservationism, which draws its sustenance from the increasing integration of rural locations into more global social formations. Our aim is not simply to pinpoint longstanding patterns of local social diversity in rural Europe but also to see how these social forms have been affected by recent societal trends. In focusing on these two broad themes we will also ask whether any differences in their occurrence or interactions carry a national identity tag.

Peasantist social tendencies

The notion that a peasant social organization exists in Europe seems to us at one level to be baffling and almost disconcerting. This discomfort might be grounded in our British background; having lost its peasantry under a wave of early agrarian capitalism Britons possibly have less empathy for what 'the peasantry' means and constitutes. But while there might be some truth in this, to accept this rationale at face value would be to deny the undoubted problematic of the idea of a peasantry. At heart, difficulty over this concept arises from distinguishing peasants from other farmers. In this we are not greatly helped by writings on peasants. To give one key example, in Eric Wolf's (1966) book *Peasants*, he devotes much more attention to making clear the factors that separate peasants from tribal societies than he does to drawing a line between peasants and small-scale farmers. We are told that a peasant unit is not merely a rural productive organization but also a unit of consumption, in which a family economy dominates. Peasants it seems are not farmers, for farming is predominantly a business enterprise, whereas a peasant principally runs a household. Despite this, and the definition provided by Ellis (1988) and referred to in Chapter 3, we are not convinced that we can distinguish between peasants and farmers. After all, in situations far removed from an appropriate description of land cultivators as peasants, we find clear patterns of social criteria dominating farmer behaviour (e.g. Gasson, 1966, 1969).[1] Indeed, we are drawn to Wright's (1964, v) warning that '... a cautious historian or social scientist would do well to avoid the quicksands of the peasant problem'. All the more so since the peasantry has been portrayed as being on the verge of extinction for some time (Mendras, 1967), with Wright (1964, 178) claiming that French peasants in the 1930s probably had more in common with their predecessors in Balzac's time that with their successors in 1963. The magnitude of sociocultural transformation that has occurred in the last 30 years suggests the prospect of a 'genuine' peasantism today is remote. Despite this, in much of southern Europe, in large segments of France, in Ireland, parts of western Germany, possibly even

1 Although the proportion of farm units which do not meet business criteria of economic viability is much higher in peasant-oriented communities; as evinced by Portugal, where 78% of farmers are defined as being a 'social type' enterprise with just 1% of national farm units considered economically viable (Avillez, 1993).

in crofting areas in Britain, rural traditions exist which owe much to a former peasantist past (e.g. Table 3.1, page 80). In some measure echoes from the peasantry provide recurring chords in the themes of traditional rurality that lie at the heart of contemporary agrarian traditions.

But what is the peasantry? In broad outline, we can think of it characteristically as a system of rural agricultural production in which smallholdings dominate, where the norm is of slight differences in the size of agricultural holdings, and where agricultural employment constitutes the mainstay of the community. How far such communities differ, say, from the occupational communities of upland Britain is a matter of debate. In characteristic descriptions of the peasantry it is usually implied or stated that relatively low incomes are characteristic of such societies, which some might associate with a weak orientation toward product sales in the market. But this could equally be explained by the small size of production units which provides little surplus for commercial sales after a household's needs are met (indeed, often there is insufficient to meet household needs). A classic investigation of this kind of community was undertaken in County Clare in the west of Ireland by Arensberg (1937). He painted a picture of a strong and stable rural community, based on principles of kinship rights and obligations, mutual aid between farming families (a practice known as *cooring*), independence from the market economy, the preservation of local traditions, patriarchy and primogeniture (in which the eldest son inherits the farm on the death or retirement of the patriarch). This image of stability, strength, homogeneity and traditional ways of life has been a key element in popular as well as academic visions of the rural past. Almost with ideological overtones, this vision has commonly drawn unfavourable contrasts with villages experiencing 'modernization', projecting the explicit or implicit message that this leads to social disintegration and a less wholesome lifestyle. Thus, in a follow-up study of villages in the west of Ireland, Brody (1973) reported that the traditional community of the 1940s had all but disappeared. The penetration of market forces, 'rapacious entrepreneurs', and the collapse of traditional systems of mutual aid, were said to lie behind this change, which had led to a 'ridiculing of patriarchs' and a high level of depressive mental breakdown as local norms and values were undermined. One explanation of this breakdown, shared poverty, coupled with strong kin obligations, had previously provided the cement that bound the system of mutual cooperation amongst cultivators. Meanwhile, with population increases exceeding the capacity of the land to feed new mouths, emigration was an economic necessity that was later to bring significant inflows of remittances as kin expressed their familial bonds by seeking to improve the living standards of those left behind. An unexpected consequence of these remittances was that they lessened reliance on mutual assistance (as services could be purchased), as well as providing the funds to buy radios and televisions, which enabled entertainment to be obtained at home rather than in neighbourhood events (Brody, 1973). As a result, individuals participated less in the neighbourhood corporate group, which went into decline. Golde (1975) noted a similar phenomenon when non-agricultural job opportunities increased in rural Baden-Württemberg, while Jenkins (1979) has described how in the mountains of southern Portugal, a system of mutual aid and close social relationships broke down in the 1960s once the village was connected via a tarmac road network to wider markets.

A number of points should be made about peasantism as a traditional form of rural society, as well as on the impact of modernization as a cause for its collapse. First, as Pongratz (1990) suggests, it is wrong to oppose traditionalism and modernity, if the former is seen as a passive and given set of rules or structures that are opposed to change. For, in West Germany at least, agricultural modernization has been conveyed through 'traditional' rules of behaviour, in which the 'traditional' goal of preserving the family farm and a 'traditional' work ethic have combined with submission to modern technical and economic demands to produce a new social reality (Pongratz, 1990). This leads to the suggestion that the notion of modernization carries within it too great a sense of rapidity, for social adjustments can occur over long periods of time, such that their incremental nature disguises the transfiguration that is occurring. This notion lies behind Wright's (1964) ideas about French peasants of the 1930s having less in common in farming techniques and social outlook with their successors than with their predecessors; albeit this recognition is couched in a manner that implicitly acknowledges the quickening pace of socioeconomic and cultural change in the twentieth century. It is also apparent in the west of Ireland, for the 'eclipse of community' that Brody identified had been occurring for more than a century. What was critical to this change was not simply extralocal interventions, but personal choice. This was visible when extra income enabled Irish cultivators to choose whether they wished to participate in *cooring* networks, just as it was evident in the decisions of Portuguese hill farmers to engage in market-oriented farm production.

The rather idyllic associations that ideas about peasant life seem to convey fail to highlight features of the reality of rural areas which are critical to any appreciation of their transformation. Very obviously there is the question of the living standards of people. Indicative of what were often harsh conditions, Shiel (1988, 15) notes how in the run-up to the Second World War more than 90% of Irish farmhouses had no piped water, with under 5% having a flush toilet and only one-in-six having a radio. Comparable indicators could be provided for this time period in much of what is now the periphery of the EC, with reports of hunger and starvation not being uncommon (e.g. Brenan, 1950; Mansvelt Beck, 1988). As Bell (1979, 7) expressed it: 'The question for peasants was not contentment but survival, not self-fulfilment but familial obligation, not advancement but stability. Within admittedly narrow constraints, however, peasants struggled to master their world or at least to function within it'.[2]

Given these constraints it is no surprise that the peasant worldview had limited horizons. Illustrative of this, Ogden (1980) found that in the eastern Central Massif 66% of adults in 1911 had been born in the commune they then resided in. But this picture of little change is exactly what we would expect of any traditional rural community. It is also an image that only fits certain places and very clearly attaches itself to agriculturalists more than others. Thus, Ogden is not alone in noting that French peasant farmers have a tradition of marrying partners who live close to them (Wylie, 1966), as well as seemingly being circumscribed in their choice of marriage partners to those from a farm background (with 70% of marriages meeting this criterion as late as 1966–1970). Moreover, this French picture is repeated in countries with less strong peasantist traditions. Benvenuti (1962), for example, found that in the Dutch community of Winterswijk some 73% of farm population marriages were between persons born within the boundaries of the municipality. Similarly, Nalson (1968) noted that 85% of marriage partners came from within 16 kilometres of the farm they then occupied in upland Staffordshire, while 90% of marriage partners were from with a 20 kilometre radius of farms in two villages in Baden-Württemberg (Golde, 1975).

Providing some explanation for this is the expectation that nonfarmers provide poor partners on farms, for marriage to someone with no farm background most commonly results in the loss of a potential farm worker (Wylie, 1966; Golde, 1975). Hence, residents of farm communities have long held to an implicit assumption that only wealthier farmers could afford to marry a nonfarmer. Frequently, this resulted in social distinctions between farmers and nonfarmers, with the latter exhibiting greater mobility, both in terms of their travel behaviour and their migration patterns (e.g. Wylie, 1966). The important point about this differentiation is that it provided a clear indication that peasant communities were not the epitome of social harmony and egalitarianism that they are implied to be in much popular culture. Indeed, what we see even in geographical dimensions of such divisions is an assertion of socially oppressive undertones. Pitt-Rivers (1971) offers a glimpse of this, when describing how those from other villages who sought to court a woman of Alcalá in Andalucía were dissuaded from pursuing their interest by being ducked in the water fountain or even being beaten-up. Yet while such an enforcement of community boundaries restricts options to internal partners, choice is not unrestricted within the locality. Quite apart from any farmer – nonfarmer divide, such communities were often divided by a status differentiation between village and open country-side. In some places, as in Ireland, this division carried political overtones, with the village being associated with external (British) control (Clark, 1975). More commonly, it emerged from a sense that country folk were 'less civilized' and more questionable characters than those who lived in the

2 Providing another perspective on this point, Pitt-Rivers (1971, 48) noted how the possession of land bestows no social virtue. Its cultivation was not seen as a sacred duty or a labour of love, but rather as an unrewarding way of making a living.

village (Pitt-Rivers, 1960; Davis, 1973; Schweizer, 1988). For Wylie (1966, 347) at least this combination of social distinctions meant that in the Anjou village he studied harmonious social coexistence only really came as a result of the crisis caused by depopulation and the problems besetting small-scale agricultural producers: 'To us it appears that the sense of community among Chanzeans grows strong as Chanzeaux turns away from its past and seeks to face the problems of modern life'.

Placing a further nail in the image of stability and social harmony as 'natural' features of peasant rural societies is the degree to which population instability characterized many places. This might seem to contradict the evidence that many residents of peasant communities were born within close proximity to their adult residence. Yet these figures refer simply to the inhabitants who remain, when these communities often lost large segments of their population through emigration. We see this occurrence in the movement of Europeans to the New World, for which the scale of outmigration was massive. Indicative of this, it is estimated that some 250 000 Germans moved to the USA between 1818 and 1828, with even more said to be moving to Brazil at that time, while one million left for the USA from 1846–1855, largely as a result of crop failures on small farms in the south west (Jones, 1960, 100, 110). Significantly these figures are for a nation that is not widely seen as a key port of emigration. When we think of Ireland, which saw its population fall from 6.53 million in 1841 to 2.82 million in 1961 (Brunt, 1988, 9), largely as a result of emigration, and Italy, which together with Ireland was a major contributor to the 15 million immigrants the USA received over the 1890–1914 period (e.g. MacDonald, 1963), we begin to get some sense of the magnitude of population change. We can link these flows to particular rural environments, with the peasant smallholdings of Galicia in Spain providing an especially strong source (Bradshaw, 1985). Thus Galicia contributed a major share of the 2 896 832 nationals that Spain lost in transoceanic movements between 1901 and 1935 (Puyol Antolín, 1989, 55-61).[3]

A degree of turbulence in population composition also existed at a local level. Thus detailed analysis of even isolated rural localities shows that population turnover was not an uncommon occurrence. In the Breton commune of St Jean Trolimon, for instance, Segalen (1991) found that around one-third of households changed their place of residence every five years as early as the nineteenth century. At a time when transport facilities were poorly developed, even quite short distance moves could reduce the social cohesion of a group. Studies of more recent decades also reveal

---

3 The peak period here was from 1905 until the outbreak of the First World War. In every year over that period at least 100 000 Spaniards made a transoceanic emigration, with 1912 recording the peak departure of 245 219 persons (Puyol Antolín, 1989, 58). As a measure of the relative magnitude of this movement, the total population on the Spanish mainland along with Islas Baleares and Islas Canarias was 18 594 405 in 1900 (Puyol Antolín, 1989, 102).

substantial turnover rates in places dominated by small agricultural landholdings (e.g. Williams, 1963; Wylie, 1966). As Wylie (1974, 352) described the situation: '... I begin to wonder what I mean when I refer to "the people of Peyrane". It seems now that this expression refers primarily to the small number of people who apparently remain in the community and transmit its culture to the many individuals who move in, remain for a few months or years, and then move away'.

The key message from all this is that peasant communities were places in which camaraderie often came about as a matter of necessity rather than choice. When given the choice or opportunity, residents either ceased to behave in a mutually supportive manner or grasped the opportunity to leave. This departure was not simply from the place but also from the occupation of farming. Most obviously we see this for the landless peasants who had to sell their labour to farmers, for often these were the first to leave (Bradshaw, 1972; Harding, 1984; Segalen, 1991). Following in their wake were those who would not inherit their parents' landholding. With few alternative means of employment, it might be thought that landlessness was the key factor behind the decision to leave. Evidence suggests that this is too simplified an interpretation. For one thing, in certain localities delays over the formal takeover of decision power into late adulthood was seen as so constraining by offspring that they would prefer to abandon their home village than wait to take on farm management (Golde, 1975; Greenwood, 1976). Whether or not farm takeover beckoned, the classic migratory pattern across southern Europe during the 1950s and 1960s was of initial moves by males to northern Europe, followed only later by women and children as part of a pattern of family reunification. Sometimes, this latter phase did not occur, with the abiding impact this had on local gender structures summarized succinctly for Portugal by Brettell's (1986) book *Men who migrate, women who wait*. And for those who had not decided on a partner, finding one became increasingly difficult. Gourdomichalis (1991) offers Greek evidence on this, with a survey of farm women in which only 3% wanted their children to become farmers, whereas some 81% had a strong desire to see that their children definitely did not go into farming (see also Davis, 1973). In many situations where inheritance passes down the male line, analysts confirm the impact of this view, with repeated reports of young women from farm backgrounds choosing to emigrate rather than marry into a farm family (Golde, 1975; Douglass, 1976; Greenwood, 1976; Wall, 1984; Breen *et al.*, 1990). In Ireland, what lay and still lies behind this pattern is a rural social system that places little value on women's work (Shortfall, 1992). Whereas in the past departing Irish women often went to work in the UK, recent demand for higher education has generated new outflows, again involving more women than men (King and Shuttleworth, 1988). Altogether, then, it comes as no surprise to find that half of all farmers in the EC are over 55 years and many have no prospective successor (CEC, 1989b; Potter and Lobley, 1992; von Cramon-Taubadel, 1993). Indeed, while the everyday life of a family on a smallholding might be hard, it is not simply the shortage of money that is problematic. We see this for return migrants who leave their communities to earn more money and have the potential to return with funds to buy a bigger landholding. From work in peripheral parts of Europe the message on return migrants is clear. With the exception of Portugal (Cavaco, 1993), relatively few return to their original localities, with many favouring cities instead (e.g. Unger, 1986). When they do return, few invest in agricultural endeavours (Lijfering, 1974; King *et al.*, 1985; Gmelch, 1986) and many come to regard their communities in a negative light and soon look to leave again (Douglass, 1976; Manganara, 1977).

The message then is that peasant communities did not form a model for the future. For many people they were certainly something to be admired, as the sheer difficulties that peasants faced in eking out a living in what were often highly unfavourable circumstances offers a vision of the imagination, resilience and adaptability of human beings. However, in glossing the peasant existence with a whole gamut of moral and idyllic images, those with real power were able to distract attention from the inequities of rural existence (Tarrow, 1977; Tracy, 1989). The utilization of an ideological gloss by power elites is not restricted to notions of the peasantry, as Newby (1975) notes for usage of the idea of a 'rural community' in England. For some neo-marxists, once they engage in market transactions, peasants are even advantageous to the needs of capital. Thus, Vergopoulos (1978, 446) has suggested that '... family farming is the most successful form of production for putting the maximum volume of surplus peasant labour at the disposal of urban capitalism. It also constitutes the most efficient way of restraining the prices of agricultural products'. This view contradicts the conclusion that '... the peasant's justified suspicion of the market system means that a subsistence orientation has survived until today' (Pina-Cabral, 1986, 10), for it points to peasants being willing to respond to new initiatives as part of a survival strategy. Evidence of this already exists for commercial initiatives in both agriculture (Eccles and Fuller, 1970) and manufacturing (Holmes, 1989). With the attraction of peasant-workers for capital lying in peasants being prepared to take relatively low renumeration, while their farm products tend to be cheap because their land is not capitalized, so ground rent does not feature as a cost in agricultural production. In addition, peasants have a crucial political significance through their support for those right-wing parties that centre on landed interests. As such it is no surprise to find that one reason for the survival of peasantist production units today is the availability of state welfare payments that support traditional production systems and make-up a significant proportion of farm income (Jansen, 1991).

Realistically, what these last points indicate is the manner in which peasants rely deeply on 'outside' opportunities for their improvement.[4] What we see in peasant communities is not an idyll but a treadmill. These were communities that were waiting to be acted upon. Even their egalitarianism weakened their ability to react to possibilities to improve their position, for they lacked the resources or connections to draw on or create new opportunities for themselves. Hence, the apparent slowness of change that they experienced owed more to an inability to do more, not from choice. When their residents were faced with choice they often selected an alternative lifestyle, and then generally away from their home locality. In this regard, despite constituting so large a proportion of the European rural past, the present for former local peasant societies is highly variable; almost being decided by whatever socioeconomic force happened upon them.

Tendencies of competition and cooperation

For many in northern Europe, what we will be examining in this section is what they would refer to as a traditional agricultural community. In many respects we would not disagree with this sentiment, at least as far as some countries or regions of northern Europe are concerned. However, as the reader will have noted, in the previous section on peasantism we provided various examples of behaviour forms that came from northern Europe. In some measure this arises because the agrarian structures of northern Europe are not uniform. Whether we look at France or Germany we see very different farm structures that epitomize dissimilar orientations in farm activity, ranging from the household-centred economy of the peasant to highly commercialized agricultural terrains (Mellor, 1978; Campagne *et al.*, 1990). However, there is another reason for this symmetry in illustrations. Since the boundaries that distinguish peasants from more competitive farmers are somewhat opaque, it is not without precedence for significant elements of their respective social systems to be shared, and even for features of them to coexist in the same location. This point is a general one that applies across the five main tendencies we have identified. Of course, the notion that two systems can coexist within the same locality might produce a sense of discomfort for those who like to see clean lines of explanatory accounting or descriptive precision. Nevertheless, its occurrence is well illustrated in one of the classic studies of rural Britain (Newby, 1977). In this modern investigation of rural farm workers, questions were asked about the reasons farm labourers believed that their workplace was the best available. The bifurcated nature of farm endeavours in this Suffolk study became clear in the responses

4 For example, when reviewing reasons for the survival of small farms in the Udine region of Italy, Saraceno (1994) argues that the abundance of small land plots for purchase and the fragmentation of holdings on inheritance underpins peasant survival. Yet this underpinning is an historical fact of postwar legislation, which encouraged the sale of small tenanted farms, with the acquisition of land having a stabilizing effect on rural populations, as it cemented a system in which former tenants became small landowners.

that were given, since for labourers on more commercially oriented farms the main attractions were high wages (32.1%) and efficient farm operations (36.9%), while for those employed on 'patriachal' farms it was having a good boss (39.6%; Newby, 1977, 311). Given the potential for such intermixing, in proximate areas especially but perhaps even in the same locality, there is need for a broad distinction between competitive and peasant social environments. For us, these largely stem from the larger landholdings and greater wealth of competitive production units. To clarify the distinction we draw here, we are still referring predominantly to family farms. However, in nations that have depended heavily on agricultural exports for some time, like Denmark and the Netherlands, prompting from state leaders helped farmers recognize that agricultural efficiency was essential for their own and national economic survival (e.g. Kamp, 1975; Rying, 1988), so inducing a notably entrepreneurial attitude towards farming (e.g. Benvenuti, 1962; Webster, 1973; de Haan, 1993).

Although perhaps less so in the past, certainly over the last 30 years production units used little bought labour. Moreover, when labour was paid for, workers were treated more as family members than employees. This provides a dividing line between competitive social orders and both conflictual and paternalistic ones. Significantly, a sense of community pervaded these competitive environments that attached itself to localities dominated by this productive mode. Offering one illustration of this, Williams (1956) reported that the practice of treating farm employees as one of the family was so entrenched in Gosforth that when one farmer sought to establish a more formal employer – employee relationship he found it difficult to get anyone to work for him. Similarly indicative of a communal spirit, we find regular reports of mutual aid, such as the borrowing of equipment between farmers (Williams, 1956; Harris, 1972). Most evidently, of course, this is seen in formalized arrangements, like the cooperatives that have so long dominated the agricultural economy of Denmark (Webster, 1973; Table 3.5, page 95). And it would generally be true to say that these cooperative systems are not the same as those which developed under peasantism, where, as in Bretagne (Brittany), the instigation and organizational impetus of the cooperative came from outside the local community, being directed by those who saw in cooperativism a particular political gain for themselves (S. Berger, 1972). What characterizes the communal elements of competitive agrarian social environments is a sense of relative equality amongst participants in social networks.

A key dimension of this is inheritance, which forms the major component of land transfer in rural areas. Thus, in Newby and associates (1978) survey of East Anglian farms they reported that 69.5% were inherited (including 60.4% of those over 400 hectares), with 80% of farmers having fathers who were also farmers. Indeed, in all EC countries except the UK and Spain, new farmers entering the sector are almost exclusively from agricultural backgrounds (CEC, 1991c). Perhaps the UK situation is slightly different, due to its more open land market and the provision of professional agricultural training which leads to 15% to 20% of new farmers coming from outside the sector. Likewise in Spain the entry of those with non-agricultural backgrounds is linked to the recent growth of capitalist farming in horticulture and other activities along the Mediterranean coast. However, over Europe as a whole, despite differences in national legislation regarding land inheritance, the transfer of agricultural land from parent to child represents a powerful force in maintaining local social cohesion. In areas with a history of commercial family farms it also emphasizes considerations of equal commonality amongst local residents.

This egalitarian tone has been identified in much writing through an emphasis on the strength of status distinctions compared with those of social class (e.g. Williams, 1963). In the *buchedd* groups of rural Wales, for instance, the dominant characteristic was extensive kin ties, a lack of occupational differentiation, and few hired workers, resulting in a status-based system of social stratification (Day and Fitton, 1975). Significantly, such social distinctions were not taken from any extralocal source, but arrived from local definitions, which placed significant stress on such activities as skill in hay cutting, ploughing or animal husbandry (Christian, 1972; Newby, 1977). Not that class was not an important factor, for while the range of class positions might have been somewhat restricted in some of these places, investigations in various locales show that even amongst farm owners, size of landholding, the number of cattle owned or some other wealth indicators were integral to social differentiation (Williams, 1956; Littlejohn, 1963; Christian, 1972). Moreover, when we examine the experience of those in different class positions we find that their life chances were different. Nalson (1968) gives a clear message on this score. He debunks the fable that farm labourers were able to move up the 'farming ladder' and take on landholdings of their own, by showing that only four of the 172 farmers in his Staffordshire study were at one time farm labourers. Although the comparison is not exact, in East Anglia the figure seems to be slightly higher, but still only 10% of farmers began their careers as either farm managers or farm labourers (Newby *et al.*, 1978). As the Keurs reported for a Drents community: 'Both renters and farm workers are conscious of their status, even while they say, "We are all alike here in Anderen" ' (Keur and Keur, 1955, 149). Nevertheless it is a common feature of such communities that genuine attempts were made to preserve social harmony. As Frankenberg (1957) found for 'Pentrediwaith' this involved seeking 'outsiders' of higher standing to act as symbolic heads of local organizations, so that if a crisis occurred they could be thrust forward to seemingly take crucial decisions, with the result that if anything went wrong acrimony would be deflected onto the 'stranger' and not disturb village unity. Williams (1963) likewise found a reluctance to accept positions in organizations, for fear that whatever actions were taken would be bound to offend someone.

Such strategies for social survival could not be as easily extended to economic survival. In terms of farm organization, the genesis of the farm structures that were associated with competitive farm environments owed much to government policies. In the case of Denmark and the Netherlands this was seen in the promotion of particular agrarian structures. In the case of Britain it resulted partly from the capitalist orientation of many large landowners, which saw the early arrival of mechanization and farm improvement in the country (Mingay, 1977). But it owed more to the 'victory' of manufacturers over landowners in national power struggles, which resulted in an emphasis on cheap food and associated large-scale imports (Crosby, 1977). This led to the breakdown of many large estates into saleable family farms or more efficient tenant operations that brought economic relief from problems generated by lower commodity prices, death duties and losses from gambling and extravagant consumption (Mingay, 1976; Beckett, 1986; Cannadine, 1990). And while the state had underwritten the creation of this social system, it was not in a position to resist continuing pressures for agrarian improvement that resulted in substantial losses in farm numbers as operators were pushed into increasing their farm size in order to create more competitive units (e.g. Table 6.1). The competitive nature of farmers in these communities in many respects led them 'naturally' towards actions that would undo their social order. In Gosforth, for example, we see the breakdown of cooperative exchanges in farm equipment as farmers were drawn to take advantage of tax benefits that accrued to those who purchased their own machinery (Williams, 1956). Indeed, Williams (1956, 20) found that almost every conversation in Gosforth was punctuated with the phrase '... that died out when t'tractor came'. Added to modernization were the pressures of urbanization and the growth of commuter dwellings in villages. In Britain the

Table 6.1 Farm holdings in Germany by area class size, 1949–1989

	Under 10 ha	10–20 ha	20–50 ha	Over 50 ha	Total
Number of holdings in 1949 (000s)	1262.5	256.2	112.5	15.6	1646.8
Number of holdings in 1989 (000s)	307.1	136.6	160.1	45.1	648.8
Percent holdings by class size in 1949	80.1	13.2	5.8	0.9	100.0
Percent holdings by class size in 1989	47.3	21.1	24.7	6.9	100.0
Percent total area by class size in 1949	39.6	26.3	24.0	10.1	100.0
Percent total area by class size in 1989	11.1	16.9	42.1	29.9	100.0

Source: based on Bergmann (1990, 50).

extensive time over which these pressures built-up is indicated by the strength of interwar political sentiments that decried the demise of the countryside under the impress of urbanism (Jeans, 1990). But their implications went beyond any changes to local social organization that commuters introduced, although these were obviously influential (e.g. Connell, 1978). Also of significance was the manner in which urban-centred influences on the countryside increased the value of land, so restricting the opportunities of farmers to enhance their productive capacity through land acquisition. As in both Denmark and the Netherlands (e.g. Kamp, 1975; Riley and Ashworth, 1975), in parts of Britain that lay within the hinterland of larger cities, this led to an intensification of production, although the fragmentation of farm holdings at times led to coping strategies which included extensification or even a de-emphasizing of farm production for other activities in places close to urban centres (Beavington, 1963; Best and Gasson, 1966). Inevitably, when associated with an influx of urbanites, this led to social change. Not surprisingly, then, in the postwar period British localities that are commonly presented as representations of competitive social tendencies were found in upland zones away from cities (e.g. Williams, 1956; Harris, 1972), with some analysts providing broad distinctions between rural societies in upland and lowland agricultural regions (Newby, 1979). Over recent decades, however, even distance from cities has not provided a sanctuary for these upland zones, which have not only been buffeted by the changing economic position of agriculture but have also been recipients of major population inflows from social groups who are new to these parts.

Clientelist social tendencies

In the same way that we can find traces of peasantist and competitive social orders unevenly distributed across past European landscapes, so too do we see the unequal incidence of clientelism. Most notably, clientelist social formations are those in which 'patrons' and 'clients' have mutually dependent social roles. Patrons are either regarded or regard themselves as some form of protector, guide or model, whose local role is most notably distinguished by their position as intermediary between the general populace and those (extralocal) agents who are believed to exercise most power over the locality (Kenny, 1960). Whether or not the patron is actually able to exert influence over extralocal power interests is neither here nor there. Indeed, there is evidence from Irish studies that patrons often do not have a great deal of influence (Bax, 1976; Sacks, 1976). What is most important is that they are believed to have the greatest prospect of exerting such influence. Commonly the clients of such patrons do not feel that they are in a position to take on a leadership role, perhaps because of local social mores about leadership

(Wade, 1971). It follows that those who avail themselves of the patron's services provide a reciprocal bond in which the client 'bestows' status and power on the patron, in return for material or intangible benefits. For the streetwise, clientelism boils down to having a 'friend' in the right place who can circumvent or has the best chance of overcoming obstacles that are thrown-up by powerful, usually extralocal, agents.

Not surprisingly, the existence of clientelism is most often identified in countries with a strong central state, or at least where there is a weak institutional structure at the local level to argue for local variations to national policies (Graziano, 1978). As such we find accounts of the southern EC nations, along with Ireland, dotted with descriptions of clientelist power structures (e.g. Kenny, 1960; Silverman, 1965; Bax, 1976). The continuing potency of clientelist forms is readily apparent in a weak response to EC measures that demand broad local support, as various analysts identify in Greece (Georgiou, 1994; Wenturis, 1994). Also distinguishing this sociopolitical process is the underlying condition of its existence, for it is characteristically found in nations with a poorly developed tradition of political democracy. There, either through deprivation resulting from colonialism or through authoritarian political rule, the populace can be seen to lack a longstanding respect for political authority, or to have become accustomed to the caprices of its rule (Giner, 1985). It is perhaps noteworthy that nations which demonstrate clientelist social forms commonly come from a Catholic tradition, given evidence that the Catholic Church has encouraged passivity amongst peasants (e.g. Furlong, 1994). But one would not want to exaggerate this point, given differences in the practices of Catholicism across nations (e.g. see Brenan, 1950, on the distinctiveness of Spanish Catholicism),[5] plus the fact that rural clientelism is not well represented in some Catholic areas (e.g. Belgium and France).

Taking these attributes as our guide, we find Murdoch and Marsden's (1994) inclusion of clientelist tendencies amongst British rural types somewhat puzzling. It is true that, over time, social formations can be transmuted, such that their innate character takes on a new form, and it becomes appropriate to view new geographical venues as demonstrators of its central elements. Certainly, social practices are not unchanging, with the evolving nature of peasantism over the last 200 years providing a clear example of this (Wright, 1964; Mendras, 1967). However, when we examine what Murdoch and Marsden mean by clientelism we find that this boils down to a dependence on state aid, primarily in remoter areas with economic bases that heavily depend on agriculture. Viewed in this manner we could extend the geo-

---

5 This distinction is important not simply on account of dissimilarities between, say, Italian and Spanish Catholicism, but also due to distinctions within Spain itself. As Brenan (1950, 94) explained: 'Generally speaking, wherever the rural population in Spain is concentrated into large villages or towns, the clergy are on the side of the middle classes and *caciques* [elite dominated political machines] and against the people, whereas when the peasantry are dispersed in hamlets and small farms, the clergy associates itself with them'.

graphical scope of clientelism very significantly. After all, the state now accounts for most of the jobs taken by women in the more urbanized portions of rural Sweden (Persson, 1992), while some 65% of state programme expenditure in rural England and Wales is estimated to go directly on income support for farmers (Hill and Young, 1991).[6] But such an extension would weaken the core characteristic of clientelism, which is a focused interaction with the outside world through the auspices of a patron. Some patrons might have few resources at their disposal, but more commonly they do exert influence. Today, for sure, that influence often comes through linkages to national organizations, which enable local agents to manipulate state resources within their locality (Littlewood, 1981). But there is a world of difference between fluctuations in the application of national guidelines resulting from bureaucratic interpretations of rules and regulations and the ability to direct resources to maintain a local position of power. The latter, for us, is clientelism, the former is not.[7] And even today clientelism can have a significant bearing on rural living conditions. The uneven power of agrarian trade unions in Italy offers one example, for the discretionary nature of welfare payments for farm workers has meant that local unions have been able to take on a patron-type role such that farm workers have effectively become clients of the state (Pugliese, 1985). More generally in Italy such clientelist practices have come to be associated with political parties (Furlong, 1994), with the distribution of work schemes as a form of unemployment relief providing an example of the same practice in southern Spain (Mansvelt Beck, 1988).

At an intra-national level, one issue that merits consideration is why

---

6 It needs to be stressed that the Hill and Young (1991) analysis focused on 170 specific state programmes. If the totality of government spending is considered the role of the state is even more evident, although the position of agriculture declines sharply. Thus, in Sweden, where some 50% of the value of agricultural production comes from government subsidies, less than 5% of state expenditure in rural areas goes on agriculture. The share of this sector is swamped by 50% of governmental expenditure going on social security transfers (unemployment, child assistance, pensions, etc.), with a further 27% allocated to the construction and maintenance of public services and infrastructure (Persson, 1992).

7 It seems to us that the main point of the distinction Murdoch and Marsden (1994, xi) draw is to focus on remoter rural areas dominated by farm activities. For us, any area that has a significant agricultural presence will also see substantial elements of its income flowing through state channels. Given the competitive nature of the markets agricultural producers operate in, state regulation and support are critical to the maintenance of a farm sector (O'Connor, 1973; Hoggart and Buller, 1987; Tarrant, 1992). Unless geographical or production spheres are 'abandoned' it follows therefore that the state will inevitably have a key role to play in regions with a significant agricultural presence. But the mere presence of high levels of governmental funding should not be equated with clientelism. Moreover, we are not convinced that a categorization for remoter agriculturally-dominated regions is particularly helpful even in understanding rural Britain. A wealth of evidence already shows that remoter areas receive substantial in-migration flows with little agricultural connection but with notable effects on local socioeconomic activities (Forsythe, 1980; Jones et al. , 1986; Williams et al ., 1989; Bolton and Chalkley, 1990; Cross, 1990), with the continuing strength of this demand apparent in nationally high rates of new housing construction in remoter areas (e.g. Hoggart, 1993).

certain locations have been more prone to clientelist social relationships than others. Clearly this has something to do with the nature of the central state itself, and the manner in which it exerts its influence unevenly across its territory. At this point to state more would be unwarranted, for the literature does not yield sufficient clues to give confidence in such a categorization (although the existence of distinct local political traditions in nations like Italy is evident; MacDonald, 1963). Fortunately, this is not required for our purposes, since our intention is to consider how a background of clientelism helps predispose localities to pressures for socioeconomic change. Here we can make pointed statements. This is because a fundamental feature of clientelism, despite its grounding in national political cultures, is that of strong local connections for the patron (e.g. Sacks, 1976). It might be that other communities in a region exhibit a similar sociopolitical practice, but these will be centred on local people of standing. This locality-specific base means that local patrons do not have a strong power base on which to resist extralocal forces of change. While they can attempt to utilize their connections at a regional or national level, it is really at this scale that power is exerted. It is also at this level that the pressures of globalization and European integration are most powerfully felt; so weakening the desire to allow specific local exclusions. Indeed, one of the characteristics of those nations that previously saw extensive clientelist practices is that they have been wholehearted in their adoption of EC initiatives (Arkleton Trust (Research) Ltd., 1992). With the national subjugated to a larger whole the local is hardly likely to be allowed to deviate. Beside which, as these states have taken a more democratic form, the range of state services on offer, and in some measure the variety of private sector openings available, has increased. As a result, the span over which extralocal forces are impacting on local communities has broadened significantly. Already, changes in the nature and intensity of clientelist relationships are documented (e.g. Bax, 1976; Furlong, 1994). The message is not simply that clientelism is in retreat but that its patrons are insufficiently powerful to divert the new forces of socioeconomic change that are engulfing rural Europe.

Rural communities of conflict

A commonly projected image of rural areas in Europe is one in which the population adheres to conservative sociopolitical values. Despite the depth and long gestation of broad agreement that sociocultural differences should not be tied to settlement forms, we unfortunately still find analysts referring to the conservative nature of rural areas (e.g. Marsden *et al.*, 1993). Perhaps as a piece of empiricism this statement fits certain nations. As a theoretical statement it should be rejected wholeheartedly. Accepting this view, we were pleased to find a category for the 'contested countryside' amongst the ideal types that Murdoch and Marsden (1994) present as characteristic of change processes in rural Britain. We must admit that we take some licence in our interpretation of their use of this category, for Murdoch and Marsden were principally concerned with contemporary change in the UK countryside, whereas we see contested rural spaces as more dominant in the past, and being weakly developed in Britain. The type of contestation that Murdoch and Marsden refer to is examined in this chapter under our consideration of recent transitions in rural Europe. What we wish to draw attention to in this section is the peculiarities of social forms that have longer traditions, even if their geographical extent is relatively limited, where the key feature of local societies comes from a political radicalism grounded in class conflict. This is a social order that is more commonly associated with urban manufacturing centres.

When we undertake a panoramic view across the present European Community, we find relatively few areas that we feel comfortable about labelling rural and radical. Amongst those that do exist, as long-term attributes of locally-grounded social systems, rather than as fleeting occurrences like the Captain Swing revolts in Britain (e.g. Charlesworth, 1994), Italy and Spain perhaps provide the most compelling examples. For both we are reminded of the strong sense of local identity that exists in these nations. In Italy the late arrival of nationhood offers some clues on the basis of this sentiment (King, 1987), while the strength of the *pueblo*, whether as an urban neighbourhood or as a village community, lies at the heart of Spanish attachments (Brenan, 1950). But not all local identities in these nations end-up with a radical sociopolitical disposition. Fortunately, we have two excellent studies which provide clues on the regional specificity of such orientations. The first of these, Gerald Brenan's *The Spanish labyrinth*, provides us with one of the most incisive and insightful studies that has been published on a European nation. The second, John MacDonald's (1963) inquiry into the links between local social structures and divergent Italian migration patterns, has a more limited focus but offers a concise account of regional differentiation. When read alongside a variety of other studies, that either explicitly focus on radical traditions (Cooke, 1985; Collier, 1987) or examine sociopolitical events in such locations (Cardoza, 1979; Snowden, 1979; Mansvelt Beck, 1988), we are left with a clear image of the basis of the distinctiveness of these regions. In a nutshell, this pinpoints the oppressive character of working conditions in areas with a strong sense of communal identity, wherein labourers or sharecroppers, such as under the *mezzadria* system, were faced with an insecure, arbitrary land tenure status, along with heavy monetary demands through rents and taxes, with the Church adopting the role of apologist for the landed elite. With spiritual leaders more prone to stir hatred than help alleviate hunger and social distress, local workers turned to political solutions that centred around a left-wing subculture which provided a theoretical rationale for changing the prevailing socioeconomic system

(Tarrow, 1977). Offering some indicators of the social framework in which such movements emerged, Brenan (1950, 116–123) reminds us that giant *latifundia* controlled 41% of agricultural land in the province of Córdoba, 50% in Sevilla and 58% in Cádiz. Three-quarters of the rural male population were *braceros* (landless day-labourers), who could generally expect to be unemployed for four-to-six months a year, with no back-up access to a personal smallholding, and an annual wage in 1930 of £15–£25. It is hardly surprising, as Brenan (1950, 122) notes, that with so many unemployed it became something of a point of honour to do as little work as possible for the landowner.

Two points should be made about this subculture. Firstly, areas like Andalucía,[8] Puglia (Apulia), Emilia and Tuscana (Tuscany) were not alone in housing large populations who lived in appalling conditions. The difference that needs to be grasped is that the populace of much of the rest of rural Europe had access to smallholdings, unencumbered by what were considered to be exorbitant rents, or else existed under more paternalistic social orders. Hence, as with Sardinian banditry (Schweizer, 1988), the political response to any sense of grievance against 'the system' had a tendency to be individualized. By contrast, as occurred when Tuscan sharecroppers came to live together in villages at the end of the nineteenth century or when labourers in Andalucía left their families to sleep in outbuildings (*cortijos*) close to the crops being harvested, there was a coming together of the aggrieved in circumstances where there was an identifiable 'enemy'. Secondly, as implied above, the existence of such subcultures relied on an interchange between the landed elite and their workers. The landed classes needed these workers, otherwise, like the highland clearances of Scotland (Preeble, 1963), they would simply have expelled them. But this dependence was not passive, for while the degree of what we would see as the callous and exploitative treatment of workers did vary from place to place, there were also efforts by these groups to forge alliances to defeat moves to better worker conditions. We see this in alliances between industrialists and landowners in the Po Delta (Cardoza, 1979), which, as in Toscana (Snowden, 1979), led to landowners supporting the fascists to defeat the working population. But perhaps most violently we see it in the causal impetus that lay behind the Spanish Civil War of 1936–1939 (Brenan, 1950).

As for the impact of this worker solidarity on the rural landscape, here we must talk about outcomes in the plural. In MacDonald's (1963) work, for instance, he is able to link worker social cohesion to a resistance to emigration in areas like Puglia and central Italy, which contrasts with mass movements that characterized areas of peasant farmholdings or even areas of large estates

---

8 We focus on Andalucía in this discussion, although this is not the only part of Spain that demonstrates a tradition of rural radicalism. However, in other regions of this ilk the historical path of protest is complicated by the presence of regional separatist sentiments. In places like Cataluña and Valencia, this provided a radical colouring even to many in the middle and capitalist classes (Brenan, 1950).

like Sicilia where worker resistance was thwarted by the Mafia. The same appears to be true in the Alentejo in Portugal, which in its consistent support for the Portuguese Communist Party can be seen as a 'radical region' (Clark and O'Neill, 1980), and where emigration again was much lower than in other 'peasant' dominated regions of the country (Almeida and Barreto, 1976). Yet in Andalucía we find that, as soon as legal restraints were removed, mass emigration to other parts of Spain and beyond into more northerly European nations became a dominant population trend (Collier, 1987; Mansvelt Beck, 1988). Also unexpected is the manner in which some 'radical' areas have proved attractive venues for decentralized industrialization. Emilia-Romagna stands out in this regard (Cooke, 1985), with Andalucía now emerging as a new growth area in Spanish manufacturing (Salmon, 1992). What these illustrate is the way in which past sociopolitical traditions, while perhaps still visible in the electoral landscape, are being acted upon by unexpected forces. As Cooke (1985) indicates, Emilia is the kind of area we would expect manufacturers to shy away from, given its worker traditions, yet it has attracted substantial new investment. From this, we get the sense that a long history of social distinctiveness might not hold the key to explaining new surfaces of socioeconomic differentiation that are being produced by globalization processes.

A paternalistic social order

Similar to contested social landscapes, the starting point for an understanding of paternalism as a form of rural social organization is the existence of sharp inequalities in household resources. This is seen especially in the presence of extensive landed estates. The key difference between contested and paternal-istic community structures arises from the strategies adopted by landed elites. As Newby and associates (1978) point out, characteristic of a paternalistic social order is recognition by the landed classes that if the labourers could be induced to identify with the system that subordinated them, then in the long-run workers would be more reliable and efficient than if they worked grudgingly. Explaining the character of the sentiments that underwrote the ethos of such landowners, Beckett (1986, 5) suggested that: 'Land was the basis of their credibility, and estates were regarded as possessions held in trust from generation to generation'. As a consequence, '... the responsibility exercised by the aristocracy in economic development was matched by a similar concept of duty and service in regard to the leadership of both locality and nation' (Beckett, 1986, 9). The gulf that separates this vision of rural leadership from that of contested social landscapes is encapsulated in the contrasting films *1900* and *The Shooting Party*.

In drawing this comparison we also highlight a particular feature of paternalistic social structures, for the very 'English' style of *The Shooting Party* gives away one of the core attributes of this social order, which is its more extensive presence in Britain than elsewhere in Europe. In some measure this resulted from the simple nature of landholding structures. As Cannadine (1990, 9–19) explains, France was unlikely to develop this system on any scale, for there were hardly any estates larger than 4000 hectares. In contrast, Britain in the 1880s had around 780 holdings of this kind, with a further 6000 owning between 400 and 4000 hectares. However, we must also recognize the peculiarities of the national social order that underscored the British landed system. For, as Mingay (1976) amongst others makes clear, the landed gentry saw themselves as pillars of social stability, such that they believed that the decline of their influence and example would inevitably be matched by a rise in lawlessness and chaos.

All this should not be read as implying that rural paternalism provided a caring and benevolent social system. This certainly was the impression that landowners wished to give, as Newby (1975, 158) recognized: 'The traditional English land owning class placed an ideological gloss on their monopoly of power within the locality through the concept of community'. Behind this facade decisions were taken, and in some parts of Britain continue to be taken, that narrowed the life-chances and quality of life of local residents (e.g. Kendall, 1963; Bird, 1982). Even in recent decades we find reports of landlords preventing their farmer tenants from improving their land in order that their own hunting or sporting is not disturbed (Shucksmith and Lloyd, 1983). Reports of deliberate attempts to restrict the availability of alternative job openings are also available (Saunders *et al.*, 1978; Wilson, 1992). There is the added restriction that workers have often had to rely on housing provided by the landowner, which increases the pressure on farm labourers to show due deference to their employers (Newby, 1977). The duality of dependency on a landowner for both employment and housing is characteristic of paternalistic rural societies. It is not uncommon for the local 'establishment' to hold powerful positions not only in the economic sphere but also in realms of culture, politics and social affairs (Johnson, 1972; Newby *et al.*, 1978). The agents of such landowners also benefit from their association with the local elite, with Cutileiro (1971) providing us with the poignant Portuguese example of the priest of Vila Velha distributing wheat flour, powdered milk and butter only to those who attended daily mass, even though these were provided by an international organization for distribution to all. Potentially, then, a sense of oppression hung over the rural community, which did not carry the violent tones of radical rural regions in southern Europe, but which did poke its head above the parapet in such guises as the landowners' agents keeping watch on polling stations at election times (Johnson, 1972).

Sustaining such a social system would always be made more difficult once the state introduced death duties on the value of property (Cannadine, 1990). But these threats were less important than the declining economic impor-

Table 6.2 Percentage change in UK farm workforce composition, 1977–1989

Farmers, partners and directors		Family workers		Hired workers	
Full-time farmers	−10.8	Full-time male	−24.3	Full-time male	−41.5
Part-time farmers	13.1	Full-time female	−47.4	Full-time female	−9.1
Spouse directors	−3.8	Part-time male	−22.1	Part-time male	−14.7
Managers	0.0	Part-time female	−26.3	Part-time female	−15.3
				Seasonal/causal	−12.5
Overall change	−4.0	Overall change	−26.7	Overall change	−26.8

Source: Munton (1992b, 69).

tance of agriculture, and the challenges to income earning capacity that arose from cheaper overseas imports. These were not simply menaces that undermined the fabric of large landed estates in the late nineteenth century. They were integral to the ongoing existence of large estates as agricultural operations, given the way in which farming was increasingly squeezed between demands for lower prices and pressures to produce greater quantities of food (Chapter 5). But once a more commercial orientation was adopted, the inherent inconsistencies in the role that the landed classes had bestowed on themselves became all too obvious. Here, on the one hand, landowners were uprooting hedgerows, ploughing out footpaths, and adopting highly commercial farming practices, yet, on the other hand, they laid claim to a stewardship of the rural landscape (and lifestyle) that saw them opposing local job creation and new housing construction (Rose *et al.*, 1976). At the same time, their workers could feel less secure about their employment, for the major losses that came from agricultural modernization were amongst ancillary workers (e.g. Table 6.2). If this did not threaten the social system sufficiently, the drawing of rural politics into mainstream national discourses (Johnson, 1972), along with the reorganization of local government England, Scotland and Wales in 1974/1975, which integrated rural and urban authorities and so diluted rural power bases, all contributed to this decline. The march of mass culture was also to have an effect but, as with many aspects of change in rural Europe, this and other changes were not simply a function of impositions on particular localities or even of a more general penetration of national or global influences. In addition, they occurred because the people who lived within these localities changed. As this happened the social character of these places also altered, with certain old practices disappearing, new ones emerging and, for elements of the new populations, a different array of restraints and opportunities that required attention.

Transformation and retransformation in rural societies

In writing on much of Europe at the turn of the twentieth century, researchers have noted the self-contained character of rural communities (e.g. Arensberg, 1937; Keur and Keur, 1955), regardless of which of the models described above might appear to apply. For the labouring classes this highly localized social existence appeared to continue for much later, as indicated by Newby's (1977) observation that farm labourers in Suffolk were only prepared to travel an average of 3.7 kilometres to work, with one-third unwilling to travel more than 1.6 kilometres. Such figures no doubt cast light on only one section of the local population, viz. those that did not emigrate, and as such are prone to contradiction from studies which note considerable population mobility (e.g. Williams, 1963; Wylie, 1966). Nevertheless, there is no doubt that substantial change has occurred in the social character of rural Europe over this century. This is readily appreciated if we focus on what many would consider the most inward looking of rural social formations, the peasant community. As Brandes (1975, 9) notes, there are three overriding symptoms of the end of the peasantry: a loss of cultural distinctiveness; a reorientation of economic identities and priorities; and, the atomization and realignment of community social relations. That the first of these has already passed is readily apparent in writings on the social organization of peasant communities. Wylie's (1974, 371) words illustrate the point: 'Peyrane has ceased to be a tight little community in which such a [noon apéritif] group plays an essential role. The Peyranais no longer feel themselves to be – in fact no longer are – a unit functioning as autonomously as possible in defence against the outside world; they have become an integral part of the world they once staunchly resisted'. According to reports on the avid acceptance of television and its messages in various peasant societies (Christian, 1972; Brody, 1973), it might even be said that peasant communities have often not simply failed to resist extralocal forces but have laid down and allowed themselves be run over by them. Certainly, with the notable exception of regionally-based ethnic separatist movements (Williams, 1980), former distinctions between urban and rural orientations towards national institutions, as in politics, have broken down even in some of the least economically advanced areas (e.g. Chadjipadelis, and Zafiropoulos, 1994).[9] In part, this owes something to the growing integration of peasant life into broader socioeconomic streams. The growth of public welfare systems offers one

---

9 We would not claim that this is a universal tendency, as the recent pattern of Norwegian opposition to joining the EC demonstrated, although the Norwegian situation has some special features that make its political divisions far more complex than a straightforward urban-rural divide (Aarebrot, 1982).

obvious example, as it has changed the nature of work in rural locales (Pugliese, 1985), broadened the range of available public services in some places at least (Sjøholt, 1988), and introduced more bureaucratically organized systems of benefit distribution (Caldock and Wenger, 1992). In the economic sphere the reorientation of economic priorities is even more sharply demonstrated. The massive abandonment of farmholdings (e.g. Table 6.1), the growth of pluriactivity on farms (Fuller, 1990) and the organization and public presentation of peasant demands for government intervention to alleviate their economic plight (Naylon, 1994), all provide transparent illustrations. As for the atomization and realignment of community relations, we need look little further than the continuing involvement of emigrants in local community affairs. This occurs through visits in vacations that change the social calendar (Mansvelt Beck, 1988; Cavaco, 1993), from the catalytic effects of remittances on community social relations (Brody, 1973; Brandes, 1975) and as a result of return migration that brings new experiences, evidence of outside wealth and (possibly) altered values into a rural area (Douglass, 1986; Gmelch, 1986).

Of course if we wish to consider the totality of social change in rural Europe our task is huge. Its description would go well beyond the tree quota the publisher is allowing us to have cut for this volume. It follows that our analysis here is inevitably restricted. In some measure this is no bad thing, as the interweaving of change processes would inevitably take us back into the realms of service sector expansion, with its new forms of behaviour and expectations of support, decentralized manufacturing production, allied as this is to the injection of new modes of labour action and regulation, and agricultural decline, with its implications for local power relationships, social status hierarchies and social class realignments. Rather than providing so broad a sweep, our intention in the second half of this chapter is to focus on two main issues: the changing composition of rural populations; and, whether these modifications have themselves led to adjustments in the socioeconomic standing of rural residents.

If we are to focus on population change, it seems logical to start with a consideration of those who are leaving rural areas. Even today, despite the wide presence of inward flows of migrants into rural Europe (Fielding, 1990), there are still parts of the continent that are experiencing depopulation (e.g. Table 2.13, page 56); and we do not include in this category regions that are recipients of inward population movements but where there is still a centralization of population into the region's more densely settled parts (Vartiainen, 1989; Persson, 1992). In the past, even the quite recent past, depopulation volumes were much greater either than they are today or, for places beyond urban commuter belts, than levels of current repopulation inflow. One example suffices to illuminate the magnitude of departure. It is taken from the predominantly rural province of Jaén in southern Spain, which was never out of the top 10 of Spain's 50 mainland provinces in terms of its net outmigration rate over the 1964–1984 period (and if a symbol of the reversal of urbanization processes over this time period is needed, it is worth recording that of these 50 provinces, Barcelona moved from having the highest rate of net in-migration in 1964 to the highest volume of outmigration in 1984; Puyol Antolín, 1989, 90). For Jaén, whose population in 1960 was some 736 391, with just 31.7% living in towns with at least 20 000 inhabitants (Marchena Gomez, 1984), the period from 1962–1984 saw the loss of 228 388 of its residents to the rest of Spain, 48 367 to other locations in Europe and 432 to places beyond Europe (Puyol Antolín, 1989). Despite an inflow of nearly 70 000 persons, this 13 year period effectively saw a loss of 37.6% of the 1960 population from outmigration, with a turnover of residents at close to 50% of its 1960 population. These changes put enormous strain on community social relations, leading to the abandonment of some settlements (Mansvelt Beck, 1988).

But the manner in which these changes affected these rural areas was not new. As we have indicated above, much of rural Europe has undergone processes of depopulation for centuries now, with massive outward movement to venues overseas as well as to internal growth centres (e.g. Saville, 1957; Jones, 1960). Those who stayed saw their community strain, creak and occasionally break as it adapted to the buffeting it was receiving, but, save for the sense of loss that accompanies friends, kin and neighbours being no longer present in times of leisure, need or joy, these communities proved remarkably resilient and adaptable (Pitt-Rivers, 1971; Bell, 1979). For us, the main threats to their souls came less from the loss of personnel than from the *introduction* of new ideas, regulations and people. As such, our descriptions of the five classes of social tendency outlined above have already built into them the reality of population losses. New forces for economic change and the pressures exerted by adjustments to government policy have likewise been touched on in the previous two chapters. What we have so far not addressed explicitly is the impact of additions to the public body. This is what we turn to now.

The return of the native

There are various ways in which we can classify in-migrants into rural areas (Forsythe, 1983), but the basic divide for us is between those who have lived in a place before and those who have not. The very obvious reason for this is that return migrants are likely to have a broader range of connections with the locality into which they move, even if, as with a major trend in French postwar retirement migration (Cribier, 1982), there might have been some gap between the time when a person left a place and the eventual return to their family roots. Signifying a key dimension of the special position of return migrants, various studies have identified family reasons as a primary motivation for return, while others point to initial departure as part of a strategy for self-improvement within their home location. In the Mezzogiorno work of King and associates (1986), for instance, two-thirds of return migrants came back for family reasons; with studies of other European regions reporting a similar importance for this reason (e.g. Cavaco, 1993). Indicative of the strategy of economic improvement, in the Italian Chieti region, 57.5% of those who bought land on their return stated that obtaining the funds required for land purchase was a key reason for their initial departure (Took, 1986). A confirmatory tone is provided by a west of Ireland study, where 76% of male return migrants and 67% of females left their community in the first place for economic reasons (Gmelch, 1986). The desire of wage labourers to utilize a 'migration strategy' to strengthen their socioeconomic position within the rural community is also recorded in Portugal (Reis and Nave, 1986).

Not unexpectedly, the intensity of return migration varies geographically, for the distribution of outmigration in earlier years or decades was far from even. In particular, if we look at the years since the 1960s, the more extensive swaths of territory characterized by rural depopulation have been found in southern Europe, with Ireland also recording high outflows (Salt and Clout, 1976). Within nations closer to the vital axis of the EC, peripherality is also important, with studies indicating that around 25% of in-migrants into Cornwall are returnees, with a companion figure of 36% for the Finistère in France (Perry *et al.*, 1986). However, it is in the more peripheral areas of Europe that regions have not seen notable incidences of so-called counterurbanization movements (Fielding, 1982, 1990), so that if return migration is to assist rural development it is here that these effects are most needed.

Certainly, as Forsythe (1983) notes, we might expect return migrants to demonstrate a more easy transition into local communities, so their positive social impacts are diffused more easily. In reality, though, we do not always find much evidence of a positive social imprint. In some instances this arises because returnees have reservations about their return, perhaps on the grounds that they find community life constraining (Gmelch, 1986), although at times disgruntlement also arises from the realization that their stay abroad has not raised their local social standing and, because they have not been involved in local social networks for some time, might actually have diminished their status (Manganara, 1977). Hence, we find indicators that return migrants feel they are very different from those who have not been abroad, as 85% of Gmelch's (1986) respondents did, with this condition leading to a reluctance to put forward new ideas that had been learnt outside the community. Elsewhere, some groups of return migrants have been found to upset established social hierarchies. Thus Wall (1984) noted that on leaving a village in the coastal Minho district of northern Portugal, only 30% of migrants were landowners, with the majority wage labourers, whereas on return 74% were landowners – although possibly this reflects the fact that those who did not own land were less liable to return. More significantly, in a village further into the Portuguese mountains, Pina-Cabral (1986) reported that returning migrants had become the socioeconomic driving force in the village, who in many respects had changed the 'peasant worldview' of villagers (also Lopreato, 1967).

However, other work within northern Portugal has challenged some of these conclusions. Thus whilst Brettell (1979) suggests that the *francês*, or villager who has worked in France, can be seen as contributing to a new social class, ultimately she argues that emigration is both the result of a rigid class system, and perpetuates this system. Measures of wealth, such as the construction of a village house (*casa francesa*) by the emigrant is seen as creating an illusion of wealth that stimulates the next generation of emigrants, rather than reflecting real social mobility. In turn, rigid class divisions are reinforced by the role such displays of wealth play in maintaining important cultural concepts such as vanity, envy and gossip. Black (1992) concurs with this view, noting that ownership of land represents a key indicator of social status, and that few returning migrants make the step from landless labourer to landowner. Meanwhile, the rigidity of the class system can be seen to pre-date the rise of emigration to France and other northern European countries, being rooted partly in a tradition of emigration to and return from Brazil that dates back to the eighteenth century.

A host of studies from Italy reach similar conclusions, with analysts noting that the vast bulk of cash that returnees have gained while abroad goes into status-oriented activities rather than productive ones (for a general overview see King, 1984). Took (1986) provides one example, noting that the first priority for investing money for 57.3% of his respondents was in a house, with a further 19.3% favouring land and just 18.3% seeing their prime investments in a business. Further indications of a less than innovative economic impact and an assertion of status considerations is seen in a marked increase in the level of self-employment after return, especially in the service sector. Thus in the Mezzogiorno study 41.6% of returnees worked in the service sector on their return, compared with 18.5% before departure and 20.1% once abroad (King *et al.*, 1986; also Lijfering, 1974). This is not to say that positive economic benefits do not result from return migration, for there is evidence of this (Lopreato, 1967; Mendonsa, 1982; King and Killingbeck, 1989), with housing construction alone having an important potential bearing on depressed regional economies (Cavaco, 1993). However, irrespective of such effects, the social geography of the community is likely to change. The trend here is apparent in other parts of southern Europe, where return migration might be less significant (e.g. Douglass, 1984), but it seems that returnees emphasize the process of restructuring the internal social geography of villages. King and his co-workers (1985, 116–117) provide the clearest statement of this process: 'What is happening in Leverano is typical of thousands of other villages in southern Europe. Like microcosms of western cities, their inner areas decay through migration and abandonment, whilst upon return these same migrants transfer the vitality of the village to the peripheral estates of villas and apartments'. All this has certain similarities with the social effects of large-scale inflows into urban commuter villages, where the complexities of social interaction and conflict are potentially even more sharply drawn and intensely felt.

Far from the madding crowd

In Britain, people not only proclaim the virtues of their countryside and submerge themselves in its cultural symbolism but also reveal a proclivity to so ennoble its character that the reality can bear little relationship to sentiment and imagery.

> Today's city dwellers main contact with the countryside is scenic and sportive; the landscape is as superficial a splendour as Madame Tussaud's, as exotic as the Elgin Marbles. We domesticate this alien presence mainly by paying it homage as heritage. The heritage landscape is less and less England, more and more "Englandland", Europe's all engulfing offshore theme park (Lowenthal, 1991, 222).

As Newby (1987, 227) points out, the countryside in Britain is a positional good (Bourdieu, 1984), something that is in fixed supply such that its consumption depends upon a person's or household's position in society. As such, at least as regards new in-migrants, possession of a rural home makes a statement about the status and cultural standing of its inhabitants (Cloke and Thrift, 1990; Shucksmith, 1990b). As Thrift and Leyshon (1992, 299) put it: 'The importance of a rural backdrop is not just about cultural mores. It also applies a quite specific set of social relations: cultural capital and social capital are interrelated'. Yet interpretations of what it means to live in a rural locality are unlikely to be shared equally by all residents (Cloke and Milbourne, 1992; Halfacree, 1993). Chevalier (1993, 178) catches the tone of such differences, if in a patronizing manner, when providing us with a French example: 'However genuine the affection of residents of renovated villages for Mediterranean art and landscapes – even if this affection is tinged with snobbery – rural people, whether farmers or not, generally do not share it, partly because they cannot afford it, but also because their cultural standards are so low that they can no longer appreciate the values of their own country'. Central to this notion is the understanding that traditional social divisions between classes of farm and nonfarm groups are being or already have been replaced by a new, overriding social division, between 'newcomers' and 'locals'.

A division of rural populations into newcomers – locals, with an associated assumption of sociocultural distinctions that are commonly subsumed under an equation reading 'newcomers means middle class' and 'locals means working class', is certainly too simplified (Cloke and Thrift, 1990). Yet this notion long had a strong magnetic effect on the ideas of those who conceptualized social change in rural areas receiving in-migrant flows. This is despite the fact that a complexity of population groups was identified for such places some time ago (e.g. Pahl, 1966); and has been confirmed empirically in a variety of studies since (e.g. Harper, 1987). In recent years a step appears to have been made toward reinvigorating our appreciation of the diversity of rural populations. Perhaps most notable in this regard are Cloke and Thrift (1987, 1990), who stress that the middle class is not unitary but contains divisions, included amongst which are locals who are middle class (as with local petit bourgeoisie); thereby demolishing any neat linkage of social class with newcomer – local standings. Incoming middle class populations are likewise conceptualized in a less generalized fashion, as they are seen to be dominated by particular fractions of the middle classes, in particular professional and managerial workers who predominantly work in service industries. For Cloke and Goodwin (1992, 328) these service classes are an 'emergent historic bloc', who take-up or develop certain styles of rural housing, dominate local politics and pursue their own sectional interest – frequently against that of the indigenous rural population. Yet behind assertions of service class sociopolitical and cultural dominance lies a sense of discomfort over the use of the notion of a service class. Providing one example of this is the uncertainty researchers express over what the service class is (Cloke and Thrift, 1990; Savage *et al.*, 1992; Murdoch and Marsden, 1994). Thus Cloke and Thrift (1990) recognize fault lines in the character of the service class, with division being conceptualized between the public and private sectors, along lines of gender and stage in the life cycle, by consumption and by place. Providing so heady a list almost leads to the conclusion that it is inappropriate to think of a service class at all, given its multifaceted diversity. Interpreted rigidly, precisely this message is embodied in interpretations of rural social relationships that point to their localized nature of their construction (Cloke and Milbourne, 1992; Murdoch and Marsden, 1994). We still see problems with this approach, as it appears to lead too readily to the implication that local social relations are dominant, when they might more accurately be the result of an interaction of supralocal sociopolitical processes.

Before turning to the complexity of in-migration and social class, it is worth noting that even a simplified newcomer – local distinction carries some weight in understanding rural social differences. We see this in studies which show how the sociocultural organization of village life changes significantly once it receives a wave of new, predominantly middle class arrivals, whether in the UK (e.g. Pahl, 1965; Forsythe, 1980), France (Jollivet and Mendras, 1971) or elsewhere. This is visible in simple indicators of the composition of rural populations. In the commuter village of Milton of Campsie, for example, Pacione (1980) found very clear distinctions between council house tenants, who had lived in the village for an average of 28 years, with 84.8%

having been born within eight kilometres of their home, and middle class
home owners, whose stay averaged four years, with just 17.8% born within
eight kilometres of their home. Viewed in this manner expectations of clear
behavioural differences between working class and middle class area residents are not difficult to predict. Conceptually, however, there is a world of
difference between visualizing this middle class group as an expression of a
nationwide category 'the middle class' and seeing it as a particular middle
class social form that is locally-grounded. For Murdoch and Marsden (1994,
11) the early work on middle class rural 'invasions', such as that by Pahl
(1965), were problematic because they sought to account for social outcomes
in terms of the interaction of national class structures. The charge was that
such studies played down the specificity of local social processes, when there
was variation across localities in social class formations. We have no problem
with this argument, as the first part of this chapter indicates. However, we are
concerned that the emphasis on local specificities carries the danger of place-
fetishism, wherein local social particularities are taken to have a causal
importance which is not merited. Put briefly, in this age of emphasizing the
'other' to the possible detriment of identifying central causal factors (Murdoch
and Pratt, 1993), we believe there is a danger that the existence of a specific
social formation will be taken as evidence of a causal local process, when it is
really little more than a contingent local variation on broader (national)
processes (Savage *et al.*, 1987).

Our concern here arises from the increasing fragmentation of the middle
classes. For, as Savage and associates (1992) make clear, the trend is toward
sharper distinctions within the middle classes, albeit at a time when national
homogenization is strengthening. As such, assuming for the moment that
there are no local deviations on class structure, particular rural areas could
quite feasibly be recipients of specific fractions of middle class in-migrants
given that there are more middle class groups nationally. Yet if there is
growing uniformity *within* class fractions, their peculiar character in one
place might not merit interpretation as a locality effect. Instead what we
should perhaps be focusing on is the processes that bring particular class
fractions to particular places. This is already being done, but in reviewing the
insights obtained analysts have not as yet gone far enough in projecting the
importance of extralocal or broader social forces. Lewis and Sherwood
(1991) offer some insight, in their emphasis on the need to contextualize in-
migrant flows in terms of local housing markets, which establish constraints
and opportunities on class formation. But perhaps the work by Cloke and
associates (1991) provides a sharper empirical manifestation of this effect,
when they note how the Welsh Office responded to fears about restrictions
on economic growth caused by limited housing opportunities for executives,
by adjusting planning regulations to enable a new, non-local class group to
gain access to countryside homes. At the same time we have to recognize the
ways in which employment markets structure opportunities for householders, such that there is an uneven geographical incidence of middle class

fractions. Boyle and Halfacree (1994) provide an obvious illustration of this, when recording the differential migration patterns of men and women in so-called service class occupations. Thus women show less inclination to move into nonmetropolitan areas, with their flow patterns being directed more forcefully toward large cities, and particularly London. As Boyle and Halfacree suggest, it is not difficult to interpret this phenomenon, given the relative absence of professional jobs in rural areas in economic sectors or occupations with a high female composition (e.g. Moseley and Darby, 1978; Little, 1994); a factor which is readily apparent when we examine the dominance of more traditional household formations in rural parts of Britain (Duncan, 1991).

For us this points to the need to avoid rather loose titles like 'the service class(es)'. Our fear is that too much attention will be directed towards certain tendencies such as middle class movements into rural areas that disguise the wider complexity of what is taking place. In the past this was seen not simply in interpretations that villages could be divided into newcomers and locals, but also in the manner in which it was assumed that newcomers were urbanites. In practice, careful analysis of such flows shows that many were and are not from urban centres. Lewis and Sherwood (1991), for instance, found that movers into relatively remote rural areas tended to come from more urbanized rural zones rather than cities, while Herington and Evans (1979) report that three-quarters of those moving into rural settlements in Charnwood did not come from large cities. Certainly there are some studies which contradict this impression (e.g. Harper, 1991), but the picture of movers coming from other rural areas is confirmed by research on British home owners in rural France. Here elderly in-migrants are disproportionately drawn from locations on the English south coast (Hoggart and Buller, 1995a), while permanent relocators of pre-retirement age were over-represented from south west England (Hoggart and Buller, 1995c).

This French work points to another question mark over interpretations of middle class in-migration into rural areas. This arises from another element of defining what the so-called service class constitutes. Certainly it does not mean people who work in service industries, many of whom are low paid and do not have white collar occupations. Clearly it does not mean managerial and professional workers in service type occupations, many of whom are self-employed petit bourgeois localists in the 'county set' mould. If we take the view that the service classes are more likely to come from monopoly or state sector institutions, as opposed to competitive sector ones (O'Connor, 1973), this still does not get around the problem, for how long does someone have to work for such large-scale organizations in order to pick-up these supposed service class attributes? As various researchers have shown, many of those who move into remoter rural areas are changing their occupation, very often into self-employment (Jones *et al.*, 1986; Williams *et al.*, 1989; Buller and Hoggart, 1994a). Even if they had their previous work experience in classic service class roles, many such people are moving to the country at a young age, while others move on retirement (Rees, 1992). So, with relative wealth an important factor in relocation decisions (Forrest and Murie, 1994), and those in managerial and professional occupations in the monopoly and state sectors best placed to engage in occupationally-inspired relocations (Fielding, 1992), the (male members of the) so-called service class should have a higher propensity to demonstrate the British 'love affair' with rural living (Boyle and Halfacree, 1994). For us, this has more to do with opportunity structures than innate attributes of a particular segment of the middle classes.

In effect, we are urging caution over too simplified a conceptualization of social change in rural areas. This is not something that we feel should be restricted to interpretations of middle class in-migration into rural locales, for it also applies to their impact once in those places. The tone of many accounts of middle class 'invasions' of rural places is that of increased social problems for local working class residents, who find that their access to housing is restricted by in-migrants' desires to stop further expansion of what they now see as 'their' rurality (Cloke *et al.*, 1991). And although their expectations of wider choice and better quality in local services can see them wanting more provision (Green, 1964; Wenger, 1980), notably, any pressures for extra services are often seen to be directed toward satisfying 'middle class sociocultural interests' (e.g. Chevalier, 1993). As Perry and associates (1986, 227) expressed it: 'Once installed in the smaller settlements, middle class "do-gooders" pressurized effectively to develop a range of arts, crafts, welfare, educational and recreational amenities, that attracted more commuters from the bigger towns. The influence of the individual actors upon the scene was thus not insignificant in local cultural structures'. Unfortunately, this change to the sociocultural flavour of the area can occur at the expense of the longstanding population, as Connell (1978) reports for Surrey, where service provision favours leisure activities more commonly associated with middle class householders and largely ignores the interests of social housing tenants. Additionally, middle class newcomers are held to disrupt established social relations (e.g. Bonnain and Sautter, 1970; Forsythe, 1980), and their desire to live out their own idealized image of rurality can lead them to resist improvements in local facilities that would help lower income inhabitants (Lowe and Goyder, 1983; Buller, 1984; Little, 19987; Murdoch and Marsden, 1994). For Newby (1979) this increases the tendency for local working class residents in English lowland villages to become 'an encapsulated community', which is isolated both on class and status grounds from professional and managerial commuters and second home owners. This sense of social isolation is increased by the physical configuration of housing, which often segregates the local working classes in social housing estates, while incomers colonize the old core of the village. Class segregation does not only occur in Britain and does not rely solely on public housing for its genesis (e.g. Wild, 1983). Thus segregation is also seen in French villages, although Belliard and Boyer (1983) suggest that here, rural newcomers shun traditional houses, seeking to create a new suburbia on vacant lots.

The picture of grumbling discontent and at times open conflict these messages project is captured in Chevalier's (1993, 179) characterization of near abandoned French villages as the only ones in which middle class incomers do not provoke problems: 'In fact it is really only when a village which used to be a peasant village is nothing more than a stone skeleton that it can be transformed into a museum-village without raising fierce opposition' (Chevalier, 1993, 179). This vision is supported by Renard (1991, 418), who accuses in-migrant 'socio-professionals' of '... destructuring the social space of rural communities' in western France, such that traditional rural features like attachment to place, autonomy, cultural homogeneity and a sense of common purpose have been lost. While in-migration into rural areas is not widespread across the remoter parts of rural Europe (Fielding, 1990; Serow, 1991), it is far from the case that the potential to generate conflict is restricted to Britain (e.g. Bonnain and Sautter, 1970), even though its intensity is likely to be strongest here, given the depth of British (particularly English) attachments to the countryside. Moreover, it is here that in-migrant production of sociocultural tensions is most prominent in discussions in the literature.

Counterposed against each of the above tendencies of new social problems, we can find documented cases that point in the opposite direction. Thus studies from across Europe can be found which show new migrants becoming integrated into recipient communities with surprisingly few problems (Anderson and Anderson, 1964; Hill, 1980; Buller and Hoggart, 1994b); although, as as Léger and Hervieu (1979) reported for a 'back-to-the-land' group in Cévennes the depth of this social integration might be paper thin (this conclusion might also be applied to the strong welcome that British in-migrants have received in rural France; Buller and Hoggart, 1994b). In addition, as Forsythe (1980) found in Rousay, it is possible that a size factor is involved, with tension heightened as the migrant influx reaches magnitude in a short time. Nevertheless, the message is that we must be careful about over-generalizing about the adverse effects of middle class in-migrants. For instance, what goes against what many see as the prevailing grain is evidence that longstanding residents lie at the heart of opposition to improved service provision. Rather than in-migrants resisting change and so restricting local residents' service access, it can be longstanding residents who oppose service change, in part because they fear that the rural character of their area will disappear (e.g. Green, 1964). In turn, in some instances local communities have been revived as social entities, and seen marked improvements in their services, almost solely due to initiatives by in-migrants (e.g. Morin, 1970). Indeed, at times incomers have provided the impetus to sustain the economic viability and redynamize the economy of their recipient communities, as noted for France by Limouzin (1980) and for Italy by Dematteis (1982).

In practice, established local power structures often lie behind the poor state of local services (e.g. Saunders *et al.*, 1978; Groenendijk, 1988). It is instructive, for example, that while in-migration and tourism are linked to rural homelessness in Britain, the procedures for deciding on who is homeless are highly subjective and offer significant scope for local officials to limit housing options (Lambert *et al.*, 1992). It is hardly surprising, then, that the willingness of 'homeless' people to approach a local council because they require housing and the propensity of councils to respond to such requests is political party-related; thus reactions in classically rural, Independent councils bear little relationship to the condition of local housing markets (Hoggart, 1995). The picture of local political elites responding weakly, if at all, to the socioeconomic needs of rural residents is something that can be related to in-migrant pressures (Connell, 1978; Little, 1987), but there is plenty of evidence that this practice is alive and kicking both in areas that receive slight in-migrant influxes or well before such inflows occurred (e.g. Sacks, 1976; Saunders *et al.*, 1978; Stone, 1990). Indeed, the potential for the distribution of service benefits to be informed by paternalism is heightened in smaller communities (Forrest and Murie, 1992). This provides a ready vehicle for the ideological biases of local elites to find expression (Rose *et al.*, 1976). Moreover, poor service provision has spin-off effects that heighten socioeconomic problems in rural areas; as David Clark (1991) indicates when noting how the majority of rural employers find an absence of low-cost housing hinders their ability to recruit needed staff for their companies.[10]

The reader should *not* come away with the impression that there is no broad trend of social disquiet generated by in-migrant flows, for a wealth of evidence signifies their strength. How far this disquiet extends beyond Britain is difficult to say, as we have found comparatively few studies which examine this issue across Europe, or at least which conceptualize it in the manner that is so dominant in the English language literature (albeit a few studies suggest similar conflicts exist elsewhere, such as Chevalier, 1993, on France, or Persson and Westholm, 1994, on Sweden). Apart from establishing its sphere of operation, a key academic problem lies in conceptualizing what is happening. In part we have presented a caricature of dominant interpretations in the literature, for the division between newcomers and locals was not as dogmatically adhered to as some of its critics portray (e.g. Cloke and Thrift, 1990). Indeed there was subtlety in many of the interpretations of the processes taking place, which is not surprising given the mixture of reasons for decisions to relocate into the countryside (e.g. Little, 1984; Perry *et al.*, 1986; Bolton and Chalkley, 1990). This complexity creates some discomfort for us when considering the notion that the attractions of rural living, and consequent power plays that restrict behaviour options for lower income residents, owe their base to a so-called service class. We have yet to be convinced that the service class is not simply another catchy academic label that carries little theoretical weight. Recognition of the diversity of the middle classes, and of conflicts between middle class fractions, is for us a very welcome addition to the rural studies literature. But our

10 As an indication of this shortage, Clark (1991) reports on a 1989 examination of 700 local housing needs surveys, which recorded a need for 376 000 low-cost rural homes, when only 34 300 were proposed and just 9000 were certain of being built.

sense is that, even for a field in which a number of excellent studies have been completed (e.g. Pahl, 1965; Connell, 1978; Forsythe, 1980), there is still a shortage of clear conceptualizations that capture what is being shown to be a highly complex process (e.g. Phillips, 1993). This complexity comes not simply from the actions of households or even of social classes, but also in the structural circumstances in which housing and migration decisions are made. These include not simply biases favouring home ownership but extend to the sentimentality that comprises an integral element of rural traditions.

Great expectations, hard times or bleak house?

So far, in terms of their impact, we have focused on in-migrants largely in terms of their consequences for host communities. This is but part of the story. In addition, we should recognize that the process of migration creates a new living environment for those who have moved. Whether or not this represents a pleasant change or a discomforting adjustment in lifestyle receives comparatively little attention in the research literature. It is as if there is an assumption that because people have moved they must have chosen to do so and will welcome the chance to live out their prior evaluation that a rural life was for them. This neglects the fact, identified more than 25 years ago by Pahl (1966), that people find themselves living in areas of low population density for very different reasons. Moreover, even if there is a positive appraisal of the decision to move to the country, if there is more than one person in a household it cannot be assumed automatically that this valuation is equally held. Neither must we presuppose that the consequences of such a move impose themselves equally across household members. How such considerations pan out is the issue to which we now turn. There are many ways in which we could examine the experience of in-migrant populations, but there is insufficient space to consider more than a limited number of population groups. Nevertheless, by including sections on gender, national minorities and the elderly, we can offer insight on a range of the issues that crop-up.

A gendered rurality

In order to appreciate the possible situation that female in-migrants might face we should start by considering the character of gender relations in rural areas. Whatever the context, in much of western society, a male-dominated ideological underpinning has largely assigned women to the domestic sphere.

For some commentators this sentiment is especially deeply-entrenched in rural areas (Little, 1990). Some indication that this is the case comes from state socialist nations, where a supposed egalitarian society has not offered much in the way of equality for farm women (First-Dilic, 1978; Repassy, 1991). Indeed, whether we look toward the early part of this century (Arensberg, 1937) or to more recent decades (Golde, 1975), the literature stresses that small-scale farms which are inherited by males would find it extremely difficult to survive without the input of a female partner. Yet farm women commonly have to work long hours, on the most repetitive tasks, without being given an equal say in decision-making (Gasson, 1981; Whatmore, 1991). The mobilization of bias against women, which is readily apparent on analysis (Shortfall, 1992), owes much to deeply-ingrained value dispositions, as Haugen (1994, 98) notes: 'Even when girls receive the practical training necessary to become farmers, their socialization still centres on constructions of womanhood as a flexible and caring gender. While the question of taking over the farm is, for girls, an evaluation of many conditions in which other people's interests play an important part, boys evaluate the situation according to their own personal interests'. More than this, young women with formal agricultural training find it more difficult to obtain farm jobs than equally qualified males (Gasson, 1981).

For some reviewers, this picture is remarkably consistent across a wide variety of cultural, religious and social contexts (Shortfall, 1992). However, if we expect national identities and traditions to carry their own weighting in social action, we should expect to find cross-national divergences in gender roles. This is what Pfau-Effinger (1994) suggests in a comparison of Finland and Germany, that is significant for the manner in which it ascribes a major part to agricultural traditions and later industrialization processes in determining dominant patterns of gender relations. In particular, a greater sense of gender equality is said to exist in Finland as a result of its slower, more decentralized processes of industrialization. This enabled a dual-earner gender model to develop and perpetuated the notion of equal contributions to the household that existed under structures of farm owner-occupation. As in rural Norway, both men and women have been able to seek and find off-farm work, resulting in the growth of 'dual career' farm households (Blekesaune *et al.*, 1993). By contrast, the faster and more concentrated industrialization of Germany, especially given its emphasis on heavy industry, promoted a more male-dominated family structure. Regionally-specific flesh is placed on this bare national skeleton by Sackmann and Häussemann (1994). In comparing regions within Germany, they concluded that there are sharp differences in gender relations amongst regions, identifying similar causal processes to those of Pfau-Effinger to account for this.

Focusing explicitly on farm contexts, we can draw further distinctions with Britain and the Netherlands, two countries in which family farming has long been highly commercialized. In the Netherlands, for instance, de Haan (1993) suggests that a functional differentiation between family and farm is underscored by an ideology specifying different spheres of activity for men and women. In this, although women might play quite a large role in farm activities, farming has always been seen as the personal enterprise of the male, which is detached from family influences and founded upon entrepreneurial skills. Hillebrand and Blom (1993) see few changes in this position in the 1980s, with a rise in husband-wife farm partnerships during that time explained principally by tax considerations. For Marsden and colleagues (1992a) meanwhile, the creation of large, integrated farming units in the UK has required a return to traditional family values, with a subordination of family members to the will of male heads of households. Similarly, Whatmore (1991) notes the impact on women of the commoditization of UK agriculture, which has led to a devaluing of their work, even though this might represent a significant share of total farm and especially household production. This is especially the case for on-farm diversification such as rural tourism initiatives, which have largely become a sphere for female activity (e.g. Neate, 1987).

In contrast, in Portugal, women have traditionally had a much greater role in agricultural tasks, with Besteman (1989) distinguishing between 'joint' farms where both husband and wife work in agriculture, and female-operated farms, where the husband is either absent or working full-time outside the farm. This gender division of labour can itself be seen as resulting from a long history of male migration (Brettell, 1986), although the strongly patriarchal system characteristic of many European farming systems, where the oldest son inherits the landholding and women are excluded, can be contrasted with the Portuguese system (which it shares with other countries where law is based on the Napoleonic code), where both daughters and sons inherit land in equal shares. It does bear remembering, then, that the seeming generality of gender relations that many studies identify are subject to important nuances and deviations at both national and regional levels. These arise from long-established social traditions that extend beyond the farm sector, although agriculture has had a key part to play in generating a specific form to gender relations (Pfau-Effinger, 1994; Sackmann and Häussemann, 1994).

Of course, today the position of women might be said to be subject to change. It was recognized some time ago that mechanization would bind farm women less tightly to on-farm work (Mørkenberg, 1978), with some recent work confirming this expectation (Siiskonen, 1988; Arkleton Trust (Research) Ltd., 1992; Hillebrand and Blom, 1993; Table 6.2, page 206). However, the picture is not universal. Thus, Pfeffer (1989) notes that as men search for off-farm employment, German women have assumed a larger share of farm tasks, including some which they may not have traditionally performed. This feminization of agricultural work is seen to be partly related to recent rural industrialization policies in the former West Germany, which had the effect of creating more work for men than for women in rural areas. Commentators on Italian agriculture also note a feminization of the workforce as males find higher paid nonfarm jobs (Douglass, 1984). Here we are probably seeing a reaction to differential regional economic performances, with relatively high income jobs being created in some regions contrasting with the move of low cost manufacturing and service employment into other parts of the continent. As Little (1990) observed, employers are not adverse to deliberately utilizing ideological expectations about female roles to manipulate job opportunities in favour of part-time work and low wages in order to raise profits. Providing one index of this she quotes figures from the UK *New Earnings Survey* showing that women who work full-time in the food industry received an hourly wage of £2.92 for manual work and £3.73 for non-manual, compared with figures for male workers of £4.03 and £6.74, respectively. As Bradley (1984) reported, in rural Britain when both male and female members of a household work, the female is likely to be better qualified, yet she will receive a lower rate of pay.

Women who migrate into rural environments are characteristically coming into an area in which their job opportunities are highly constrained by their relative inaccessibility (Moseley and Darby, 1978). They are confronted by an environment in which there are few jobs for those with qualifications, so that work which is available is commonly part-time and requires a relatively low level of skill (Little, 1994). In a nutshell, they enter an environment in which the prospects of social mobility are much more limited than in urban areas (Chapman, 1986). And while the decentralization of production and the creation of new service jobs might help alleviate this in the future, the beneficial effects of such changes will be uneven (see Chapter 5). Moreover, to take-up such opportunities women will find their path is strewn with difficulties laid down by poor support systems, with the almost invisible provision of childcare facilities in much of rural Europe being one example (e.g. Stone, 1990). This is something that even the Scandinavian countries appear not to have addressed outside more urbanized country areas (Persson, 1992). At the same time, there are social pressures arising from the expectation that a 'rural woman' should utilize her caring skills to benefit the community as a whole. As Pescatello (1976) notes for Iberia, a consistent theme in this 'rural ideology' is that kinship, neighbourhood and village social relations are feminine concerns (also Davis, 1973). Significantly, in one British study it was found that middle class families identify especially with this vision of a woman's role (Little, 1994). Faced with this combination of restraints, newly arriving women might well find their rural community a less than benevolent environment, in which they are nonetheless expected to devote time to promoting social harmony. How far in-migrant women actually find such environments constraining could depend upon initial expectations and (if relevant) a partner's activities, for irrespective of sex many of those who migrate into remoter rural areas are prepared to accept lesser job openings, more insecure employment and lower rates of pay in order to live in what they see as a better environment (Jones *et al* ., 1986; Williams *et al.*, 1989; Bolton and Chalkley, 1990; Buller and Hoggart, 1994a).

The rural world of national minorities

When writing *Rural development* one of our abiding memories was of a lack of writing on ethnic and racial minorities in the European countryside (Hoggart and Buller, 1987). Save for work in North America, where large populations of blacks, hispanics and native indians are found in nonmetropolitan areas, it is probably fair to state that little work has been completed on minority populations in rural zones in the northern hemisphere. Certainly, within the English language at least, issues of ethnicity and race appear to be habitually ignored in rural texts. This is perhaps understandable, given that immigration – the principal source of ethnic diversity in western Europe – has largely been directed towards urban-industrial agglomerations (Schmitter-Heisler, 1986), with rural areas generally regarded as being more ethnically (and culturally) homogeneous. Nonetheless, ethnic diversity is both a longstanding and a growing feature of many of Europe's rural areas.

We should not downplay the importance of the volume of immigrants that some nations receive, and the potential impact this has on interpretations of the importance of different socioeconomic processes. For Kontuly and Vogelsang (1988), for example, large-scale immigration into cities was the main reason why counterurbanization was not observable in German demographic statistics until the mid- to late-1970s, when government policy severely restricted new immigration. But while most German immigrants went to cities, with only 1% of foreign workers employed in agriculture, by 1982 they nonetheless made up 8% of total employment in the sector. Recognizing the importance of imported labour for this sector, the Federal Labour Office (*Bundesanstalt für Arbeit*) started recruitment in 1955 aimed specifically at agriculture and construction, with foreign workers for manufacturing only being targeted later (Castles *et al.*, 1984). In Switzerland also the system for recruiting foreign labour was targeted at the needs in agriculture, along with construction, hotels and catering. France was little different, with labour shortages in the Mediterranean regions producing the employment gaps that immigrants came to fill. Thus, Castles and co-workers (1984) note that from its inception in 1945, the French *Office National d'Immigration* specifically recruited agricultural workers, so that by 1970, the sector was employing around 100 000 seasonal foreign labourers each year, most of them from Spain (see Mansvelt Beck, 1988, for an account of this from the viewpoint of Spanish villagers). Many of these workers, as with many who came to France before the *Office National d'Immigration* began its work, settled in local communities and are reported to be well integrated socially. This process was eased by the intimate social atmosphere of the places they settled in, along with the fact that they contributed to the economic wellbeing of their new home region (Toujas-Pinède, 1990; Berlan, 1995; Noiriel, 1995).

In the past, such migrant labourers largely came from the poorer agricultural regions of southern Europe. However, a more recent development in southern Europe has been associated with an inflow of foreign labour from the southern shores of the Mediterranean. Again, urban centres appear to be most affected, although significant numbers have been employed in capitalist-oriented and rapidly intensifying agricultural areas. In the case of the Spanish province of Zaragoza, for example, Escalona Orcao and Escolono Utrilla (1993) noted that 29% of migrants included in the regularization campaign of 1990 were issued with work permits for agriculture, the largest single occupational group in this exercise. Meanwhile, the involvement of foreign workers in agriculture, and in rural areas more generally, extends elsewhere in southern Europe. King and Rybaczuk (1993), for example, note the importance of foreign labour in areas of specialized agriculture in Italy, such as west and south east Sicilia (vineyards), Puglia (vineyards and other tree crops) and the Neapolitan plains (tomatoes, tobacco, market gardening), as well as the presence of a Tunisian fishing community in rural Mazara del Vallo. These rises in the number of foreign workers are linked to the availability of alternative jobs for local people in some regions, with pressures to lower production costs being important elsewhere. Both of which are not unrelated to the feminization of farming noted above. At present these immigrant workers are most numerous in the southern regions of Campania (8935 in 1991), Sicilia (7500) and Calabria (2920), with the northern regions of Veneto (2600), Trentino (2000) and Emilia-Romagna (1690) some way behind, but with a trend of increased numbers in all regions save Alto Adige and Sardegna (Istituto Nazionale Economia Agraria, 1992). Even the development of intensive agricultural production in the Attiki plain of central Greece has received a boon from the influx of Albanian migrants since 1991 (King and Rybaczuk, 1993).

Across the EC as a whole, the extent to which foreign workers in agricultural or other rural occupations has engendered social change or conflict is difficult to gauge, not least because studies of such populations are rare. Certainly in Epirus in northern Greece, the porosity of the border with neighbouring Albania since the fall of communism has led to a large increase (over 100 000) in a partly itinerant foreign population, a process that has been widely reported in the Greek press as leading to an increase in crime, inflation and social conflict (Black, 1994). However, the situation here is complicated by the fact that many (though by no means all) of the newcomers are ethnically Greek, coming from a province of Albania that many Greeks still refer to as 'northern Epirus', so there is some ambivalence in the response both of Greek citizens and the Greek state to these new arrivals (Kokkinos, 1992). Meanwhile, reports of racially-motivated attacks, and discrimination at various levels of society and economy, continue to be reported throughout Europe, but the extent to which distinctive rural and urban trends can be identified is limited (Wrench and Solomos, 1993).

Of course, ethnic divergence in European society is not limited to questions of immigrant minorities, for the so-called 'national minorities' arguably have a much higher profile and significance for rural areas (Williams, 1980). For example, there are an estimated 16 million people in Europe who speak 'single minority languages' or languages which are not the official language of a state in Europe, whilst all western European states have national minorities if those speaking other national languages are included. In turn, although two high profile 'minority' languages, Catalan and Euskadi (Basque), are spoken mainly in urban centres, at least eleven more are primarily spoken in rural areas (Occitan, Galego, Sardinian, Breton, Welsh, Friulian, West Frisian, Corsican, Scots Gaelic and Irish), where the speaking of the minority language can be seen as an expression of cultural, ethnic or national difference, in opposition to the mainstream social world of the state.

At this point, then, the best we can do is point to our limited knowledge on the behaviour and impact of minority groups in rural areas. That they can have a bearing on rural social relationships that merits research attention cannot be doubted. Highlighting one potential effect on rural social stratification, Cloke and Thrift (1987) argue that as cultural and linguistic cleavages cut across social class divisions, ethnicity, along with retired migrants and those seeking alternative lifestyles, points to the need for more research on intra-class social divisions within rural localities. Meanwhile, the movement of 'national minorities' within the EC may produce effects akin to those of in-migrant foreign workers. Redclift (1973) provides one illustration with the breakdown of community relations in a Spanish Pyrenean village. This was spurred by the in-migration of *gallegos* and *andaluces* – workers from the Spanish provinces of Galicia and Andalucía – who came to construct a dam for hydro-electric power, with inputs from tourism and changed agricultural practices hastening communal disintegration.

The elderly in the country

Analysts of elderly migration in a variety of countries are finding an increasing tendency for moves into rural locales on retirement, with the predilection that formerly held in many countries for a return to one's family roots breaking down in favour of a more general quest to find a home in an amenity rich environment (Rees, 1992; Cribier, 1993). Hervieu (1991) offers a flavour of the growing importance of rural places of residence for the aged when noting that in the 1970s those over 50 years old were no more numerous than those under 20 in French rural areas, yet by 1988 the ratio had grown to 1.7 to 1, with figures as high as 2.5 to 1 in certain rural *départements*, such as Limousin. Indeed, in much of Europe there is an over-representation of the elderly, and especially those in their early retirement years, in net flows from cities to the countryside (Serow, 1991). Perhaps only southern European regions that were experiencing urbanization processes long into the era when northern Europe was seeing counterurbanization gain strength offer an exception to this rule (e.g. Table 2.1, page 35).

Whatever the particular reasons for this in-migration process, the ageing of rural populations can be seen to have had a number of implications both in the economic and the social spheres. However, while evidence on any of these is not plentiful, on examination there appears to be less to worry about negative consequences than surface impressions would suggest. At one time, for example, concern was expressed in national policies about the ageing of the agricultural workforce and the likelihood that older farmers would be less innovative and open to new technologies (Black, 1992). However, with today's agricultural emphasis on more environmentally friendly farming, one wonders if these fears will be so loudly put (Potter and Lobley, 1992). Likewise, there are concerns that numerous elderly residents in rural areas will create problems for service provision, for example in the fields of housing, health and social services. Distance, inaccessibility and low population densities are held to compound difficulties in providing such services, especially in the context of the progressive centralization and rationalization of many state services. Once again, however, rural-urban differences are less than might be expected. Indeed, Caldock and Wenger argue that a focus on rurality tends to divert attention from the more fundamental variable in service provision quality, which is effective organization and management of provision: 'Continual reference to geographical factors may prove simply to be the perfect scapegoat for explaining away failures and deficiencies' (Caldock and Wenger, 1992, 67). So, while distance does add to the problems of the centralized bureaucratic provision systems that have commonly been devised for urban areas, and extended unchanged to rural places, adjusting to this extra problem might be largely an administrative issue. As background to understanding this point we can point to a limited number of case studies, for research that specifically focuses on the rural elderly in Europe is not well developed. Nonetheless, from the suggestions that are available it seems that the standard of health and mobility of elderly people in rural areas is no different from that in urban places (Wenger, 1981). In addition, the rural elderly appear to have as much contact with their families as retired people in urban settings. They are more likely to belong to voluntary organizations. They are also more inclined to be part of a religious institution which provides further opportunities for informal support (Grant, 1984). The notes of caution we would add are twofold. First, in former peasant communities there appears to be an assumption that traditions of close family kinship will provide a solid base for supporting elderly relatives. However, as Triantafillou and Mestheneos (1988) note for Greece (as a whole), the capacity and willingness of family networks to cope with the health problems of the elderly is often overestimated. Second, we recognize that any problems which do arise can be heightened if migrants decide to spend their retirement in another country. Age Concern England (1993) has highlighted this problem for Spain (although much of the coastal strip of

Spain might not be regarded as 'rural'), and we find some parallels in rural France (Hoggart and Buller, 1995a), where similar problems of fluctuating exchange rates, the death of a partner and oscillating house prices can create economic and personal distress. But more generally those who retire to country areas appear to welcome their new lifestyle (e.g. Hoggart and Buller, 1995a). Indeed, in so far as many of these in-migrants are home owners, they have often traded-down in house prices so that they have a significant capital stock to draw-on to support their retirement (Buller and Hoggart, 1994a; Forrest and Murie, 1994).

Social stability and social change

In this chapter we have explored the divergent kinds of social organization that are found today in rural Europe, then examined the changing composition of rural populations that is dismembering, reshuffling and recreating longstanding social traditions. Although it might be possible to provide a more finely-grained categorization of rural social tendencies, we believe that the ideal-types we have identified pick out the principal dimensions of distinction. Indeed, they overlap and intermingle in some measure, which suggests to us that a proposal to add more classes would need very careful thought and justification. As far as national impressions on these ideal-types are concerned, these are fairly clear. They are not necessarily one-to-one relationships, as some states are characterized by more than one social tendency and their trace elements can be identified in a variety of national settings. Whether each national context provides its own peculiar twist to the more general social characterization we have provided we cannot evaluate, for we would need to be sure that considerably more evidence was available before feeling comfortable about undertaking such an interpretation. Yet the prospects of being able to conduct such an analysis effectively are not high. Principally this is due to the major restructuring of the European population that has taken place in the last 40 years. The European countryside has lost significant portions of its population in the past, but the dramatic changes of recent decades have seen the disappearance of many rural settlements, major transformations in rural economic bases which have introduced populations with different views on production practices and workplace social relationships, along with a more consumerist population infusion that potentially has even more divergent values and behaviour patterns. How these forces of change merge, intersect, mutually disturb or conflict, and what geographic fixes and fluidities are associated with each, is a matter that is too little understood or researched. It certainly seems that the place of national identities and traditions in rural social tendencies is likely to change but how is still an open question. The issues are there but more research is needed to try to find answers.

|7|

Environmental challenges for rural areas

It is fitting that the main body of this book should close with a chapter on the European rural environment for that environment is becoming the central defining feature of rural areas themselves and of distinctive economic and social trends that are currently emerging in the European countryside. Although it might seem paradoxical to speak of the greening of rural Europe, it is nonetheless true that the importance of rurality in economic, political and social processes is increasingly a consequence of a growing prioritization of its environmental qualities.

The physical environment constitutes the natural bedrock of rural areas. Traditionally, this environment has been interpreted, principally by the indigenous population, in two ways: first as constraint, a source of hardship to be overcome or endured (e.g. Bell, 1979); and, second, as a source of 'natural' richness to be exploited and rendered subservient to human needs (Berger, 1979). Both interpretations have played a major role in influencing human socio-spatial organization (Dion 1934; Hoskins 1955; Dolfus, 1979), as well as in defining the 'quality' of rural areas, as Arthur Young's famous observations on French openfield landscapes of the 1780s demonstrate (Braudel, 1990). Within this viewpoint, a pleasing rural scene was one that was both productive and exemplified human mastery of nature's caprices. Thus, whereas rural activities, and particularly agriculture, had historically had major environmental impacts, these were seen as essentially formative and civilizing rather than destructive or heretical (Short, 1991; Ferkiss, 1993).

As science and technology gradually replaced God and theology in determining humanity's relationship to the natural world, that relationship shifted dramatically. The anthropomorphism introduced by the Renaissance, based more on technological aptitude than on moral imperative, provided the philosophical and scientific justification for ever-greater changes to the natural world and threw into focus a different set of issues concerning social organization and economic development upon which the industrial revolution was ultimately predicated. Since the industrial revolution in the northern European states, which polarized urban and rural differences, the relationship of human society to the countryside has seen a shift in perception, first, from contempt to nostalgia, then from a belief in boundless resource availability to fears over resource scarcity (Thomas, 1983). The development of rural policy in the nations of western Europe closely mirrors these shifts, as well as reflecting an important process of rural redefinition.

If rural areas have always been to a greater or lesser extent areas for which the *proximity* of the physical world constitutes a central defining feature, they have only relatively recently become environmental areas for which the *quality* of the natural environment is a key element in their wider conception and representation. This shift in nomenclature reflects a series of profound changes that take us from a notion of rural areas as a venue of primary production to one in which rurality is associated with a multitude of consumption and production activities, in which landscape and its physical qualities occupy a central position. Thus, agriculture, as the traditional organizing and defining activity in the European countryside, has been replaced by the wider and less monofunctional concept of 'rural' (Bontron, 1994). 'Rural' has itself become 'environment' (Mathieu and Jollivet, 1989), while 'nature' has shifted, following a parallel path, first, to 'landscape' then also to 'environment' (Cadoret, 1987; Robic, 1992). These transitions are fundamental for they underlie both general tendencies within society and a series of transformations related specifically to the vocation and perception of rural space. On the one hand, the broad package of socio-demographic and economic phenomena that we label, for the sake of convenience, restructuring, including the urban-rural manufacturing shift, counterurbanization, and changing patterns of consumption, is contributing to a fundamental redefinition of the role and function of rural space. This redefinition is significantly centred on the environmental qualities of this space (Buttel, 1992). On the other hand, within rural areas, a demographic and spatial retreat from the farming sector is occurring, in which land abandonment and agricultural modernization are altering both the relationship of farmers to the 'natural' space they farm (Billaud and de la Soudière, 1989) and the role accorded to farmers by society as a whole (Laurent, 1994). A third set of issues relates to the increasing prioritization of green issues both in social discourse and in political action. As Buttel (1992, 24) points out:

> Environmentalization portends a decisive reversal of the inconsequentiality of matters rural and nonmetropolitan. Much of the green agenda brings rural societies and their environments centre stage.

In short, the rural areas of Europe are being redefined according to environmental principles and criteria. In turn, this is having a major impact not only upon the activities undertaken in rural areas but also upon the sociopolitical composition and behaviour of people within and beyond the countryside. Buttel again (1993, 24) provides helpful insight: 'I cannot help but wonder whether greening and environmentalization might not over time lead to a fundamental shift in how rural spaces are symbolized, and accordingly how we define and deal with rural problems'. In this chapter, we identify and trace the emergence and development of a rural environmental paradigm and examine its consequences for rural Europe by focusing on three thrusts of change: the threat to the environment posed by technological advances and the crisis of agricultural production; the growing commodification of the rural environment as a key element in social consumption; and, lastly, the evolving position of rural areas as a focal point in the institutional and social processes of environmentalization and greening. Understandably, we hope, our treatment of these three themes is not uniform. The environmental impact of contemporary agricultural activities forms the major part of this chapter due to the primordial place this sector has in popular definitions and representations of the European rural environment. 'Nothing could be more obvious', claims Thompson (1994, 1), 'than agriculture's importance for environmental quality'.

Production and environmental change in rural Europe

' "How beautiful nature is". Say this every time you are in the country', extolled Flaubert in the mid-nineteenth century (Flaubert, 1976), who referred to an association that for many underlies the basic difference between urban and rural areas. Yet while the eighteenth century rural commentator Marshall claimed that '... wherever cultivation has set his foot, Nature has become extinct' (quoted in Fairbrother 1970, 11), it is in the countryside, in the largest sense of that word, that people feel closest to nature (Viard, 1990). It is also arguably in the countryside that contemporary society is brought most frequently and easily into contact with the realities of environmental change; whether this be catastrophic, as with recent floods in Europe, or diffuse, as in the removal of hedgerows or the ineluctable physical extension of towns and cities. As Chapter 3 demonstrated, awareness of such changes can play a critical part in the development of national rural traditions.

It has become almost a banality to refer to the transformation of the European rural environment. Protection, preservation and conservation have been the leitmotifs of rural policy throughout the later part of the twentieth century, particularly in northern Europe (Richez, 1991), although increasingly also in the south (Baker *et al.*, 1994). Textbooks on rural geography now contain obligatory chapters on such issues (e.g. Robinson, 1990), which places environmental issues alongside other typically 'rural' subjects such as agriculture, migration and settlement. The seeming omni-presence of rural environmental concern indicates, however, that we should tread carefully in presenting broad generalizations. For while rural environ-mental problems seem almost ubiquitous, given the generality of forces like agricultural modernization, urban sprawl and rural depopulation at various points in time, there are important regional and national differences in the nature and perception of rural environmental change (Jollivet, 1994). Fur-thermore, in a number of cases, the emergence of environmental issues is closely linked to processes of European integration (Baldock and Long, 1987: Montenari, 1991). For the development of EC environmental policy has led to a necessary Europeanization of environmental issues, which has created new environmental agendas in a number of member states, at the same time as the normative and regulatory tone of EC environmental policy has provoked member states to recognize new environmental problems (Buller *et al.*, 1992).

Just as the European continent is subdivided into political divisions, so too, claims Lavoux (1991), can it be broken down into environmental divisions. In broad terms, we can identify a southern model, where the dominant concern is retaining environmental quality while encouraging industrial growth and agricultural modernization; a Scandinavian model, where natural resource protection is the principal concern; and, a north west European form, characterized by land-use conflict and an essentially con-sumption-oriented conception of nature and the environment. Significantly, the nations that compose these groupings view the growing intervention of the European Community in the field of environmental policy in differing lights. While Germany and the Scandinavian states see EC environmental legislation as considerably weaker than their own (Bruckmeier and Glaeser, 1995), in southern nations it has been interpreted as a north European agenda with little relevance to them (La Spina and Sciortino, 1993; Kousis, 1994; Pridham, 1994). Britain, for Rose (1990) the 'dirty man of Europe', remains perhaps uniquely equivocal, resisting increasingly onerous European environmental legislation while at the same time undergoing a fundamental shift in its own style of environmental management (Ward *et al.*, 1995). Significantly, the combination of governmental reluctance to act and a highly informed and active environmental lobby (Lowe and Goyder, 1983), means that Brussels receives more complaints relating to Britain's poor performance in implementing EC environmental legislation than any other state (see Table 7.1, page 232).

Nevertheless, despite considerable variety in rural environmental preoc-cupations, and in the relative position adopted by member states with respect to EC policy, there is a set of common rural environmental problems which are emerging in all European states. The most intrinsically rural amongst these has been environmental degradation linked to agricultural practices, with other transnational concerns including the extension of urban areas and activities into rural zones, and the ecological and landscape implications of

Table 7.1 Complaints received by the European Commission relating to failures of member states to implement EC environmental policy, to 31.12.1990

Country	Number of complaints
Belgium	17
Denmark	3
France	47
Germany	56
Greece	40
Ireland	19
Italy	33
Luxembourg	3
Netherlands	7
Portugal	19
Spain	111
United Kingdom	125

Source: Collins and Earnshaw (1992, 232).

modern tourism. All three issues are essentially human-induced, although they can have significant bearing on processes within the physical environment. Commonly in fact, processes of environmental change, whether diffuse or catastrophic, are fundamentally mediated through existing practices of human land-use and land transformation (Veryet and Pech, 1993). Nevertheless, we must be careful not to ignore the power of nature. As Ward and Munton (1992, 130) warn: 'The environment is socially constructed but not to the point of suggesting that its bio-physical characteristics are unimportant'.

The three environmental issues examined in the remainder of this chapter largely focus on processes of change that occur within rural areas. However, it would be remiss of us to proceed without laying down a marker to signify the importance of largely 'urban' industrial activities for rural environmental change. Our emphasis in this chapter is principally on environmental issues as they are generated by those in rural areas. In this regard some important environmental issues are not considered, primarily because their genesis, and the need to address their causation, owes much to actions in cities or more broadly to actions and attitudes at a national or even international scale. One example is acid rain, which is estimated to have damaged nearly two-thirds of Britain's forests, along with around half of those in Germany, Italy and Norway (Cole and Cole, 1993). The source of such pollution may be far removed from the rural environments that are affected, as past friction between EC12 countries, in which sulphur emissions are concentrated, and

Scandinavian countries, where affected forests are a significant rural re-source, readily demonstrates. Similarly, the eastern *länder* of Germany are affected by a range of pollution problems of mainly urban origin, such as the use of lignite for power generation or motor vehicle emissions, whose impact spreads well beyond urban areas into rural zones. Issues such as these constitute a serious environmental concern for rural areas, and reinforce the need for Europe-wide solutions to environmental problems. However, given their primarily urban nature, they are not discussed in detail in this chapter, save to note the general point that environmental pollution respects neither international borders nor the rural-urban divide.

Agricultural intensification and environmental decline

At a scientific and ecological level, the environmental consequences of modern agriculture are increasingly known and documented (Conway and Pretty, 1991), as are its land-use implications (e.g. Brouwer *et al.*, 1991). What we seek to stress is that these consequences are linked to two distinct and opposed trends. In certain parts of the EC the tendency is toward an increasingly technologically-based agriculture, whereas in other areas a gradual retreat to less intensive forms of farming has equally significant outcomes. These trends not only find a differential spatial expression within Europe but also engender variable sociopolitical consequences for the farm sector, as well as altering nonfarm attitudes toward farmers as protectors of the rural environment.

The environmental impact of modern agriculture is in large part a product of the growing importance of technological approaches to primary produc-tion, which have repeatedly earned the epithet the 'postwar agricultural revolution'. In Clark and Lowe's (1992, 11) terms: '... the inherently unreliable and refractory processes of nature have come more under agricul-ture's control through the intervention of science and technology'. Specifically, this has meant increased mechanization and technological innovation, along with the accompanying structural changes this has required. An increased usage of artificial inputs, ranging from pesticides, herbicides and fertilizers to plastic sheeting, has accompanied these production shifts, which have more recently spread to biogenetic engineering. Taken together these changes have pushed agriculture toward increasing specialization and monocultivation (e.g. Bowers and Cheshire, 1983; Goodman *et al.*, 1987; Bowler, 1992). Encouraged and sponsored both by individual governments and, for much of the postwar period, by the CAP, the technological transformation of farming has been predicated upon three assumptions regarding the environment. The first is that what technology has done wrong, technology can put right. The second is that farmers are cushioned against direct responsibility for environmental problems by financial and technological support systems that encourage increased production and compensate for losses due to environmental factors. With support systems not simply permitting but encouraging the increased subjugation of natural influences, this assumption boils down to economic risks taking precedence over environmental risks. The third supposition is that the farming community is ultimately the best custodian of the rural/agricultural environment and, as such, should not fall under the same regulatory controls that regulate the impact of other industries on the natural environment.

Of course, the technological farming revolution is primarily a lowland phenomenon, and as such is more evident in broad urban hinterlands. Nonetheless, it reaches all the way from Sweden (Anderberg, 1991) down to Spain (MOPU, 1989). Meanwhile, although intensive monoculture and industrial husbandry are largely confined to these areas (Kostrowicki, 1991), the environmental effects of technological innovations are felt more widely, with agricultural pollution being a major concern in all European states (OECD, 1989a). Thus Jenkins (1979) describes the development of a 'potato-fertilizer syndrome' in the mountains of southern Portugal, whereby small producers have become increasingly locked into more intensive cultivation practices that disrupt the traditional symbiosis between agriculture and the environment, and are seen to lead to the physical and chemical degradation of soils. Similarly, Black (1992) notes how in the north of Portugal, one aspect of agricultural change has been a shift in the type of material used as bedding for cattle, with a consequent decline in the quality of manure that had been crucial in maintaining soil quality under the traditional farming system.

Critical differences between states or regions do not, however, lie in the form or extent of environmental pollution and degradation that is caused by intensive agriculture. There are some specific regional problems, such as the overgrazing of thin, dry Mediterranean soils (Margaris, 1987) or the effects of acid rain on northern European timber and crop production (Park, 1987), but they share the environmental stage with a wide range of seemingly ubiquitous environmental issues that are provoked by agricultural intensification. Thus excessive concentrations of nitrates in surface and ground waters, habitat destruction and species contraction, and the presence of pesticides in drinking water are common problems of intensive agricultural regions be they East Anglia, Brabant or the Po Valley (Louloudis, 1985; Bruckmeier, 1987; Larrue, 1988a; Frouws, 1989; MOPU, 1989; Beopoulos, 1991; Osti, 1992; Pitman, 1992). Similarly, significant national and regional differences do not seem to be found in regulatory responses to such issues (see Larrue, 1988b; Lavoux *et al.*, 1988; Bodiguel, 1989). The bulk of European states possess a wide range of broadly similar regulatory, contractual, and fiscal measures which are designed to restrict agricultural pollution and encourage environmentally friendly farming (e.g. Larrue, 1988b; Bodiguel, 1989; Bruckmeier and Teherani-Krönner, 1992). Where critical differences do arise between states is in the reasons why agricultural pollution has been raised as an public issue, and the more general manner in which it has been addressed. Viewed in this light, the agricultural-environmental issue is demonstrated to encompass much more than ecological degradation, for it has implications for a variety of concerns other than ecological ones.

Reasons for the emergence of environmental concern in agriculture reveal nine distinct themes:

1 an ecological concern for the impact of intensification upon ecosystems
2 a concern for the public health implications of resource contamination
3 a concern for the farming sector's relative freedom from regulatory constraint
4 a concern for the amenity value of the countryside
5 a concern for access to rural areas threatened by the intensification of agricultural land-use
6 a concern that local economic development could be hindered by agricultural monopolization
7 a concern to limit agricultural surpluses generated by intensification
8 a concern to improve farm incomes by diversifying farm activities into landscape protection
9 an ethical concern for a declining sense of environmental responsibility that is considered inherent in contemporary agricultural policy.

These concerns are not universally present in rural Europe. Indeed, it is here that we might begin to identify distinct trajectories which mirror varied national and local conceptions of the role of agriculture within contemporary rural space and, *inter alia*, differing patterns of rural economic and social change. Significantly, such trajectories bear witness to an overall sense of decline in the position of agriculture within Europe's rural regions.

It has long been apparent that in many north European regions, particularly those that lie relatively close to major urban centres, an incoming middle class population is increasingly ready to challenge not only the environmental effects of contemporary agriculture, which often accord badly with their chosen vision of rural arcadia, but also the local political hegemony that sustains the power of local farming lobbies (Saunders *et al.*, 1978; Lowe *et al.*, 1986). Typifying this phenomenon, agricultural pollution and environmental degradation have been seized upon as effective political battlegrounds in political disputes in the shire counties of Britain. As with elsewhere in Europe, conflicts have developed in remoter areas (Chevalier, 1993), when the role of the farmer as custodian of the rural environment is challenged by incoming nonfarm populations, who demand, on the one hand, a more environmentally sustainable agriculture and, on the other, stricter environmental regulation of agriculture and related activities (Shoad, 1987). The link between this challenge and the changing socioeconomic

composition of rural areas has been long-established in lowland Britain (e.g.
Newby, 1979). It is more recent in other European states, partly because of
their dissimilar rural traditions (as evoked in Chapter 3), and partly due to a
lesser pace and scale of agricultural modernization and middle class in-
migration. It is noticeable, for example, that in rural France the emergence of
agricultural pollution issues reflects less the actual geography of agricultural
pollution than recent urban-to-rural migration trends (Bodiguel and Buller,
1989). By contrast, the relatively slow development of an organized challenge to intensive farming in Germany reflects the public's late awareness of
this as a local issue, along with attempts by the agricultural lobby to manage
problems through voluntary control and diversification into organic methods
(Bruckmeier, 1994). Yet, despite this encouragement, it has been more from
the relatively small number of women farmers that Germany's 'green'
challenge to intensification has come (Schmitt, 1994).

Rural social transformation has been a primary stimulus for a developing
concern for the ecological, amenity and access implications of modern
agricultural practices, and has thereby raised agricultural pollution onto the
political agenda. However, where the agricultural population remains nu-
merically and economically important, more direct intra-rural and rural-urban
conflict between economic interests has been generated. This has led to calls
for state intervention to regulate the environmental consequences of modern
agriculture (Lowe, 1992). Critical to these calls is the fact that the effects of
agricultural pollution reach far beyond the confines of an individual farm and
an immediate rural zone, both in space and time (e.g. Pitman, 1992). Of
course, farmers themselves can be the first economic actors to be affected by
farm-based pollution. This is well recognized in the Nordic countries and
Switzerland, where the environmental fragility of primary sector industries
traditionally induced a high degree of awareness of the environmental limits
of certain agricultural practices (Chabert, 1993; Pettersson, 1993). Elsewhere, where primary production is less of a central rural vocation, the
pollution of shellfish waters by slurry run-off, the contamination of water
sources used by the bottled water industry, the proliferation of algae in
bathing sites and, within the food industry itself, the contamination of water
supplies, all threaten the sustainability of rural economic development
(Bodiguel and Buller, 1989). On a larger scale, just as sulphur emissions from
power stations lead to the loss of distant forests in northern Europe, so too
the contamination of water supplies by pesticides and nitrates threatens the
public health of city-dwellers many miles removed from the sources of farm
pollution. This shift from agricultural pollution as a localized amenity
concern to a broad issue of public health and rural economic sustainability
has profound consequences for policy responses and the regulation of
agricultural production (CEC, 1990a). Throughout Europe moves to adopt
a more regulatory approach to farm pollution are emerging (CEC, 1989c;
OECD, 1989a). In Belgium such calls have come primarily from the urban
industrial sector (Mormont, 1994b), while they have come from individual

communes charged with supplying clean water to consumers in France (Bodiguel and Buller, 1989). The longstanding treatment of agriculture as a 'special case' in the regulation of pollution is now increasingly challenged. As Baldock (1992) demonstrates, this 'special case' has been founded upon an entrenched belief in northern Europe that family-farms are guardians of the integrity of their local environments. Yet the inevitable logic of intensification is removing primary production from its 'natural' environmental milieu.

Another growing tendency, especially amongst newer members of the EC, is for issues of agricultural pollution to be raised as a direct consequence of Community policy. This takes two forms. First, the imposition of EC environmental regulations and directives has undoubtedly made policy-makers and perhaps the public more aware of environmental problems, such that it has 'created' new environmental issues (Buller *et al.*, 1992). As an example, EC Directive 80/778 on the quality of drinking water has been fundamental in raising both public and governmental awareness on the problems of nitrate and pesticide pollution that agricultural practices cause, and has led to a substantial shift in styles of environmental regulation on farms (Ward *et al.*, 1995). Equally, playing the rural environmental card has emerged as an important means of gaining EC funding for rural development projects. Negri (1993) illustrates this by showing how traditional forms of economic development aid to the marginal uplands of Lombardia have proved relatively inconsequential in comparison with aid for risk management and environmental improvement. Similarly, agri-environmental measures in rural Spain need to be seen as part of a broad spectrum of rural aid mechanisms and not simply as solutions to particular environmental problems (Mendoza-Gomez, 1994).

Agricultural extensification and the environment

The environmental consequences of modern agriculture are not solely linked to processes of intensification and technological investment, although these may be the dominant environmental issues in regions most suited to intensification. However, in less favoured environments, production intensification, socio-demographic change in farm populations and the differential effects of agricultural policy, are having fundamentally different effects. In the upland agricultural regions of Europe (Figure 2.3, page 34), as well as in lowland zones with unfavourable climatic or soil fertility bases, socioeconomic change has more likely prompted rural decline, net outmigration and land abandonment. In such places, the effects of agricultural extensification and the end of land tendering can be equally catastrophic, provoking soil erosion, desertification, forest fires and increasing the chances of major floods (Comolet, 1989; Mata-Porras, 1994). Here we find important national variations. In Britain a dominant issue is the maintenance of countryside heritage and amenity, with the spectre alongside agricultural withdrawal being insidious residential growth (Watkins and Winter, 1988). In France, land abandonment is viewed more philosophically, as nature regaining what humanity has fashioned; a return to the ancient struggle between civilization and wilderness (Terrasson, 1988). But in Mediterranean regions fears revolve more around the risks of forest fires and flash floods. Thus a media debate is held each year in Portugal over the spread of forest fires in eucalyptus and pine plantations during the summer, although this debate cannot be divorced from heated political questions over who should own and manage forests on land traditionally held in common by local communities (Brouwer, 1993).

A pertinent statement of distinctions in national responses is found in the adoption of set-aside policy (Briggs and Kerrell, 1992). This was introduced formally in the EC in Community Regulation 1094/88, which was later considerably enlarged after the CAP reforms of 1992. Rates of adoption of set-aside, prior to it becoming mandatory for larger cereal growers in 1992, show considerable variation across member states (see Table 7.2). While British, German and Italian farmers have tended to place a significant proportion of arable land under voluntary set-aside, though as Ilbery and Bowler (1993) note this concerns principally the marginal cereal growing regions and not the major cereal-producing areas, the French agricultural

Table 7.2 Set-aside within Community member states, 1988–1991

	Land set-aside (ha)	Percent of EC land area set-aside	Percent of arable land
Belgium	740	0.04	0.2
Denmark	5520	0.29	*
France	166 575	8.73	0.9
Germany (West)	293 384	15.38	4.1
Germany (Ex-East)	599 243	31.41	*
Greece	250	0.01	*
Ireland	1766	0.09	0.2
Italy	608 705	31.91	6.7
Luxembourg	85	0.01	0.2
Netherlands	14 606	0.77	1.7
Spain	84 087	4.41	0.5
UK	132 622	6.95	1.9

Source: Ilbery and Bowler (1993, 19). Note: Portugal has been exempted from set-aside until 1994; * no data.

community continues to demonstrate resistance to this measure (Monteux, 1993). Although part of the explanation for uneven rates of adoption comes from differential rates of grant-aid, another reason lies in the standing of this measure in its broader rural policy context. In Britain, for example, set-aside policy has been incorporated in the wider 'countryside' ethic that is so dominant. Illustrative of this, in December 1991 *The Sunday Times* published a front-page story headed, 'Wealthy land-owners get handouts for doing nothing' (Rufford *et al.*, 1991), while the Ramblers Association has criticized the Countryside Commission's policy for providing grant-aid for set-aside when land is not open to public access (Gillie, 1993).[1] The issue in Britain has been seen essentially as one of justifying paying farmers and landowners, whereas in France, where all such payments are generally welcomed, the issue is a more deep-rooted one relating to the ethic of productivism. Set-aside land is widely described as a 'Community wound' on the once healthy *corpus agricole* of rural France. The receiving of grant-aid for abandoning land, however temporarily, is portrayed as an insult both to the farmer's *raison d'être* and to a long French history of agricultural development (Buller, 1992).

Quite different issues are raised by set-aside in Spain. Here, as Mata-Porras (1994) outlines, although the lower use of inputs such as herbicides and fertilizers may decrease land pollution, the prospect that land which is set-aside will be left without vegetative cover significantly increases the likelihood of wind and water erosion. This is already a serious concern in Spain, as significant areas of land were lost to cultivation in the decades prior to the implementation of EC set-aside measures. Thus Papadimitriou (1994) points out that across southern Europe, land under cultivation declined by 14% between 1966 and 1990, with concern raised at an EC level that this could produce long-term environmental decline and 'desertification'. This view has not gone unchallenged. Grove (1994), for example, argues that Mediterranean land degradation is no more serious now than at other times in the last century, and that there are far more important environmental issues such as pollution by mining wastes, salinization linked to the over-use of aquifers and salt-water incursion along the Mediterranean coastline.

These words of caution noted, it is nonetheless the case that postwar agricultural policy has tended to emphasize major divisions between upland and lowland farms, between irrigated and non-irrigated zones and, under CAP, between northern and southern Europe. To some extent, such divisions have always existed, and in policies for Less Favoured Areas (LFA) and under Integrated Mediterranean Programmes (IMP) the EC has taken the steps toward addressing potential environmental problems that arise in areas

---

1 The Countryside Commission could point with real justification to very low levels of public access to rural areas in Britain compared with its European partners. Millward (1993) provides a ready indication of this, but rather tamely posits that this might have something to do with landholding structures. Other analysts have been much less reserved in condemning the generally deleterious effects that large landholders have had in restricting public access to their land (e.g. McEwen, 1977; Norton-Taylor, 1982; Wilson, 1992)

facing land abandonment (Clout, 1984). For certain, the EC has recognized for some time that a failure to tender the land in such environments can be problematical.

> Sufficient numbers of farmers must be kept on the land. There is no other way to preserve the natural environment, traditional landscapes and a model of agriculture based on the family farm as favoured by society generally (CEC, 1991b, 9–10).

However, recent production-limiting reforms have substantially altered the framework in which decisions about farming are likely to be made. At the European periphery, as well as within the core, the decoupling of price support and farm income support, lower commodity prices and agri-environmental measures all, arguably, alter the balance of CAP support in favour of the less intensively farmed regions (Robinson, 1993; Whitby, 1994). Hence, while agriculture in the Po Valley is threatened by production limitation measures and stricter environmental controls, the traditionally marginal regions of Italy are seeing their 'original' products favoured, owing to their less intensive organization and their role in sustaining rural economies and populations (De Meo, 1991). Similarly, in Ireland, the intensive dairy industry of the east, which was strongly grant-aided until the onset of EC milk quotas in 1984 and CAP reforms of 1992, is now amongst the most menaced sectors in the country (Reid, 1992).

What is of interest here is that the consequences of farm and land abandonment are long-term and not easily resolved by technical solutions. What is in question is the very function of agriculture, which has prompted debate about the viability of a non-productive role for European farming (Muller, 1991). If agriculture is to be undertaken only where production is profitable in European and world markets, then much of the present farmland in Europe will no longer be productive agriculturally (Brouwer and Chadwick, 1991; Bolsius, 1993). In its place a service agriculture is perhaps destined to emerge; one that possibly provides for local specialist markets or is funded by the public purse for environmental reasons rather than supported for its productive role (Le Meur, 1993).

Both technological advances in farming and, for those unable or unwilling to benefit from such advances, farming withdrawal, are sides of the same coin. This coin is minted in a productivist ideology. But within Europe both have negative environmental effects. How then can contemporary farming and environmental protection be reconciled, apart from by stringent regulation? For most the answer is environmentally-sensitive farming, although this is anathema for those who see farming as essentially destructive or see it as implicitly environmentally friendly. Whatever the viewpoint, there is no single approach to environmentally-sensitive farming, for in Europe such practices have taken two very different trajectories. The first has adopted specific agri-environmental measures to reduce the environmental impact of agricultural practices through subsidies and voluntary adherence to less intensive methods of production. The second has sought alternative forms of agricultural production, most notably organic, biological or biodynamic farming.

For the first of these the notion that has emerged most strongly in policy circles is of a contractual guardian or renumerated steward. This role is expected to provide a user-friendly means of retaining farmers on the land, as well as producing environmental management gains. Although described by Thompson (1994, 89) as '... an ethic of use, if not exploitation' and as such is '... in opposition to an ethic of preservation', the voluntary approach has a number of advantages for farmers. It safeguards the agricultural community's hegemony in land-use decision-making, it provides income support in areas threatened by farm income decline, it builds upon farmers' know-how in rural management, and thereby reinforces a sense of moral responsibility, and it could produce durable environmental gains (Cox *et al.*, 1990). Britain has been a principal sponsor of this stewardship approach. In part this arises from its history of management agreements, such as those fostered under the 1949 National Parks and Access to the Countryside Act, the 1968 Countryside Act and the 1982 Wildlife and Countryside Act. Its received experience has also been informed by the actions of the Farming and Wildlife Advisory Group, which was set up by the National Farmers' Union alongside the Royal Society for the Protection of Birds in 1969. Most recently, Environmentally Sensitive Areas (ESAs), which were initially established in 1985, have created and built upon a now firm tradition of voluntary stewardship which recently achieved Europe-wide prominence, first in EC Regulation 797/85 and latterly in Regulation 2078/92 (Whitby, 1994).

Although other European states, notably France and Germany, have similar, if more recent and less widely adopted, voluntary schemes (Buller, 1991; Europe Environnement, 1993; Wilson, 1994), the rapidity with which ESAs multiplied in Britain suggests their particular adaptability to the British context. While Britain currently boasts some 1 472 958 hectares under 29 ESA designations, France can claim only 772 000 hectares split amongst 62 operations, with many still at the 'projected' stage. In part, this is a consequence of the former's voluntary tradition in countryside management. In part too, it reflects a greater readiness on the part of British farmers to accept a custodianship role with respect to their traditional domain, one that French farmers continue to resist (Deverre, 1995; Boisson and Buller, 1995). The growth of ESA policy within Europe has nonetheless coincided with a wider debate over the rationale of the CAP and the role of farming in modern society (Courtet *et al.*, 1993). ESAs, as they are known in Britain, thus form only one element in an increasingly complex set of Community inspired and national agri-environmental measures (CEC, 1990a; Wilson, 1994). Under Community Regulation 2078/92, the British territorial approach to contractual guardianship is augmented by a more operational thrust which seeks to reduce chemical inputs, promote extensification, preserve rare breeds and take farmland out of cereal production. Agri-environmental measures are thereby seen as an increasingly central element of EC agricultural policy and, as a consequence, are set to account for a larger proportion of the EAGGF guarantee fund. Significantly, the shift away from a production-focused CAP to one that seeks a greater level of agricultural sustainability, is contributing to the resurgence of rural and agricultural issues in national policy priorities. Thus both ESA designation and the adoption of agri-environmental measures following Regulation 2078/92 are confirming and reinforcing rural policy agendas, whether involving upland landscape management in the UK or combating rural depopulation in France (Boisson and Buller, 1995).

The second thrust in environmentally-sensitive agriculture is centred on organic and biological farming, which offers a more holistic conception of the relationship of agricultural activity to the 'natural' world. In distinction to the agri-environmental measures described above, organic farming combines images of an alternative lifestyle and ethical behaviour (as in the German biodynamic movement). It also offers a practical means of addressing unintended consequences of postwar agricultural policy. It is largely this practical side to organic and biological farming that led to its current elevated status. Certainly, the 1980s saw a rapid growth of organic farming in Europe. This is particularly the case in Germany (see Table 7.3), while in other states organic farming is often practised by migrant Dutch or German farmers who wish to escape the productivist rationale of their home nations (Hoetjes, 1993; Clunies-Ross and Cox, 1994). For a long time considered a fringe and

Table 7.3 Organic farming in the European Community, 1989

Country	Producers	Area (ha)
Belgium	150	1,000
Denmark	500	7,000
France	3,000	2,000
Germany	2,685	54,300
Greece	–	–
Ireland	97	1,500
Italy	800	9,000
Luxembourg	11	450
Netherlands	410	6,200
Portugal	34	420
Spain	350	2,800
United Kingdom	620	15,000
European Community	8,655	117,670

Source: CEC (1992c, 59).

alternative activity, particularly by the established farming community and their policy-making representatives, organic farming has emerged as an important component in the agriculture-environment debate. The reasons for this shift are fourfold: first, because organic farming methods are offered as a viable means of reducing agricultural surpluses; second, because there are clear environmental benefits to be derived from this form of production; third, because there is a growing consumer demand for pesticide-free and 'naturally' produced farm commodities; and, finally, because the low intensity of this production style can contribute to the maintenance of an *in situ* agricultural population. Thus, Community Regulation 2078/92, which sought to encourage environmentally friendly agricultural practices, specifically named organic farming as a potential grant-aided activity. Prior to that date, the Commission sought, in Regulation 2092/91, to standardize organic production and product labelling, hitherto the major stumbling block to an effective international commercialization. This 1991 Regulation has had a major impact on the organization of organic farming in member states. It necessitated a certification of producers and distribution networks and, in doing so, permitted assessments of the spread of organic farming within the Community (Table 7.3). Moreover, what growth has occurred in organic farming cannot be accounted for by the availability of grants for conversion to organic production methods. For in the majority of EC nations, such grants have yet to be formally announced, other than as limited pilot schemes. So rather than looking to public policy initiatives as an explanation for the expansion of organically-produced food items, it is more appropriate that we look to the consumer and, through their response to profitable consumer interests, the stores that sell these farm products.

Such a focus draws our attention to the fact that the evolving nature of agricultural production processes in Europe are raising new questions about the relationship of farming to contemporary society. Fundamental amongst these are questions that pertain to the rural environment. If agriculture has 'produced' nature within Europe, then what we are seeing today is that nature is increasingly being 'consumed' by non-agricultural interests within and beyond the countryside. It is to this consumption of nature, and its implications for what is meant by 'nature', that we now turn.

The consumption of rural space

Throughout this book, we have striven to show how restructuring is contributing to a geographically differentiated redefinition of rural vocation and organization in Europe. Earlier in this chapter, we maintained that the rural environment is, or has the potential to be, a key component in the restructuring process. The desire for a rural setting whether it is for economic activity, residential location or simply recreation is a fundamental spatial choice in modern social consumption (Kayser *et al.*, 1994). Similarly, the growing prioritization of environmental concerns within civil society and government places rural areas, as *de facto* environmental areas, in the forefront of current debates. In the second part of this chapter we seek to put flesh on the bones of these suppositions by looking more closely at the consumption of rural environments and its implications for broader process of rural restructuring. To illustrate the issues raised, three specific items are addressed: the protectionism afforded to national parks; the environmental consequences of residential expansion; and the implications of rural tourism.

The emergence of national parks

The generalized use of rural space for non-production activities might be said to have started toward the end of the last century when an essentially urban population actively sought, largely through membership of voluntary associations, such as the Dutch *Vereniging tot Behoud van Naturmonumenten*, the Touring Club of France or the British National Trust, to protect rural areas of natural beauty. Motivated by a desire to maintain and preserve national heritage, to escape, however temporally, from the trammels of urban life, and to protect and appreciate natural beauty that was represented by the countryside, various national protectionist movements sprang-up and contributed to the establishment of national parks and other forms of protectionist designation. Today, national parks are found in virtually all European states (see Table 7.4, page 245), although the precise bases of their definitions vary considerably (Richez, 1991).

The European national park experience differs fundamentally from the US model, which predates the first formal European national park by some 50 years. For one thing, with one or two exceptions, European national parks have not emerged from a wilderness tradition but reflect a concern for the protection of environments that are strongly marked by human activities. This is so even though the first European national park, which was created in Sweden in 1909, was essentially uninhabited 'wilderness' in public ownership. Since then, European national parks have been characterized more by a requirement that they be reconciled with the needs of their resident population; one consequence of which has been policy variability to cover their distinctive circumstances. Behind the common terminology, different social practices and conceptions of the very notion of a national park are found. This is revealed in a highly variable chronology. Despite the relative temporal synchronicity of landscape and nature protectionism in the late nineteenth century, only in a few states were national parks created before the middle of the twentieth century. These tended to be in sparsely populated areas, as with

Table 7.4 National Parks in the EC as defined by individual member states, 1988

	Number	Surface area (000s ha)
Belgium	0	0
Denmark	*	*
Germany	3	234.0
France	6	345.7
Greece	10	72.1
Ireland	3	20.4
Italy	5	269.0
Luxembourg	*	*
Netherlands	8	33.3
Portugal	1	52.0
Spain	9	122.7
United Kingdom	10	136.4

Source: Richez (1991); Department of the Environment (1991); Ministère de l'environnement (1994). Note: * no data.

Engadine in Switzerland (1914), Covadonga and Ordesa in Spain (1918) and Gran Paradiso in Italy (1922). Where land occupation was denser, economic problems (as in the 1930s) tended to favour policy emphases on rural economic expansion rather than landscape protection, with the added difficulties of reaching agreement over suitable levels of regulatory control and the potentially inflammatory issue of the relationship been landscape protection and the rights of land ownership, all placing national park issues within the remit of broader rural policy considerations. This delayed policy development and meant that some effective but geographically limited local voluntary efforts were not readily transformed into national government action (MacEwen and MacEwen, 1982; Duffey, 1983; Leynaud, 1985).

It also has to be said that rural protection policies emerged in very different national contexts. Although we might identify the key role of internationally-informed nineteenth century artistic elites, and can state unequivocally that policies were inspired by those living in towns rather than country dwellers, the development of an appreciation for 'the countryside' displayed great social variety. In France, the *maisons de campagne* of the Parisian lower bourgeoisie constituted an important first step in establishing an enjoyment of the non-productive use of the rural landscape (Green, 1990). However, this petit bourgeois desire contrasted with two further interests that revealed different goals for the rural environment. One of these came from the state forestry corps who, throughout the nineteenth century, were engaged in a 'crusade' of tree planting (Kalaora and Savoye, 1985, 8). Their concerns for maintaining viable rural communities were linked to a desire to prevent the degradation of the forestry stock. Yet this more functional view ultimately clashed with the aesthetic goals that the urban upper classes exhibited (Kalaora and Savoye, 1985). The difficulties of reconciling these competing interests, plus a lack of broad public interest in protected area policies (Poujade, 1975), undoubtedly contributed to the lateness of national park designations in France. Thus the first national park, Vanoise, was not formally established until 1963. The French experience of class differentiation over the most appropriate uses for particular rural territories, also finds parallels in Spain (Ojeda Rivera, 1989). In Britain, by contrast, the class concerns that retarded national park development in the early decades of the century were those of the wealthy; with landowning interests on one side seeking to block potential restrictions on the usage of their land while an often urban-based upper middle class preservationist movement pressed for the retention of national heritage (MacEwen and MacEwen, 1982; Jeans, 1990). An urban impetus was also party to conflict in the Netherlands, where clashes occurred between the desires of a 'rural agrarian folk culture' and the preferences of a town-based leisured society (Volker, 1989). As Green (1990, 120), reflected on France, '... investment in nature had much to do with the materiality of space and those historically-defined ways of looking grounded in urban life'.

In recent years of course as well as being venues of rural consumption, national parks have emerged as equally important elements in rural economic development and restructuring. Thus French parks play a dual role as conservation sites and rural growth poles (Leynaud, 1985), while Italian parks, particularly in the Alpes, and despite criticism of their questionable environmental gains, have permitted controlled ski and tourism facilities (Richez, 1991). In similar vein, certain Spanish parks have become central elements in broad regional development planning (Gomez-Mendoza, 1994). The same is true for the recent establishment of national parks in the Republic of Ireland, although here, in certain guises, the role of urban elites and rural interests are reversed. Thus recent controversy over the designation of a small national park on Mullach Mor in County Clare has seen a coalition of the state (in the form of the Ministry of Public Works) and local entrepreneurs (seeking tourist development possibilities) being opposed in their attempt to establish a park by a largely professional and urban-based elite. This opposition group is supported by a nascent local environmental movement which focuses its criticisms on the damage tourism could pose for an environmentally-sensitive and scientifically important site. The reality then is that uneven and shifting priorities have been seen in responses to national parks since the inception of this idea around the turn of the last century.

Residential growth and landscape quality

On a broader front, in the bulk of European states rural protection policies predate the mass growth of residential preferences for rural locations. Indeed, the latter are a relatively new phenomenon for most European countries. Only in Britain was the emerging rural protectionist movement of the late nineteenth century and early twentieth century partly motivated by concern for the environmental effects of urban sprawl (e.g. Lowe, 1989; Jeans, 1990). Inversely, whereas the urban-rural residential shift is often forwarded as an explanation for the development of local environmental considerations, in Britain these sentiments have directly fuelled counterurbanization processes (e.g. Newby, 1979; Murdoch and Marsden, 1994). Certainly, the urban elites of France, Italy and other states have historically occupied rural *demesnes* in Lombardia, Sologne and elsewhere. However, counterurbanization processes of the 1970s and 1980s substantially altered the spatial pattern of housing growth within these nations (e.g. Dematteis and Petsimeris, 1989; Costa and Formentini, 1990; White, 1990; Troufleau, 1992). In Chapter 6 we considered the implications of urban-rural migration for the social and class composition of rural areas. In this chapter, we seek rather to assess the environmental implications of this shift in residential preferences. To some extent, the separation of these different strands is to create distinctions where none exist. As Murdoch and Marsden (1994, 17) point out, '... it is action over the environment that provides a useful illustration of how the processes of class formation may be part and parcel of activities which ostensibly have no class complexion'. Nevertheless, it is important to grasp the impacts that rural in-migration processes have on general understandings of the 'the rural environment'.

With the installation of a non-agricultural population, which has often had essentially urban origins, adjustments to public conceptions of what was meant by the rural environment underwent three changes. The first was a greater prioritization of those rural attributes that attract nonfarmer, possibly urban groups into rural zones in the first place. In some nations this contributed to a shift from dominant agrarian ideologies to those linked with naturalist traditions (Chapter 3). More generally, this prioritization points toward a reformulation or reconstitution of what is meant by rurality, as well as different priorities over what local residents value about their own rurality (Berger, 1994; Murdoch and Marsden, 1994). Second, it introduced a new set of environmental service issues, many of which are more traditionally associated with urban zones (like refuse collection and water quality). Third, it ushered a new set of economic actors into the rural sphere. This particularly involved those who were associated with house building and the property market, but also those in rural tourism, both of whom have an interest in 'producing' and 'commodifying' aspects of rurality so cherished by in-migrants or holiday-makers (Cloke and Goodwin, 1992). Within all three of these inputs, the potential for conflict or at least challenge to established rural practices and conceptions is high.

Certainly, where in-migrants are predominantly from urban backgrounds, the antagonism that can develop between newcomers and farmers is generally a conflict over an appropriate vision for the future of a rural area (e.g. Chevalier, 1993). This is linked to different conceptions and definitions of the production of public goods (the countryside) and the consumption of private goods (space and property). Notably, the intensity of such conflicts tends to be greater in nations with a more urbanized rurality. For certain, what we see in the strength of counterurbanization trends and non-agrarian notions of rural potentialities in such nations is a view that rurality offers something 'personal' to the individual (or household). In this regard it merits note that the desire of many British property-buyers in rural France to distance themselves from fellow Britons is regarded by the indigenous population as distinctly odd (Buller and Hoggart, 1994b). It seems that where the English seek castles, the French prefer *lotissements*. But Britain (or most accurately England) is perhaps an extreme case, for its naturalist rural tradition so heightens attachments to 'the countryside', at the same time as its highly urbanized rurality thwarts immediate experiences of the spacious rurality that is found in much of the rest of Europe; and these sentiments are especially intense in south east England, where the highest incomes are found, but so too is the highest sense of urbanism and the longest distances from the closest Britain can claim to any wilderness (the national parks and the Scottish highlands). Given this background, it is not surprising that there is a significant pool of research on the prioritization of consumption-led expectations about rurality on the part of in-migrant householders, as well as well-researched pieces on the effects this has on both rural policy (e.g. Buller, 1984; Murdoch and Marsden, 1994) and the economic and social composition of rural space (e.g. Shucksmith, 1990a). Such inflows have certainly reinforced the British rural preservationist ethic, although this predates the massive urban-rural population shift that began in earnest in the 1960s (Lowe, 1989; Jeans, 1990; Lowenthal, 1991). But what these in-migrant flows have also prompted is a transformation of the broad underpinnings of that ethic, for this was founded on agricultural sustainability as much as protection against urban sprawl. Today, amongst many of those British middle class residents of rural zones, naturalist tendencies are increasingly being viewed largely in terms of an exclusionary localism (e.g. Cloke and Thrift, 1987). This involves objectives that generally prioritize self-interested consumption goals, while making the productive functions of rural space increasingly difficult to perform. The rural environment it seems is for personal rather than general consumption, and decidedly for consumption rather than production (Lowenthal, 1991).

Elsewhere in Europe, it has been less easy to point a finger at so direct a relationship between shifts in rural vision, population change and reorientations in rural policy (though see Guyomard, 1981; Noque I Font,

1988; Volker, 1989; Chevalier, 1993; Kayser, 1994; Mormont, 1994a; Fourny, 1995). In part, the particularly British amenity movement is a middle class response to the participatory and policy-influencing opportunities offered by the restraint-oriented post-1947 planning system (Lowe and Goyder, 1983). Where land-use planning systems have had fundamentally different objectives, as with enabling tools for rural development in France (Buller, 1993) or mandatory zoning mechanisms for the spatial resolution of land vocations in the Netherlands (Groenendijk, 1988), neither opportunities for middle class political mobilization nor specific exclusionary policy outputs have generally been available.

The entry of environmental service demands into rural areas marks a more paradoxical transition in the conception and consumption of the rural environment. In many European rural regions, unanticipated population increases, following decades of rural population decline, have placed unprecedented strains on services like drinking water distribution, household waste disposal and waste water treatment. This has exacerbated variations in the supply and treatment of water across Europe (see Table 7.5). Coincidentally, these strains have emerged at a time of growing EC intervention in regulating environmental quality, and the two together are fast creating a sense of crisis in rural environmental management (Ward *et al.*, 1994). The consequences of this include the diversion of investment priorities away from housing and economic growth towards improved environmental services to meet Com-

Table 7.5 Water supply and treatment in EC nations

	Water supply per capita (m^3)	Population served by waste water treatment, percent	Date of national statistical source
Belgium	917	23	1980
Denmark	228	98	1988
France	781	52	1989
Germany (West)	728	90	1987
Greece	720	1	1986
Ireland	233	11	1979
Italy	977	60	1987
Luxembourg	156	91	1989
Netherlands	993	92	1988
Portugal	125	11	1987
Spain	1186	48	1989
UK	254	84	1987

Source: Eurostat (1992).

munity norms (Ward *et al.*, 1995), the replacement of a traditionally concili-
atory means of environmental policy-making with a more confrontational
style and shifts in institutional responsibility or culpability (as the 1994
incarceration of three French mayors for failing to supply drinking water of
an acceptable quality to their commune demonstrates). Taken together, these
different elements suggest a wide-ranging process under which the traditional components of the rural environment, the fields, the hedges, the rivers
and streams of Abercrombie's England, for example, are being transformed
from visual and amenity components of a rural experience to variables in a
more empirical notion of environment linked directly to human health.

The tourist product and the rural environment

A further component in the consumption of the rural environment is rural
tourism in all its diverse forms, from the ski resorts of the Italian Alps to
cream teas in Devon farmsteads. It is not our intention here to consider in
detail the evolution of rural tourism in Europe. Rather, we focus specifically
upon how changes in the perceived role of the rural environment is contributing to the promotion of rural tourism. Two trends in particular are worthy
of note. First, the adoption of farming operations to the new exigencies of
increasingly market-led EC policy is leading to a widespread diversification
of income generation amongst farmers and their families, for which farm-
based tourism and the sale of farm buildings for tourist or residential use are
an important element (e.g. Arkleton Trust (Research) Ltd., 1992; Evans and
Ilbery, 1992). The second trend is the development and growth of green
consumerism, which has induced a search for product authenticity. The
economic need to match more immediately accessible holiday experiences of
mountain and coastal destinations, and thus to create a viable and unique
rural tourism, has intensified moves toward the 'invention' and marketing of
rural products, whose chief selling points are that they are 'locally' produced
and of traditional fabrication (this includes the massive increase in bottled
spring water, although some of this is linked to health concerns). Thus the
internationalization of the food industry and the globalization of manufacturing have been accompanied by an increasingly high profile for regional
product specificity (although often, but far from always, these are produced
by large corporations; as with so many regional British draught beers). The
principle of the French *appellation d'origine controlée* or the Italian
denominazione di origine controllata, which are generally associated with
wine production, are being extended to a wide range of agricultural products
throughout Europe (e.g. Moran, 1993).

   To a certain degree, both these trends might be seen as responses to the
growing breadth of national and Community agricultural policies, in that

these are increasingly having to take on board, through structural and other measures, different conceptions of agricultural activity and, thereby, of rural vocation. Hence, in Austria, Fuller (1990) identified a shift from part-time farming to pluriactivity as part of an intentionally more holistic approach to rural management (Sieper, 1993). Nonetheless, for the most part, diversification, whether through pluriactivity or plurifunctionality, is a novel though growing European-wide adjustment to agricultural retreat, declining farm incomes and the rationalization of farm holdings (Jollivet, 1988; Ilbery and Bowler, 1993); although its incidence, partly due to uneven opportunity structures, is less marked in southern Europe (Arkleton Trust (Research) Ltd., 1992). As one part of these changes, on-farm tourism and the sale of redundant farm buildings for residential or productive use have become increasingly attractive strategies where the demand exists or can be promoted (e.g. Hoggart and Buller, 1994). Not surprisingly, such strategies are promoted by the governments of individual member states (for example the British government's ALURE initiative and the Ministry of Agriculture's subsequent Farm Diversification Grant Scheme), as well as by the EC (as in the LEADER programmes and the Rural Zone Development Programmes under Objective 5b). The common emphasis in these aid schemes is on a holistic approach to rural development and the incorporation, maintenance or reinsertion of the agricultural sector within that approach. This has alarmed British countryside protectionist bodies who see the relaxing of planning controls on the use of farm buildings as a threat to the ethos of landscape preservation (see Marsden *et al.*, 1993, 122), and in the Netherlands it has led to calls for an extension of planning controls to new on-farm activities (van den Berg, 1993). However, in nations with less urbanized rural belts these steps have been broadly welcomed, especially in regions where economic development in specifically rural locales is regarded as essential to rural sustainability (Laine, 1993).

The marketing of rural space and rural commodities, be they former farm buildings in Devon, distinctive regional products in Bavaria, luxury dwellings in Toscana, gîtes in the Peleponese, villas in the Algarve, cottages in Herefordshire or tax-free greenfield industrial estates in Picardie, is the *fin de siècle* growth industry in Europe (Marsden, 1992). Basic to all the above examples is recognition of the marketable value of the rural environment: 'if the British are buying up rural France, it is because the French are selling it to them' claimed a French estate agent in a recent survey (Buller and Hoggart, 1994c). British house builders have long recognized this value and have consequently become a major interest group in rural policy-making, seeking favourable housing land allocations through the planning system and fighting both farming lobbies and amenity interests in the process (Elson, 1986; Rydin, 1986). 'The local planning system', claim Marsden and colleagues (1993, 24), 'by means of its control over land availability fulfils a mediating role between the accumulation strategies of the housebuilding industry and the consumption norms of the new rural middle class'. What we should stress

here is that the commodification and marketing of rural spaces, and specifically their visual attractiveness and their sense of nostalgia which is
accompanied by a sense of exclusiveness, is neither a universal nor a
permanent phenomenon. As one illustration, British house buyers in rural
France are highly selective both over the regions in which they choose to buy
and in the type of houses they acquire (Buller and Hoggart, 1994a). Their
favoured localities are determined by a combination of aesthetic sensibility,
exclusivity and the desire to rediscover a sense of rural community they feel
has been lost in Britain. Thus, while Périgord becomes the lost Eden of
southern Englanders, central France remains empty and emptying. For
Marsden and colleagues (1993, 10), these different forms of rural experience
are desired attributes of an urban existence. However, this is perhaps too
Anglo-Saxon an emphasis. Elsewhere in Europe, where family and cultural
attachments to rural areas remain strong, yet urbanization continues to draw
people away from the countryside, the rural experience becomes a necessary
part of maintaining identity (Friedl, 1959; Burgel, 1981; Abrahams, 1991).
Yet even this is spatially focussed. As Cribier (1993) has demonstrated,
certain French rural regions, notably the coastal and more attractive regions,
have a stronger pull on their émigrés than others. What is clear from these
various reports is that the attractiveness of rural regions for new forms of
consumption is highly uneven. At the same time, as more and more actors
seek to join the rural commodification bandwagon, such forms of consumption become increasingly banal, as the multiplicity of heritage trails, villages
of character, or regional and local products of varying degrees of authenticity
demonstrate. Effectively, like other aspects of consumption, in a rush to
accumulate desired objects, there is a failure to recognize that the object itself
is being changed, potentially losing the essence it was first sought for, and
becomes 'fundamentally inauthentic' (Gane, 1991, 77). Certainly, this might
lead to a new sense of what authenticity is (Cohen, 1988), but this so often
fails to meet the expectations of 'purists', who move on for a more 'genuine'
experience.[2]

Not surprising therefore that new forms of consumption are also found to
be fickle. The regions that are sought as residential or tourist locations today
are not necessarily going to retain that interest as fashions change or
residential and tourist developments lose their attractiveness (see Butler,
1980, for an account of this process for tourist destinations). The evolving

---

2 We can provide two suggestive indicators of this from our work on British property
  purchases in rural France. The first is the greater propensity for elderly in-migrants into
  rural France to come from locations on the English south coast (Hoggart and Buller,
  1995a). The second is for a disproportionate number of those of pre-retirement age who
  moved their main home to France to be from south west England (Hoggart and Buller,
  1995c). These indications are, however, only suggestive, as we did not inquire into
  previous residential histories in Britain to establish whether these people had first moved
  from other parts of the UK into these desired British destinations (e.g. examine Williams
  et al., 1989 or Bolton and Chalkley, 1990, on movement into the south west).

geography of French second homes is a revealing example. While the Paris Basin has traditionally been a strong focus for second home acquisition (Clout, 1977), recent trends show not only that would-be second home owners are increasingly looking further afield but also that existing owners within the region are selling-up (INSEE, 1992). In a more general sense, the boom and bust cycle that so often seems to be a part of development based on the residential property market does not augur well for the long-term prospects of sustained growth in rural regions.

The message of this last section is that the shift towards a more consumption-oriented exploitation of rural space, based upon the undoubted environmental assets of the European countryside, should not be considered an unqualified panacea for rural survival and sustainability. Undeniably, the growing environmental preoccupations of modern society are bringing a new pertinence to rural areas and, as we said above, are emerging as a basis for a new rural definition. But the consequences for rural areas themselves should not be considered as uniformly or permanently positive. The changing context of economic production is having important and divisive effects upon the role and function of European rural zones. The same could be said for the consumption process. The broadly symbiotic spatial organization of the past, in which distinct rural areas performed a set of independent functions in respect to towns, has been irrevocably lost. The process of globalization is, paradoxically, a process of profound division (Urry, 1981).

The greening of rural Europe

Regional components of rural Europe distinguish themselves first and foremost by their physical characteristics. Yet centuries of human occupation have dulled the immediate imprint of these characteristics, while transnational economic processes have tended to bring this physical diversity under a homogenizing yoke. Even so, environmental characteristics remain a powerful and pertinent basis for rural distinctiveness, whether in terms of relations between rural and urban areas (Buttel and Flinn, 1977), with respect to individual and collective representations of rurality and rural identity (Burgess, 1982), or in the biological resource that non-urban zones represent. Additionally, they play a fundamental role in defining rural policy responses throughout Europe. What we are seeing today, and indeed have seen in varying degrees over the last 20 or so years, is a gradual shift in the thrust and direction of rural policy both within member states and within the CEC as new rural preoccupations are taken into account and at times come to dominate policy responses. Foremost amongst these preoccupations is a new 'environmental paradigm' (Billaud, 1994). In rural policy terms, authors refer to a process of 'greening' (Clout, 1991; Buttel, 1993; Harper, 1993), whereby environmental considerations become a central focus of rural policy. In this chapter, we have already examined a number of individual elements of this greening process. These include the challenges, not only on environmental but also on economic grounds, to intensive farming and its CAP support, the commodification of rurality and residential amenity protectionism. In this final section, we seek to draw these elements together and consider the differential evolution of rural policy as it has responded to the 'greening' process. The starting point for any such analysis must be the identification of 'pre-greening' rural policy. For, as we see it, rural policy as a whole contains two central thrusts characterized both by different objectives and by differential capacities to adapt to the 'greening' process.

At the risk of oversimplification, specific rural policies, operating at the national level, initially emerged as a consequence of the urban-rural divide that processes of modernization fostered (Murdoch and Pratt, 1993). As Buller and Wright (1990, 6) maintain, rural policies have been required:

> ... firstly, to ameliorate the disadvantages and deprivations experienced by residual populations as a consequence of the uneven rates and distribution of development and, secondly, actively to promote growth in areas of economic decline or stagnation as a means of overcoming these social and economic difficulties.

Throughout Europe, rural policy has traditionally focused upon an essentially residual population. At least in stated intent, it has sought to redress the imbalance between urban and rural areas in services, employment opportunities, quality of life and so on, in order to maintain viability for those activities that are essentially the domain of rural space, viz. agriculture or forestry (House of Lords, 1990). As such, an intention of rural policy has been to overcome the negative effects of distance or space. Policy has thus been a response to rural depopulation and agricultural retreat. Rural development policy has sought to minimize the negative economic and social consequences of distance and isolation in rural life and, in many European states, these persist as the dominant rural policy issues (Kayser, 1990).

Parallel to this central objective is a second thrust of rural policy where the focus is not so much on a specific rural population as on the state as a whole. Here, rural policy is a component of national enterprise. This second policy thrust is fundamentally different to the first. First, this is because it is framed in the context of general national goals, be they the maintenance of a viable agricultural population, self-sufficiency in agricultural produce or landscape and environmental protection. As such, this second policy thrust is built around the notion that the recipients of rural policy are engaged in the provision of a 'public good'. At the same time, the provision and maintenance of this desired 'public good' has commonly involved rural policy in seeking to retain distance distinctions and low density human activities (including farming); one obvious illustration being in the quest to preserve the sanctity of the rural landscape (Jeans, 1990). This second thrust is thus implicitly opposed to many of the solutions that are proposed or sought under the first thrust, like rural industrialization, settlement growth, urban-style service provision and so on.

All states are faced with the problem of reconciling these two thrusts. For some, this has been achieved by seeking to transform the needs of rural inhabitants into a public good. Where agrarian policy dominates rural policy, the social wellbeing of the agricultural population becomes the *sine qua non* of rural policy. This was one of the basic principles of the 1942 Scott Report in Britain and still finds pride of place in French and Danish rural policy (Just, 1992; Kayser, 1994). For other states, rural planning has emerged as a means of tying specific rural policy objectives to defined spatial areas, be they national parks, rural development zones or urban fringes.

The point is that changes in rural areas brought on by such processes as counterurbanization, rural industrialization and the general weakening of rural economic and social distinctiveness, seem to call into question the bases upon which these two thrusts of rural policy have hitherto been reconciled. Clearly, the need to address the inability of market mechanisms to assure a just distribution of services and allow for the redistribution of benefits remain key themes of rural policy (Cavailhès *et al.*, 1994). Yet in many rural regions such a policy now operates in a context of social and economic diversification rather than shrinkage. Owing to the emergence of new populations (of higher socioeconomic standing) the targets are less the general rural population than an increasingly minor and encapsulated proportion of it. As Lowe (1995) has argued, *real disadvantages* may *increase* for certain groups as the *relative disadvantages* of rural life *decrease* for the rural population as a whole. It is, however, in the second thrust of rural policy, and in the allocation of public goods to rural areas, that we observe the most substantial change. A broad set of essentially consumption-oriented environmental principles, together with shifts towards post-productivist objectives, are increasingly guiding rural policy. At one level, the reform of the CAP and Community structural policies are replacing agricultural productivity with rural sustainability as the dominant goal of European rural policy (CEC, 1991a). At another level, the expanding use of European rural space for residential and recreational use is raising the profile of amenity and environmental service considerations (CEC 1988). Increasing environmental awareness and more demand for environmental constraint on human activities is permeating economic and industrial policy. Mandatory impact assessments are being lined up alongside the spread of environmental auditing. Institutional adaptation is also required in response to new requirements in trade and environmental policy, as well as shifts in the prioritizing of environmental issues within nations. Each of these is part of a wider greening process. 'Were we to represent Europe by a colour', reaffirms a recent Commission document (CEC, 1992b, 7), 'that colour would undoubtedly be green'. In terms of national rural policies, these diverse elements are having the effect, intended or otherwise, of forcing a reorientation of objectives that privilege rather than negate general rural distinctiveness. The definition or the identification of the national good that rural areas represent is altering. Self-sufficiency in food production, a concern prompted by the menace of global conflict and climatic variability, has been superseded as the current public good that rural areas produce by concerns for residential access and amenity, for recreational and aesthetic provision, and for biological and ecological resource preservation. These are set to emerge as dominant considerations, not only in future rural policy-making but also in national modernization and post-modernization strategies (Gevaert, 1994; Hervieu and Viard, 1994). As such, they call into question many of the former cornerstones of rural policy. The shifting nature of the public good that is being demanded of farmers, from food supplies to sustainable rural communities, and thence to a maintained ecological re-source, is not always easily to follow, as experiences in France and Italy have shown (Bell *et al.*, 1993). At the same time, the traditional role of the state as the *de facto* rural policy-maker and regulator is being eroded by localized rural policy agendas (Stöhr, 1992), by privatization (Cavailhès *et al.*, 1994) and by deregulation (Marsden *et al.*, 1993). In France, at least, the current rural debate has become a focal point for defining the future role of the central state in territorial planning (Kayser, 1993; Hervieu and Viard, 1994). The EC is a central player in this transformation. Examination of the sectoral and territorial destinations of Structural Funds show that although agriculture and rural development remain the central thrusts of Objective 5b, environmental and tourism activities constitute an important part of Community expenditure (see Table 7.6). The distribution of funds under Objective 1 (Table 4.8, page 139), which specifically targets less developed zones, reveals a substantial divide between those nations where agricultural investment is proportionally high and environmental investment low (France, Greece, Portugal) and those where environmental investment is high and agricultural investment low (Britain, Italy and Spain).

Table 7.6 Percentage budget allocations for Objectives 2 and 5b Structural Funds to different sectors, 1989–1993

	Objective 2	Objective 5b
Business investment	61	21
Human resources	4	16
Communications infrastructure	13	1
Agriculture and rural development	–	37
Environment	14	12
Tourism	8	11
Miscellaneous	–	2

Source: CEC DG-XVI (1992, 3).

The point on which we choose to end is a simple one. The greening or the environmentalization of rural policy within Europe is becoming a pan-European phenomenon; a regional response to global concerns, a public response to changing consumption demands and an institutional response to changing international (or Community) policy contexts. Whereas traditional rural policy, whether it seeks to address rural disadvantage or encourage rural activities, operates essentially in a national framework, responding to national priorities and thus reflecting national concerns, the greening process is an increasingly international one. The danger here is that the targets of the former become ignored by the latter. The clients of this new thrust of rural policy are not those disadvantaged by their rural location but increasingly those who are able to choose a rural lifestyle on their own terms. As Buttel (1992, 23) asks: 'Will we ... witness a further erosion of commitment to improving the livelihoods of the rural poor and to rural development? Can we think meaningfully of "sustainable development" in nonmetropolitan contexts of the advanced countries?'

|8|

Postscript

When we started work on this project we had no clear idea of our conclusions. Had we, the work would have been no real challenge and it probably would not have been started. What led us to this work was the identification of what we saw as a gap in the ideas that were being put forward in rural studies. This concerned national-specificities in structures and processes affecting the character and trajectories of change in the rural areas of Europe. Starting a project on this issue was not undertaken lightly, for prevailing trends clearly favour economic globalization, the internationalization of mass consumption, supranational political organization and the worldwide integration of transport and communication networks. Indeed, we were persuaded some time ago by the argument that the nature of '... the capitalist system makes of national boundaries only matters of political formality' (Grant, 1965, 43). Yet our reading of the literature on rural Europe, and on country areas beyond this continent, kept bringing home the importance of processes and structures that are specific to one country but find weak expression elsewhere. This was much more than recognition of distinctive types of rural region. Even official reports are now regular harbingers of the message that rural space embodies distinctions that separate predominantly urbanized regions, significantly rural regions and predominantly rural regions (CEC, 1988; OECD, 1994d). What interested us was that we saw dissimilarities across these categories that required deeper analysis and interpretation. This implied going beyond the surface impressions of socio-economic indicators. Certainly, these provide evidence of a high degree of disparity within rural Europe (Chapter 2); as in restricted opportunity structures in the south that limit engagement in pluriactivity (Arkleton Trust (Research) Ltd., 1992) or the strength of depopulation at the periphery of the continent during the 1960s and 1970s (Salt and Clout, 1976). However, on close examination, within such patterns some nation-specific forces are at play. We see this at the regional level, where socioeconomic standing and performance distinguish areas in the same nation from regions in other countries. Pompili (1994) makes this observation when examining socioeconomic problems in the EC's Objective 1 areas (Chapter 2). Rodríguez-Pose (1994) provides confirmatory evidence when noting how patterns of economic change in the 1980s had divergent regional trajectories in Italy and Spain compared with their nations' performances in the continent as a whole (Chapter 5).

But ironically what consolidated our interest in the national sphere was the emphasis that the literature seemed to place on locality and locally-informed theorization of socioeconomic processes. This gave us some unease, since our reading of the literature left us with a sense of little consideration of any notion between the local and the global. Indeed, there almost seems to be a tendency to conflagrate the national and the global, when the two are clearly not the same. If we wanted empirical evidence to feed our discomfort over the national being ascribed too little priority, it was not hard to find. In Greece and Portugal, for instance, there is no strong tradition of local community initiative, with intrusive (and often unsympathetic) national government agents and agencies limiting local or regional involvement in development efforts (Georgiou, 1994; Syrett, 1994). In Britain too, the growing centralization of state authority has worked against the assertion of local or regional autonomy (Gamble, 1988). Then rural programmes are subjected to a mixture of authorities in Germany and Spain, so the array of initiatives that impose themselves on local development come from a variety of geographical scales (e.g. Arkleton Trust (Research) Ltd., 1992; Wilson, 1994). Certainly local issues can place marked constraints on state actions, as seen in Belgium and Spain, where national governments must progress with caution, given the strength of regionalist sentiments (Williams, 1980; Pijnenberg, 1989; Conversi, 1990), while Chapter 5 indicated that new forms of economic organization, like industrial districts, appear to have few links with the specifics of nationhood. But what this points towards is our starting point, which was an underlying sense that there is greater complexity in causal processes than is often acknowledged. Our fear, that finds particular expression in rural studies, is that research in social science runs the risk of being lulled by the charms of new research approaches and theorizations, which do not penetrate the complexity of causation at different structural 'levels', but package material as if all we need to examine is global structures and contextualized local 'deviations' around them. Our sense is that we need to be more conscious of processes and structures which are 'organized' at different geographical scales. In rural geography this certainly means raising our eyes above the local scale, so attention to global processes is particularly welcome (e.g. Le Heron, 1994), but we also require a focus on the nation. In expressing this view we take significant cues from Michael Mann (1986), who provides in the first pages of *The sources of social power* a cogent, exhilarating exposition of the merits of greater attention to the intricacies of dissimilar sources of power and the manner in which the interweaving of their force complicates, while making more intriguing, processes of social transformation.

As Mann (1986, 17) expressed it: 'To conceive of societies as confederal, overlapping, intersecting networks rather than as simple totalities complicates theory. But we must introduce further complexity'.

Positioning our start point for this book, we can again call on Mann (1986, 2): 'State, culture and economy are all important structuring networks; but they almost never coincide'. Such ideas directed us deliberately not to impose a national specificity to the structures and processes we examined. We wanted to undertake an exploration of whether these forces existed, not present a bold characterization that positions them in the forefront of explanation. As we have made clear in various parts of the text, there are large areas that we know too little about; where we did not come across research output or theoretical ideas that helped us formulate a clear image of either geographical distinctions or causal importance. It follows that we have not sought to package the material we have presented into one coherent conceptualization. Readers will note, for example, that there is no direct overlap between the rural traditions we identified in Chapter 3 and the social tendencies we described in Chapter 6. There are none because we do not see a direct meshing of these schemas. The rural traditions we described were clearly informed by social tendencies in their base areas, but the emergence of a rural tradition is conditioned by compromises between power interests and dominant forces which change over time. In part this is because new interests have become involved as the political organization of space has changed. The uniting of Germany and Italy in the late nineteenth century provide obvious illustrations of this, especially in the Italian case (Graziano, 1978; Davis, 1979), but we also see the process operating in the tempering of national interests in the formation of the European Community (Tracy, 1989; Milward, 1992). In addition, we see change in traditions as new groups thrust themselves into dominant power circles. The nineteenth century 'victory' of manufacturers over the landed aristocracy of Britain provides an illustration, even if the success of manufacturers was diluted by their desire to mimic aristocratic lifestyles (which has much to do with the depth of naturalist rural traditions in Britain), with real power retained by finance capital as it was able to utilize the successes of manufacturers and the country's empire builders for their own gratification (Wiener, 1981; Ingham, 1984). This brings us to two points that need stressing.

The first concerns the distribution of power and determining forces in rural traditions. Here we see dominant trends in social mores being decided by power elites, even if in the present world of swelling middle classes others often carry the banner. This is not to deny the existence of inputs from the so-called average person, as we showed in identifying the significant role of farm workers in generating conflictual social formations in Italy and Spain (Chapter 6). Nor do we wish to deny the importance of advertising and the mass media as pseudo-independent impositions on the modern psyche. What we would say is that core values take time to develop and the most powerful have the means, and find it easier to organize themselves, to project their preferred social model (Urry, 1986). This point we hope is apparent from our analysis of social tendencies in Chapter 6. It is certainly evident in the way the landed classes used the imagery of the honest peasant to garner support for their own brand of agrarianism in France (Tarrow, 1977; Tracy, 1989). As Mann (1986, 7) points out, it is almost inevitable that '[t]he masses comply because they lack the collective organization to do otherwise, because they are embedded within collective and distributive power organizations controlled by others'.

This brings us to our second point, which is that traditions of rurality, and ultimately the socioeconomic processes that affect rural areas, are deeply permeated by broader power structures. At one level this is manifest in the way national economic performance penetrates the viability and sustainability of rural modes of life. An obvious example is the stultifying impact of the Franco and Salazar regimes on rural development in Spain and Portugal in the middle of the twentieth century (e.g. Harding, 1984; Pinto *et al*., 1984; Mansvelt Beck, 1988). A further example comes from the forces that have propelled Britain toward relative economic decline over the last century, which affect not simply the magnitude and character of economic activity but also its spatial organization (Massey, 1986; Taylor, 1989). As pointed out in Chapter 5, for instance, it is instructive to note that the decentralization of producer services from large cities follows different paths in Britain and Germany. The dominance of City finance in Britain is notably allied to a deconcentration process that clings to the South East and produces yet more pressures on its heavily urbanized rurality (Marshall, 1988; Barlow, 1989; Thrift and Leyshon, 1992). By contrast, under more decentralized structures of power in Germany, the trend in service deconcentration is for moves away from the core into peripheral areas (Jaeger and Dürrenberger, 1991). Further illustrations of national specificities are visible in state policies toward private sector companies. Thus Guesnier (1994) suggests that the emphasis on large corporations that is part and parcel of the French adherence to Fordist principles has restricted the potential of SMEs in that nation, while mistrust of large business organizations by Italian governments, albeit alongside strategies to distribute political patronage effectively, means that its governmental policies are advantageous for SMEs (Furlong, 1994). If further signification of national inputs is required, it bears noting that the often illegal or semi-illegal practices that characterize flexible accumulation strategies in some states inevitably draws succour from governments turning a blind eye (Hadjimichalis and Papamichos, 1990). For us what all this points to is the merits of more cross-national analyses that investigate links between national power structures and processes of socioeconomic differentiation in rural areas (and of rural areas within the broader context of their national economies, polities and societies).

This being said we must not fall into the trap of assuming a national causal force is operating just because national differences can be identified. Consider the point made by Thomas (1983, 243) that the industrial revolution created an urban-rural divide that has subsequently underscored civilization's attitudes towards nature and the environment. In that the pace and scale of industrial expansion across Europe was uneven, it is possible that dissimilar industrialization experiences led to different conceptions of the environment, which became integral to rural traditions. Certainly such rural-urban distinctions might have been barely apparent when market towns saw their *raison d'être* in the economies of surrounding rural zones. But since industrialization took place within national bounds this provided ample opportunity for national ideologies to incorporate images of rurality into mainstream ideas (Anderson, 1983). Today we can see this effect in the divergent attitudes of Britain and France toward environmentally-sensitive farming proposals (Chapter 7). Moreover, as Pfau-Effinger (1994) identifies in a study of Finland and Germany, we can link processes of industrialization, when aligned with the agricultural base prior to their genesis, to national divergences in gender differentiation. However, whilst it would be easy to interpret this as a nation-specific force, in reality it might be no more than a particular manifestation of a general process, albeit with some local contextualization. The issue behind this point is that geographical concentrations can provide a particular flavour to a nation, without the causal impetus being national as such. This thought is provoked by comparing Pfau-Effinger's study with that of Sackmann and Häussermann (1994). This latter work does take one of Pfau-Effinger's case study states, namely Germany, but explores regional dissimilarities in gender roles rather than their dominant national characteristics. Yet they reach very similar conclusions to those of Pfau-Effinger concerning the respective roles of farm organization and the nature of later industrialization processes. Our question is whether Pfau-Effinger identified a point of distinction between Finland and Germany that was genuinely nation-specific or if the mainstream of German gender relations is different from those in Finland largely because it happened to be a venue in which a particular 'global' process was played out (i.e. compared with Finland, Germany experienced concentrations of a particular kind of industrialization that owed little to the fact of nationhood).

If question marks about nation-specific influences can be levied against processes of the past, they are even more pertinent today, given that the dynamism and complexities of socioeconomic change have made our understanding of the rural world more uncertain.

> The decline of the postwar certainties, most notably agricultural productivism and its corporatist structures has opened the way for the emergence of a more differentiated countryside, one whose trajectory is no longer determined to the same degree by the fortunes of a single industry ... (Marsden *et al.* , 1993, 185).

Placed in this light, the certainties of the rurality of industrial societies are fast disappearing in the face of post-industrialism and globalization (Kayser *et al* ., 1994; Lévy, 1994). Certainly there is a wealth of evidence pointing to the intermixing of rural traditions, with the production-related mass migrations of European populations in the 1960s and 1970s away from rural areas now seeing a counterpart consumption-oriented movement into the same spaces. While the latter might not have had a major effect on transforming images of rurality, more recent trends do appear to be having this effect (e.g. Forsythe, 1980; Buller and Hoggart, 1994a). At the same time we see an enhancement of transnational structures of regulation that are forging higher levels of standardization over the utilization of rural territories, both in their prospects for income generation (Chapter 4) and in the utilization of their environment (Chapter 7). For both of these, there have been very rapid changes in the circumstances of rural life. This perhaps is true of the postwar years in general, although the pace of change has increased over time. This leads us to note that ideas about nation-specific effects need to recognize distinctions between expediency in the shorter term and decisions that are drawn from deeply-entrenched traditions. In Chapter 4 this issue was explored in the context of transnational policy, which clearly reveals a depth of sentiment towards kinds of rurality that differ across the states of Europe. These dissimilar visions underwrote the responses of national governments to the formation, management and reform of the CAP. Yet what we also see in these processes is the hand of expediency or context that thwarts the automatic imposition of rural traditions on critical policy forms. The juxtaposition of a European economic recession and Britain's recent accession to the EC in the mid-1970s provides one illustration, for this undoubtedly restricted efforts to reform the CAP at an earlier time. In the same way, do we really imagine that the CAP would have taken its present form had the UK not been so aloof to European integration in the early 1950s and become a founding member of the Community? When you join late you need to accept the rules that have already been worked out, even if, as for Britain, Greece, Portugal and Spain, this involved compliance with a vision of agriculture that did not mesh too closely with their longstanding national values.

Where this brings us to is the inherent condition that has underwritten so much of this book, which is that there is little certainty in the picture that we are able to paint. We have tried to provide a guide to the issues and cajole the reader into thinking about connections and causations. What we have not sought to do is impose an explanatory position in a dogmatic way. No doubt this will not please some readers. When *Rural development* was published (Hoggart and Buller, 1987), a few reviewers criticized us for not providing a single theoretical framework that would explain the processes and structures we described. As it was the fashion amongst some rural analysts at that time, we were not surprised that the call was for a structuralist mode of interpretation. We smile wryly to find the same analysts now calling for intermediate level analyses and decrying the rigidities of top-down structuralist accounts. Mind you, one of us has had the misfortune to have one of his books described as postmodern. To see this comment made about the present volume would make us cringe; which at least would be a repetition of what happened last time. This would indeed be a low blow, for we are not seeking to chase the fashionable, nor extol the need to examine yet one more 'other'. What we seek to emphasize is the need for quality research that examines the complexities of causal processes and their outcomes. This will require careful work that examines processes at each geographical level in a detailed but structured context. Where we still fall short in rural studies is in analyses of structuration processes. We already have excellent studies of individual localities, for which we must largely thank anthropologists and sociologists of the 1950s and 1960s, although a few geographers also made their mark (e.g. Connell, 1978). Of course, this work in general has rightly been criticized for its weak theoretical base. Yet what was provided was detailed, highly informed accounts of rural existence. Today, we fear, too much research is falling between two desirable goals; quality research of local level processes and theoretical contextualization of the structured circumstances of activities in rural areas. Our hope is that this book, aimed largely at the latter goal, will at least provoke researchers to think more deeply about the structures that inform local social action. For theoretical advance we need at least to debate and explore the role of nation-specific forces and not take them for granted, or even worse ignore them.

Bibliography

AAREBROT, F.H. 1982: Norway: centre and periphery in a peripheral state. In S. Rokkan and D.W. Urwin (eds.) *The politics of territorial identity*. London: Sage, 75–112.

ABRAHAMS, R. 1991: *A place of their own: family farming in eastern Finland*. Cambridge: Cambridge University Press.

ABRAHAMSSON, K., editor, 1994: *Work and leisure: the national atlas of Sweden*. Stockholm: National Library.

AGE CONCERN ENGLAND 1993: *Growing old in Spain: older British people resident in Spain*. London.

AGNEW, J.A. 1981: Political regionalism and Scottish nationalism in Gaelic Scotland. *Canadian Review of Studies in Nationalism* 8(1), 115–129.

AGNEW, J.A. 1987: *The United States in the world-economy*. Cambridge: Cambridge University Press.

AGNEW, J.A. 1993: The United States and American hegemony. In P.J. Taylor (ed.) *Political geography of the twentieth century*. London: Belhaven, 209–238.

ALAPURO, R. 1982: Finland: an interface periphery. In S. Rokkan and D.W. Urwin (eds.) *The politics of territorial identity*. London: Sage, 113–164.

ALMEIDA, C. and BARRETO, A. 1976: *Capitalismo e emigraçao em Portugal*. third edition. Lisboa: Prelo.

AMIN, A. 1989: Flexible specialization and small firms in Italy: myths and realities. *Antipode* 21, 13–34.

ANDERBERG, S. 1991: Historical land-use changes: Sweden. In F.M. Brouwer, A.J. Thomas and M.J. Chadwick (eds.) *Land-use changes in Europe*. Dordrecht: Kluwer, 403–426.

ANDERSON, B. 1983: *Imagined communities: reflections on the origins and spread of nationalism*. London: Verso.

ANDERSON, R.T. and ANDERSON, B.G. 1964: *The vanishing village: a Danish maritime community*. Seattle: University of Washington Press.

ANDERSON, R.T. and ANDERSON, B.G. 1965: *Bus stop for Paris: the transformation of a French village*. Garden City, New York: Doubleday.

ANDRETSCH, D.B. and FRITSCH, M. 1994: The geography of firm births in Germany. *Regional Studies* 28, 359–365.

ANDRIKOPOULOU, E. 1987: Regional policy and local development prospects in a Greek peripheral region: the case of Thraki. *Antipode* 19, 7–24.

ANDRLIK, E. 1981: The farmer and the state: agricultural interests in West German politics. *West European Politics* 4, 104–119.

ANGEL, D.P. 1994: Tighter bonds? customer-supplier linkages in semiconductors. *Regional Studies* 28, 187–200.

ARENSBERG, C.M. 1937: *The Irish countryman: an anthropological study.* 1968 edition. Garden City, New York: American Museum of Natural Science.

ARKLETON TRUST (RESEARCH) LTD. 1992: *Farm household adjustment in western Europe 1987–1991.* two volumes. Luxembourg: Office for Official Publications of the European Communities.

ARROYO ILERA, F. 1990: *El reto de Europa: España en la CEE.* Madrid: Editorial Síntesis.

ARTER, D. 1993: *The politics of European integration in the twentieth century.* Aldershot: Dartmouth.

ASHFORD, D.E. 1986: *The emergence of the welfare states.* Oxford: Blackwell.

ASHFORD, S. and TIMMS, N. 1992: *What Europe thinks: a study of western European values.* Aldershot: Dartmouth.

ASSOCIATION DES RURALISTS FRANCAIS 1986: Les études rurales sont-elles en crise? *Bulletin de l'ARF* 41–42.

AUDRETSCH, D.B. and FRITSCH, M. 1994: The geography of firm births in Germany. *Regional Studies* 28, 359–366.

AUTY, R.M. 1993: Emerging competitiveness on newly industrializing countries in heavy and chemical industries. *Tijdschrift voor Economische en Sociale Geografie* 84, 185–197.

AVDELIDIS, P. 1981: Le rôle de la coopération dans le développement rural. In S. Damaniakos (ed.) *Aspects du changement social dans la campagne grecque.* Athens: Greek Review of Social Science special edition, 42–48.

AVERYT, W.F. 1977: *Agropolitics in the European Community: interest groups and the Common Agricultural Policy.* New York: Praeger.

AVILLEZ, F. 1993: Portuguese agriculture and the Common Agricultural Policy. In J. da Silva Lopes (ed.) *Portugal and EC membership evaluated.* London: Pinter, 30–50.

AYDALOT, P. and KEEBLE, D.E. 1988: High technology industry and innovative environments in Europe: an overview. In P. Aydalot and D.E. Keeble (eds.) *High technology industry and innovative environments.* London: Routledge, 1–21.

BACALHAU, M. 1993: The image, identity and benefits of the EC. In J. da Silva Lopes (ed.) *Portugal and EC membership evaluated.* London: Pinter, 182–188.

BAGNASCO, A. 1977: *Tre Italie.* Bologna: Il Mulino.

BAKER, S., MILTON, K. and YEARLEY, S., editors, 1994: *Protecting the periphery.* London: Frank Cass.

BALABANIAN, O., BOUET, G., DESLONDES, O. and LERAT, S. 1991: *Les états méditerranéens de la CEE.* Paris: Masson.

BALDOCK, D. 1992: The polluter pays principle and its relevance to agricultural policy in European countries. *Sociologia Ruralis* 32, 49–65.

BALDOCK, D. and LONG, I. 1987: *The Mediterranean environment under pressure: the influence of the CAP on Spain and Portugal and the Integrated Mediterranean Programmes in France, Greece and Italy.* London: Institute of European Environmental Policy.

BANDARRA, N.J. 1993: The Community structural funds and post–Maastricht cohesion. In A.W. Gilg (ed.) *Progress in rural policy and planning: volume three.* London: Belhaven, 221–242.

BARKE, M. and NEWTON, M. 1994: A new rural development initiative in Spain: the European Community's Plan Leader. *Geography* 79, 366–371.

BARKER, M.L. 1982: Traditional landscape and mass tourism in the Alps. *Geographical Review* 72, 395–415.

BARLOW, J. 1989: Regionalisation or geographical segmentation? developments in London and South East housing markets. In M. Breheny and P. Congdon (eds.) *Growth and change in a core region: the case of South East England.* London: Pion, 182–202.

BARRAL, P. 1968: *Les agrariens français: de Méline à Pisani.* Paris: Colin.

BARRY, N.P. 1987: *The new right.* London: Croom Helm.

BAUER, G. and ROUX, J. 1976: *La rurbanisation ou la ville éparpillée.* Paris: Seuil.

BAUWENS, A.L. and DOUW, L. 1986: Rural development: a minor problem in the Netherlands? *European Review of Agricultural Economics* 13, 242–366.

BAX, M. 1976: *Harpstrings and confessions: machine-style politics in the Irish Republic.* Assen: Van Gorcum.

BEAVINGTON, F. 1963: The change to more extensive methods in market gardening in Bedfordshire. *Transactions of the Institute of British Geographers* 33, 89–100.

BEAVIS, S. and HALSALL, M. 1992: Britain told to pass on EC funds for pits. *The Guardian* 26 September, 38.

BECKETT, J.V. 1986: *The aristocracy in England 1660–1914.* Oxford: Blackwell.

BELL, F. and others 1993: The role of policy influencing farm household behaviour in European mountain areas. *Revue de Géographique Alpine* 81, 101–127.

BELL, P. and CLOKE, P.J. 1989: The changing relationship between private and public sectors: privatisation in rural Britain. *Journal of Rural Studies* 5, 1–15.

BELL, R.M. 1979: *Fate and honor, family and village: demographic and cultural change in rural Italy since 1800.* Chicago: University of Chicago Press.

BELLIARD, J.C. and BOYER, J.C. 1983: Les nouveaux ruraux en Ile de France. *Annales de Géographie* 512, 433–451.

BENENATI, A. 1981: *Le développement inégal en Italie.* Paris: Economica.

BENNEMA, J.W. 1978: *Traditions of communal cooperation among Portuguese peasants.* Amsterdam: University of Amsterdam Anthropology-Sociology Centre Paper on European and Mediterranean Societies 11.

BENNETT, R.J. and KREBS, G. 1994: Local economic development partnerships: an analysis of policy networks in EC-LEDA local employment development strategies. *Regional Studies* 28, 119–140.

BENVENUTI, B. 1962: *Farming in cultural change.* Assen: Van Gorcum.

BEOPOULOS, N. 1991: L'intensification de l'agriculture grecque et les problèmes de l'environnement. Paper to RAFAC seminar, CHIEAM, Montpellier, October 1991.

BERGER, A. 1994: L'espace rural: les perspectives d'une recomposition. *Revue de l'Économie Méridionale* 42, 5–26.

BERGER, A. and ROUZIER, J. 1977: *Ville et campagne: la fin d'un dualisme.* Paris: Economica.

BERGER, J. 1972: *Ways of seeing.* Harmondsworth: Penguin.

BERGER, J. 1979: *Pig earth.* London: Hogarth Press.

BERGER, S. 1972: *Peasants against politics: rural organization in Brittany 1911–1967*. Cambridge, Massachusetts: Harvard University Press.

BERGMANN, T. 1990: Socio-economic situation and perspectives of the individual peasant. *Sociologia Ruralis* 30, 48–61.

BERLAN, J-P. 1995: Dynamique d'intégration dans l'agriculture méditerranéenne de la France. *Etudes Rurales* 135/136, forthcoming.

BERLAN-DARQUE, M. and KALAORA, B. 1992: The ecologization of French agriculture. *Sociologia Ruralis* 32, 104–113.

BERMEO, N. 1986: *The revolution within the revolution: workers control in rural Portugal*. Princeton, New Jersey: Princeton University Press.

BERRY, B.J.L. 1972: Hierarchical diffusion: the basis of developmental filtering and spread in a system of growth centres. In N.M. Hansen (ed.) *Growth centres in regional economic development*. New York: Free Press, 108–138.

BEST, R.H. and GASSON, R.M. 1966: *The changing location of intensive crops: an analysis of their spatial distribution in Kent and the implications for land-use planning*. Ashford: Wye College Studies in Rural Land-Use Report 6.

BESTEMAN, C. 1989: Economic strategies of farming households in Penabranca, Portugal. *Economic Development and Cultural Change* 37, 129–143.

BÉTEILLE, R. 1981: *La France du vide*. Paris: Litec.

BILLAUD, J-P. 1994: Environnement et questions rurales: vers un nouveau paradigme? In Association des Ruralistes Français (ed.) *Le monde rural et les sciences sociales : omission ou fascination?*. Paris.

BILLAUD, J-P. and DE LA SOUDIERE, M. 1989: La nature pour repenser le rural. In N. Mathieu and M. Jollivet (eds.) *Du rural à l'environnement*. Paris: L'Harmattan, 180–191.

BIRD, S.E. 1982: The impact of private estate ownership on social development in a Scottish rural community. *Sociologia Ruralis* 22, 43–55.

BLACK, R. 1992: *Crisis and change in rural Europe: agricultural development in the Portuguese mountains*. Aldershot: Avebury.

BLACK, R. 1994: Asylum policy and the marginalization of refugees in Greece. In W.T.S. Gould and A.M. Findlay (eds.) *Population migration and the changing world order*. Chichester: Wiley, 145–160.

BLACKSELL, M. and FUSSELL, C. 1994: Barristers and the growth of local justice in England and Wales. *Transactions of the Institute of British Geographers* 19, 482–495.

BLAKE, D.H. and WALTERS, R.S. 1992: *The politics of global economic relations*. fourth edition. Englewood Cliffs, New Jersey: Prentice–Hall.

BLEKESAUNE, A., HARVEY, W.G. and HAUGEN, M.S. 1993: On the question of the feminization of production on part-time farms: evidence from Norway. *Rural Sociology* 58, 111–129.

BLOCK, F. 1977: *The origins of international economic disorder*. Berkeley: University of California Press.

BLOCK, F. 1987: *Revising state theory*. Philadelphia: Temple University Press.

BOCOCK, R. 1993: *Consumption*. London: Routledge.

BODENSTEDT, A. 1990: Rural culture – a new concept. *Sociologia Ruralis* 30, 5–47.

BODIGUEL, M., editor, 1989: *Produire et préserver l'environnement: quelles règlements pour l'agriculture européenne?*. Paris: L'Harmattan.

BODIGUEL, M. and BULLER, H. 1989: Agricultural pollution and the environment in France. *Tijdschrift voor Sociaal Wetenschappelijke Onderzoek van de Landbouw* 3, 217–239.

BODIGUEL, M,. BULLER, H. and LOWE, P.D. 1990: Concepts, definitions and research traditions. In P.D. Lowe and M. Bodiguel (eds.) *Rural studies in Britain and France*. London: Belhaven, 37–52.

BODSON, D. 1993: Les *villageois*. Paris: L'Harmattan.

BODY, R. 1991: *Our food, our land: why contemporary farming practices must change*. London: Rider.

BOISSON, J-M. and BULLER, H. 1995: The French experience. In M.C. Whitby (ed.) *The European experience of policies for the agricultural environment*. Wallingford: CAB International, forthcoming.

BOLSIUS, E.C.A. 1993: Possible changes in land-use in rural areas in the European Community. In E.C.A. Bolsius, G. Clark and J.G. Groenendijk (eds.) *The retreat: rural land-use and European agriculture*. Utrecht: Nederlandse Geografische Studies 172, 42–51.

BOLTON, N. and CHALKLEY, B. 1990: The rural population turnaround: a case study of north Devon. *Journal of Rural Studies* 6, 29–43.

BONNAIN, R. and SAUTTER, G. 1970: Société d'ici, société d'ailleurs. *Études Rurales* 74, 23–50.

BONNAMOUR, J. 1993: *Géographie rurale: position et méthode*. Paris: Masson.

BONTRON, J-C. 1989: Equipement et cadre de vie. In A. Brun (ed.) *Le grand atlas de la France rurale*. Paris: de Monza, 88–92.

BONTRON, J-C. 1994: Le concept de ruralité à l'épreuve du changement social. In C. Courtet, M. Berlan–Darqué and Y. Demarne (eds.) *Territoires ruraux et développement*. Paris: CEMAGREF, 162–166.

BONTRON, J-C. and MATHIEU, N. 1977: *La France des faibles densités*. Paris: SEGESA.

BORCHARDT, K. 1991: *Perspectives on modern German economic history and policy*. Cambridge: Cambridge University Press.

BOUCHER, S., FLYNN, A. and LOWE, P.D. 1991: The politics of rural enterprise: a British case study. In S.J. Whatmore, P.D. Lowe and T.K. Marsden (eds.) *Rural enterprise*. London: David Fulton, 120–140.

BOUQUET, M. 1986: Domestic economy and welfare. In P.D. Lowe, T. Bradley and S. Wright (eds.) *Deprivation and welfare in rural areas*. Norwich: Geo Books, 85–96.

BOURDIEU, P. 1984: *Distinction: a social critique of the judgement of taste*. London: Routledge and Kegan Paul.

BOUSSARD, I. 1990: French political science and rural problems. In P.D. Lowe and M. Bodiguel (eds.) *Rural studies in Britain and France*. London: Belhaven, 269–285.

BOWERS, J.K. and CHESHIRE, P.C. 1983: *Agriculture, the countryside and land-use: an economic critique*. London: Methuen.

BOWLER, I.R. 1985: *Agriculture under the Common Agricultural Policy*. Manchester: Manchester University Press.

BOWLER, I.R. 1986: Intensification, concentration and specialisation in agriculture: the case of the European Community. *Geography* 71, 14–24.

BOWLER, I.R. 1992: The industrialization of agriculture. In I.R. Bowler (ed.) *The geography of agriculture in developed market economies*. Harlow: Longman, 7–31.

BOYLE, P.J. and HALFACREE, K.H. 1995: Service class migration in England and Wales 1980–1981: identifying gender-specific mobility patterns. *Regional Studies* 29, 43–57.

BRADLEY, T. 1984: Segmentation in local labour markets. In T. Bradley and P.D. Lowe (eds.) *Locality and rurality*. Norwich: Geo Books, 65–90.

BRADSHAW, R. 1972: Internal migration in Spain. *Iberian Studies* 1(2), 68–75.

BRADSHAW, R. 1985: Spanish migration: economic causes, grave social consequences. *Geography* 70, 175–178.

BRAITHWAITE, M. 1994: *The economic role and situation of women in rural areas.* Luxembourg: Office for Official Publications of the European Communities (Green Europe 1/94).

BRANDES, S.H. 1975: *Migration, kinship and community: tradition and transition in a Spanish village.* New York: Academic Press.

BRANDES, S.H. 1976: The impact of emigration on a Castilian mountain village. In J.B. Aceves and W.A. Douglass (eds.) *The changing faces of rural Spain*. Cambridge, Massachusetts: Schenkman, 1–16.

BRASSLOFF, W. 1993: Employment and unemployment in Spain and Portugal: a contrast. *Journal of the Association for Contemporary Iberian Studies* 6(1), 2–24.

BRAUDEL, F. 1973: *The Mediterranean and the Mediterranean World*. London: Collins.

BRAUDEL, F. 1990: *The identity of France*. London: Collins [originally published in 1986 as *L'identité de la France*. Paris: Artaud].

BREEN, R., HANNAN, D.F., ROTTMAN, D.B. and WHELAN, C.T. 1990: *Understanding contemporary Ireland: state, class and development in the Republic of Ireland*. Dublin: Gill and Macmillan.

BRENAN, G. 1950: *The Spanish labyrinth.* second edition. Cambridge: Cambridge University Press.

BRETTELL, C.B. 1979: Emigrar para voltar: a Portuguese ideology of return migration. *Papers in Anthropology* 20, 1–20.

BRETTELL, C.B. 1986: *Men who migrate, women who wait: population and history in a Portuguese parish.* Princeton, New Jersey: Princeton University Press.

BRIASSOULIS, H. 1993: Tourism in Greece. In W. Pompl and P. Lavery (eds.) *Tourism in Europe*. Wallingford: CAB International, 285–301.

BRIGGS, D.J. and KERRELL, E. 1992: Patterns and implications of policy-induced agricultural adjustments in the European Community. In A.W. Gilg (ed.) *Restructuring the countryside: environmental policy in practice*. Aldershot: Avebury, 85–102.

BROADY, M. 1980: Rural regeneration: a note on the mid-Wales case. *Cambria* 7(1), 79–85.

BRODY, H. 1973: *Inishkillane: change and decline in the west of Ireland*. London: Allen Lane.

BROUWER, F.M. and CHADWICK, M.J. 1991: Future land-use patterns in Europe. In F.M. Brouwer, A.J. Thomas and M.J. Chadwick (eds.) *Land–use changes in Europe*. Dordrecht: Kluwer, 49–78.

BROUWER, F.M., THOMAS, A.J. and CHADWICK, M.J., editors, 1991: *Land–use changes in Europe*. Dordrecht: Kluwer.

BROUWER, R. 1993: Between policy and politics: the Forestry Service and the commons in Portugal. *Forest and Conservation History* 37(4), 160–168.

BROWN, C. 1988: *Price policies of the CAP: retrospect and prospect.* København: Statens Jordbrugsøkomiske Institut Rapport 41.

BRUCKMEIER, K. 1987: Local nitrate policy in West Germany. Berlin: WZB Working Paper 87–15.

BRUCKMEIER, K. 1994: République Fédérale Allemagne. In M. Jollivet (ed.) *Bilan des recherches en sciences sociales sur les problèmes d'environnement en milieu rural dans les pays européens.* Paris: CNRS Groupe de Recherches sur les Mutations des Sociétés Européennes, 111–121.

BRUCKMEIER, K. and GLAESER, B. 1995: Integrating EC environmental directives in Germany. Berlin: WZB report to CEC DG-XII.

BRUCKMEIER, K. and TEHERANI-KRONNER, P. 1992: Farmers and environmental regulation: experiences in the Federal Republic of Germany. *Sociologia Ruralis*, 32, 66–81.

BRUNET, P. 1992: *L'atlas des paysages ruraux de France.* Paris: de Monza.

BRUNT, B. 1988: *The Republic of Ireland.* London: Paul Chapman.

BRUSCO, S. 1986: Small firms and industrial districts: the experience of Italy. In D.E. Keeble and E. Wever (eds.) *New firms and regional development in Europe.* London: Croom Helm, 184–202.

BUCK, N., DRENNAN, M. and NEWTON, K. 1992: Dynamics of the metropolitan economy. In S.S. Fainstein, I. Gordon and M. Harloe (eds.) *Divided cities.* Oxford: Blackwell, 68–104.

BULLER, H. 1984: Amenity societies and landscape conservation: an analysis of political efficacy of local groups in West Sussex. London: unpublished PhD thesis, King's College London, University of London.

BULLER, H. 1991: *Agricultural structures, policy and the environment in France.* Paris: Université de Paris X CNRS Groupe de Recherches Sociologiques, report to CEC DG-VI.

BULLER, H. 1992: Agricultural change and the environment in Western Europe. In K. Hoggart (ed.) *Agricultural change, environment and economy.* London: Mansell, 68–88.

BULLER, H. 1993: The French planning system. In Welsh Affairs Committee of the House of Commons *Rural housing: volume III – minutes of evidence and appendices.* London: HMSO, 217–228.

BULLER, H. and HOGGART, K. 1994a: *International counterurbanization: British migrants in rural France.* Aldershot: Avebury.

BULLER, H. and HOGGART, K. 1994b: Social integration of British home owners into French rural communities. *Journal of Rural Studies* 10, 197–210.

BULLER, H. and HOGGART, K. 1994c: Vers une campagne européenne: les Britanniques en France rurale. *Espace Géographique* 1994/3, 263–273.

BULLER, H. and WRIGHT, S. 1990: Concepts and policies of rural development. In H. Buller and S. Wright (eds.) *Rural development: problems and practices.* Aldershot: Avebury, 1–26.

BULLER, H., LOWE, P.D. and FLYNN, A. 1992: National responses to the Europeanization of environmental policy. In J. Liefferink, P.D. Lowe and A. Mol (eds.) *European integration and environmental policy.* London: Belhaven, 175–195.

BUNCE, M.F. 1994: *The countryside ideal: Anglo-American images of landscape.* London: Routledge.

BUREAU, J.C., BUTAULT, J.P., HASSAN, D., LEROUVILLOIS, P.H. and ROUSSELLE, J.M. 1991: *Generation and distribution of productivity increases in European agriculture 1967–1987.* Luxembourg: Office for Official Publications of the European Communities.

BURGEL, G. 1981: La Grèce rurale révisitée. In S. Damaniakos (ed.) *Aspects du changement social dans la campagne grecque.* Athens: Greek Review of Social Research special edition, 11–17.

BURGESS, J. 1982: Selling places: environmental images for the executive. *Regional Studies* 16, 1–17.

BURTON, R.C.J. 1994: Geographical patterns of tourism in Europe. In C.P. Cooper and A. Lockwood (eds.) *Progress in tourism, recreation and hospitality management: volume five.* Chichester: Wiley, 3–25.

BUTLER, R.W. 1980: The concept of a tourist area cycle of evolution: implications for the management of resources. *Canadian Geographer* 24, 5–12.

BUTTEL, F.H. 1992: Environmentalization: origins, processes and implications for rural social change. *Rural Sociology* 57, 1–27.

BUTTEL, F.H. 1993: Environmentalization and greening: origins processes and implications. In S. Harper (ed.) *The greening of rural policy.* London: Belhaven, 12–26.

BUTTEL, F.H. and FLINN, W.L. 1977: The interdependence of rural and urban environmental problems in advanced capitalist societies: models of linkage. *Sociologia Ruralis* 17, 255–281.

BUTTON, J., editor, 1976: *The Shetland way of oil: reactions of a small community to big business.* Sandwick: Thuleprint.

BYMAN, W.J. 1990: New technologies in the agro-food system and US-EC trade relations. In P.D. Lowe, T.K. Marsden and S.J. Whatmore (eds.) *Technological change and the rural environment.* London: David Fulton, 147–167.

CABOURET, M. 1984: Paysages agraires et sociétés. Paris: SEDES.

CABRAL, A.J. 1993: Community regional policy towards Portugal. In J. da Silva Lopes (ed.) *Portugal and EC membership evaluated.* London: Pinter, 133–145.

CADORET, A. 1985: *Protection de la nature – histoire et idéologie: de la nature à l'environnement.* Paris: L'Harmattan.

CAILLE, M–T. 1986: *Images des paysans.* Paris: Musée d'Orsay.

CALDENTEY ALBERT, P. 1993: Industria agroalimentario. In M.M. Rodríguez (ed.) *Estructura económica de Andalucía.* Madrid: Espasa-Calpe, 375–395.

CALDOCK, K. and WENGER, G.C. 1992: Health and social service provision for elderly people: the need for a rural model. In A.W. Gilg (ed.) *Progress in rural policy and planning: volume two.* London: Belhaven, 55–72.

CAMAGNI, R. 1988: Functional integration and locational shifts in new technology industry. In P. Aydalot and D.E. Keeble (eds.) *High technology industry and innovative environments: the European experience.* London: Routledge, 48–64.

CAMPAGNE, P., CARRÈRE, G. and VALCESCHINI, E. 1990: Three agricultural regions of France: three types of pluriactivity. *Journal of Rural Studies* 6, 415–422.

CANNADINE, D. 1990: *The decline and fall of the British aristocracy.* New Haven, Connecticut: Yale University Press.

CAPPELLIN, R. 1992: Theories of local endogenous development and international cooperation. In M. Tykkyläinen (ed.) *Development issues and strategies in the new Europe.* Aldershot: Avebury, 1–20.

CARDOZA, A.L. 1979: Agrarians and industrialists: the evolution of an alliance in the Po Delta 1896–1914. In J.A. Davis (ed.) *Gramsci and Italy's passive revolution.* London: Croom Helm, 172–212.

CARRIERE, J-P. 1989: *Les transformations agraires au Portugal.* Paris: Economica.

CARTER, I. 1974: The Highlands of Scotland as an underdeveloped region. In E. de Kadt and G. Williams (eds.) *Sociology and development.* London: Tavistock, 279–311.

CASTELLS, M. 1980: *The economic crisis and American society.* Princeton, New Jersey: Princeton University Press.

CASTLES, S., BOOTH, H. and WALLACE, T. 1984: *Here for good: Western Europe's new ethnic minorities.* London: Pluto.

CAVACO, C. 1993: A place in the sun: return migration and rural change in Portugal. In R.L. King (ed.) *Mass migration in Europe.* London: Belhaven, 174–191.

CAVAILHÈS, J., DESSENDRE, C., GOFFETTE–NAGOT, F. and SCHMITT, B. 1994: Analyse des évolutions récentes de l'espace rural. *Économie Rurale* 223, 13–19.

CAZORLA PEREZ, J. 1989: *Retorno del sur.* Cádiz: Oficina de Coordinación de Asistencia Emigrantes Retornados.

CEC 1987: *Regional development programmes for Denmark 1986–1990.* Luxembourg: Office for Official Publications of the European Communities.

CEC 1988: *The future of rural society.* Report of the Commission to the Parliament, COM (88) 371. Luxembourg: Office for Official Publications of the European Communities.

CEC 1989a: *Eurobarometer/Eurobarometre 31: public opinion in the European Community.* Luxembourg: Office for Official Publications of the European Communities.

CEC 1989b: *A Common Agricultural Policy for the 1990s.* Luxembourg: Office for Official Publications of the European Communities.

CEC 1989c: *Intensive farming and the impact on the environment and rural economy of restrictions on the use of chemicals and animal fertilizers.* Luxembourg: Office for Official Publications of the European Communities.

CEC 1990a: *Environmental policy in the European Community.* fourth edition. Luxembourg: Office for Official Publications of the European Communities (European Documentation Publication 5/1990).

CEC 1990b: Written reply to the European Parliament 10 April 1990: QXWO5224/90FR. Brussels.

CEC 1991a: La *Communauté Européenne et le développement rural.* Luxembourg: Office for Official Publications of the European Communities (Dossiers de l'Europe 5).

CEC 1991b: *The development and future of the Common Agricultural Policy.* Luxembourg: Office for Official Publications of the European Communities.

CEC 1991c: *Farm take-over and farm entrance within the EEC.* Luxembourg: Office for Official Publications of the European Communities.

CEC 1992a: *Farm incomes in the European Community 1990/1.* Luxembourg: Office for Official Publications of the European Communities.

CEC 1992b: *Agriculture in Europe: development, constraints and perspectives.* Luxembourg: Office for Official Publications of the European Communities.

CEC 1992c: *L'état de l'environnement dans la Communauté Européenne.* Luxembourg: Office for Official Publications of the European Communities.

CEC 1993a: *Farm incomes in the European Community in the 1980s.* Luxembourg: Office for Official Publications of the European Communities.

CEC 1993b: *The challenge of enlargement: Commission opinion on Austria's application for membership*. Luxembourg: Office for Official Publications of the European Communities (Bulletin of the European Communities Supplement 4/92).

CEC 1993c: *The challenge of enlargement: Commission opinion on Sweden's application for membership*. Luxembourg: Office for Official Publications of the European Communities (Bulletin of the European Communities Supplement 5/92).

CEC 1994a: *The agricultural income situation in less favoured areas of the EC*. Luxembourg: Office for Official Publications of the European Communities.

CEC 1994b: *Community structural funds: guide to the Community initiatives 1994–1999*. Luxembourg: Office for Official Publications of the European Communities.

CEC 1994c: *Competitiveness and cohesion: trends in the regions*. Luxembourg: Office for Official Publications of the European Communities.

CEC 1994d: *Basic statistics of the Community 1994*. thirty first edition. Luxembourg: Office for Official Publications of the European Communities.

CEC 1994e: *Eurobarometer/Eurobarometre: public opinion in the European Community: trend variables 1974–1993*. Luxembourg: Office for Official Publications of the European Communities.

CEC 1994f: *The agricultural situation in the Community: 1993 report*. Luxembourg: Office for Official Publications of the European Communities.

CEC DG-XVI 1991: *Europe 2000: outlook for the development of the Community's territory*. Luxembourg: Office for Official Publications of the European Communities.

CEC DG-XVI 1992: *Structural policies: mid-term assessment and outlook*. Luxembourg: Office for Official Publications of the European Communities.

CEC DG-XXIII TOURISM UNIT 1994: *The evolution in holiday travel facilities and in the flow of tourism inside and outside the European Community: part II main report*. Luxembourg: Office for Official Publications of the European Communities.

CEC LONDON 1994: The enlargement of the European Union. London: Commission of the European Communities Background Report ISEC/B19/94.

CEC LONDON 1995: This week in Europe: 2 February 1995. London: Commission of the European Communities Report WE/4/95.

CENTRAL STATISTICAL OFFICE 1994: *Regional trends*. London: HMSO.

CENTRO DE ESTUDIOS DE ORDENACION DEL TERRITORIO Y MEDIO AMBIENTE 1981: *Analisis territorial definición de un sistema nodal de referencia*. Madrid: Ministerio de Obras Publicas y Urbanismo.

CEPFAR 1990: *L'état des lieux de la ruralité: les réalités nationales face à l'approche communautaire*. Brussels: CEPFAR.

CHABERT, L. 1993: L'aménagement de la montagne Suisse. *Revue de Géographie Alpine* 83, 51–83.

CHADJIPADELIS, T. and ZAFIROPOULOS, C. 1994: Electoral changes in Greece 1981–1990: geographical patterns and the uniformity of the vote. *Political Geography* 13, 492–514.

CHADWICK, J.W., HOUSTON, J.B. and MASON, J.R.W. 1972: *Ballina: a local study in regional economic development*. Dublin: Institute of Public Administration.

CHAPMAN, T. 1986: Class and social mobility: patterns of employment in rural and urban Scotland. In P.D. Lowe, T. Bradley and S. Wright (eds.) *Deprivation and welfare in rural areas*. Norwich: Geo Books, 175–192.

CHARLESWORTH, A. 1994: The spatial diffusion of riots: popular disturbances in England and Wales 1750–1850. *Rural History* 5, 1–22.

CHASE-DUNN, C. 1989: *Global formation: structures of the world economy.* Oxford: Blackwell.

CHESHIRE, P.C. 1993: European integration and regional responses. London: National Institute of Economic and Social Research Single European Market Initiative Working Paper 1.

CHESHIRE, P.C. and HAY, D.G. 1989: *Urban problems in western Europe.* London: Unwin Hyman.

CHESHIRE, P.C., CAMAGNI, R., DE GAUDEMAR, J-P. and CUADRADO ROURA, J.R. 1991: 1957 to 1992: Moving toward a Europe of regions and regional policy. In L. Rodwin and H. Sazanami (eds.) *Industrial change and regional economic transformation.* London: Harper Collins, 268–300.

CHEVALIER, M. 1993: Neo-rural phenomena. In *Two decades of l'Espace Géographique: special issue in English.* Paris: Reclus, 175–191 [first published in French in 1981 in *l'Espace Géographique* 10, 33–47].

CHOSSON, J-F. 1990: *Les génerations de développement rural 1945–1990.* Paris: Librarie Générale de Droit.

CHRISTIAN, W.A. 1972: *Person and god in a Spanish valley.* New York: Seminar Press.

CHRISTIANS, C. 1993: La croissance démographique dans les campagnes Belges et la minorisation des agriculteurs. In V. Rey (ed.) *Géographies et campagnes.* Paris: Ecole Normale Supérieure, 139–143.

CHRISTIANS, C. and DAELS, L., editors, 1988: *Belgium: a geographical introduction to its regional diversity and its human richness.* Liege: Bulletin de la Société Géographique de Liege 24.

CHURCH, A. and STEVENS, M. 1994: Unionization and the urban-rural shift in employment. *Transactions of the Institute of British Geographers* 19, 111–118.

CIACCIO, C. 1990: L'urbanisation touristique des espace ruraux en Sicile. In P. Korcelli and B. Galczynska (eds.) *The impact of urbanization upon rural areas.* Warsaw: Institute of Geography, Polish Academy of Sciences Conference Paper 7, 135–143.

CLARK, D.M. 1991: *Rural housing supply: a 1990 survey of affordable new homes.* Salisbury: Rural Development Commission.

CLARK, G. 1991: People working in farming: the changing nature of farmwork. In A.G. Champion and C. Watkins (eds.) *People in the countryside.* London: Paul Chapman, 67–83.

CLARK, G.L. 1993: Global interdependence and regional development: business linkages and corporate governance in a world of financial risk. *Transactions of the Institute of British Geographers* 18, 309–325.

CLARK, S. 1975: The political mobilization of Irish farmers. *Canadian Review of Sociology and Anthropology* 12, 483–499.

CLARK, S.M. and O'NEILL, B.J. 1980: Agrarian reform in southern Portugal. *Critique of Anthropology* 15, 47–74.

CLARKE, J. and LOWE, P.D. 1992: Cleaning up agriculture: environment, technology and social science. *Sociologia Ruralis* 32, 11–29.

CLEARY, M. 1989: *Peasants, politicians and producers: the organization of agriculture in France since 1918.* Cambridge: Cambridge University Press.

CLOKE, P.J. 1989a: Rural geography and political economy. In R. Peet and N.J. Thrift (eds.) *New models in geography: volume one.* London: Unwin Hyman, 164–197.

CLOKE, P.J. 1989b: Deregulation and New Zealand's agricultural sector. *Sociologia Ruralis* 29, 34–48.

CLOKE, P.J. and GOODWIN, M. 1992: Conceptualizing countryside change: from post–Fordism to rural structured coherence. *Transactions of the Institute of British Geographers* 17, 321–336.

CLOKE, P.J. and MILBOURNE, P. 1992: Deprivation and lifestyles in rural Wales: rurality and the cultural dimension. *Journal of Rural Studies* 8, 359–371.

CLOKE, P.J. and THRIFT, N.J. 1987: Intra-class conflict in rural areas. *Journal of Rural Studies* 3, 321–334.

CLOKE, P.J. and THRIFT, N.J. 1990: Class and change in rural Britain. In T.K. Marsden, P.D. Lowe and S.J. Whatmore (eds.) *Rural restructuring*. London: David Fulton, 165–181.

CLOKE, P.J., PHILLIPS, M. and RANKIN, D. 1991: Middle class housing choice: channels of entry into Gower, south Wales. In A.G. Champion and C. Watkins (eds.) *People in the countryside*. London: Paul Chapman, 38–52.

CLOUT, H.D. 1968: Planned and unplanned change in French farm structures. *Geography* 53, 311–315.

CLOUT, H.D. 1971: *Agriculture*. London: Macmillan.

CLOUT, H.D. 1972: *Rural geography*. Oxford: Pergamon.

CLOUT, H.D. 1977: Résidences secondaires in France. In J.T. Coppock (ed.) *Second homes: curse or blessing?*. Oxford: Pergamon, 47–62.

CLOUT, H.D. 1984: *A rural policy for the EEC?*. London: Methuen.

CLOUT, H.D. 1986: *Regional variations in the European Community*. Cambridge: Cambridge University Press.

CLOUT, H.D. 1991: The recomposition of rural Europe: a review. *Annales de Géographie* 561/562, 714–729.

CLOUT, H.D. 1993a: The revival of rural Lorraine after the Great War. *Geografiska Annaler* 75B, 73–91.

CLOUT, H.D. 1993b: *European experience of rural development*. Salisbury: Rural Development Commission.

CLOUT, H.D., BLACKSELL, M., KING, R.L. and PINDER, D.A. 1993: *Western Europe: a geographical perspective*. third edition. Harlow: Longman.

CLUNIES-ROSS, T. and COX, G. 1994: Challenging the productivist paradigm: organic farming and the politics of agricultural change. In P.D. Lowe, T.K. Marsden and S.J. Whatmore (eds.) *Regulating agriculture*. London: David Fulton, 53–74.

COCCOSSIS, H. 1991: Historical land-use changes: Mediterranean regions of Europe. In F.M. Brouwer, A.J. Thomas and M.J. Chadwick (eds.) *Land-use changes in Europe*. Dordrecht: Kluwer, 441–461.

COHEN, A.P. 1982: Belonging: the experience of culture. In A.P. Cohen (ed.) *Belonging: identity and social organisation in British rural cultures*. Manchester: Manchester University Press, 1–17.

COHEN, E. 1988: Authenticity and commoditization of tourism. *Annals of Tourism Research* 15, 467–486.

COLE, J.P. and COLE, F. 1993: *The geography of the European Community*. London: Routledge.

COLLARD, A. 1981: The inequalities of change in a Greek mountain village. In S. Damaniakos (ed.) *Aspects du changement social dans la campagne grecque*. Athens: Greek Review of Social Research special edition, 208–220.

COLLIER, G.A. 1987: *Socialists of rural Andalusia: unacknowledged revolutionaries of the Second Republic*. Palo Alto, California: Stanford University Press.

COLLINS, K. and EARNSHAW, D. 1992: The implementation and enforcement of European Community environment legislation. *Environmental Politics* 1, 213–249.

COMOLET, A. 1989: *Prospective à long terme de la déprise agricole et environnement*. Paris: Institut pour une Politique Européenne de l'Environnement.

CONDRADT, D.P. 1982: *The German polity*. second edition. London: Longman.

CONNELL, J. 1978: *The end of tradition: country life in central Surrey*. London: Routledge and Kegan Paul.

CONVERSI, D. 1990: Language or race? the choice of core values in the development of Catalan and Basque nationalisms. *Ethnic and Racial Studies* 13, 50–70.

CONVERY, F.J. and BLACKWELL, J. 1988: The land-use planning system of the Republic of Ireland. In M.C. Whitby and J. Ollerenshaw (eds.) *Land-use and the European environment*. London: Belhaven, 108–112.

CONWAY, A. 1991: Agricultural policy. In P. Keatinge (ed.) *Ireland and EC membership evaluated*. London: Pinter, 42–59.

CONWAY, G. and PRETTY, J. 1991: *Unwelcome harvest*. London: Earthscan.

COOKE, P. 1985: Radical regions? space, time and gender relations in Emilia, Provence and south Wales. In G. Rees, J. Bujra, P. Littlewood, H.E. Newby and T.L. Rees (eds.) *Political action and social identity*. Basingstoke: Macmillan, 17–41.

COOKE, P. 1992: Regional innovation systems: competitive regulation in the new Europe. *Geoforum* 23, 365–382.

COOKE, P. and DA ROSA PIRES, A. 1985: Productive decentralization in three European regions. *Environment and Planning* A17, 527–554.

COPPOCK, J.T., editor, 1977: *Second homes: curse or blessing?* Oxford: Pergamon.

CORBIN, A. 1988: *Le territoire du vide*. Paris: Aubier.

CORBRIDGE, S., THRIFT, N.J. and MARTIN, R.L., editors, 1994: *Money, power and space*. Oxford: Blackwell.

COSTA, M. and FORMENTINI, U. 1990: Urbanisation et ruralité en Toscane. In P. Korcelli and B. Galczynska (eds.) *The impact of urbanization upon rural areas*. Warsaw: Institute of Geography, Polish Academy of Sciences Conference Paper 7, 25–44.

COULOMB, P. and DELORME, H. 1989: French agriculture and the failure of its European strategy. In D. Goodman and M. Redclift (eds.) *The international farm crisis*. Basingstoke: Macmillan, 84–112.

COUNCIL OF EUROPE 1988: *Rural tourism in Europe*. Strasbourg: European Campaign for the Countryside Study 2.

COUNCIL OF EUROPE 1993: *European Social Charter*. Strasbourg.

COURT, Y. 1989: Denmark: towards a more deconcentrated settlement pattern. In A.G. Champion (ed.) *Counterurbanization*. London: Edward Arnold, 121–140.

COURTET, C., BERLAN-DARQUE, M. and DEMARRE, Y., editors, 1993: *Un point sur agriculture et société*. Paris: Association Descartes.

COX, G., LOWE, P.D. and WINTER, M. 1985: Changing directions in agricultural policy: corporatist arrangements in production and conservation policies. *Sociologia Ruralis* 25, 130–153.

COX, G., LOWE, P.D. and WINTER, M. 1986: Agriculture and conservation in Britain: a policy community under siege. In G. Cox, P.D. Lowe and M. Winter (eds.) *Agriculture: people and policies*. London: Allen and Unwin, 181–215.

COX, G., LOWE, P.D. and WINTER, M. 1988: Private rights and public responsibilities: the prospects for agricultural and environmental controls. *Journal of Rural Studies* 4, 323–337.

COX, G., LOWE, P.D. and WINTER, M. 1990: *The voluntary principle in conservation*. London: Packard.

COX, R.W. 1987: *Production, power and world order: social forces in the making of history*. New York: Columbia University Press.

CRIBIER, F. 1982: Aspects of retired migration from Paris. In A.M. Warnes (ed.) *Geographical perspectives on the elderly*. Chichester: Wiley, 111–137.

CRIBIER, F. 1993: La *fonction d'accueil des retraités citadins en milieu rural et son évolution probable*. Paris: Université de Paris VII CNRS Equipe de Géographie Sociale et Gérontologie.

CROCI–ANGELINI, E. 1992: Agricultural policy. In F. Francioni (ed.) *Italy and EC membership evaluated*. London: Pinter, 31–50.

CROSBY, T.L. 1977: *English farmers and the politics of protection 1815–1852*. Brighton: Harvester.

CROSS, D.F.W. 1990: *Counterurbanization in England and Wales*. Aldershot: Avebury.

CROUZET, F. 1962: Les conséquences économiques de la révolution. *Annales Historiques de la Révolution Française* 34, 182–217.

CRUZ VILLALÓN, J. 1987: Political and economic change in Spanish agriculture 1950–1985. *Antipode* 19, 119–133.

CURBELO, J.L. and ALBURQUERQUE, F. 1993: The lagging regions of the EC in the face of economic and monetary union. *European Planning Studies* 1, 275–298.

CURTIN, C. and VARLEY, T. 1991: Populism and petit capitalism in rural Ireland. In S.J. Whatmore, P.D. Lowe and T.K. Marsden (eds.) *Rural enterprise*. London: David Fulton, 97–119.

CUTANDA, A. and PARICIO, J. 1994: Infrastructure and regional economic growth: the Spanish case. *Regional Studies* 28, 69–77.

CUTILEIRO, J. 1971: *A Portuguese rural society*. Oxford: Clarendon.

DANIELS, S. 1993: *Fields of vision: landscape imagery and national identity in England and the United States*. Cambridge: Polity.

DANISH FARMERS' UNION 1990: *Agriculture in Denmark*. København.

DAUMAS, M. 1988: L'évolution récente des structures agraires en Espagne. *Annales de Géographie* 542, 419–443.

DAVIDSSON, P., LINDMARK, L. and OLOFSSON, C. 1994: New firm formation and regional development in Sweden. *Regional Studies* 28, 395–410.

DAVIES, T.M. 1978: Capital, state and sparse populations. In H.E. Newby (ed.) *International perspectives in rural sociology*. Chichester: Wiley, 87–104.

DAVIS, J.A. 1973: *Land and family in Pisticci*. London: Athlone Press.

DAVIS, J.A. 1979: The South, the Risorgimento and the origins of the 'southern problem'. In J.A. Davis (ed.) *Gramsci and Italy's passive revolution*. London: Croom Helm, 67–103.

DAWSON, A.H. 1993: *A geography of European integration*. London: Belhaven.

DAY, G. and FITTON, M. 1975: Religion and social status in rural Wales: 'buchedd' and its lessons for concepts of stratification in community studies. *Sociological Review* 23, 867–892.

DE HAAN, H. 1993: Images of family farming in the Netherlands. *Sociologia Ruralis* 33, 147–166.

DE HAAN, H. and NOOIJ, A. 1985: Rural sociology in the Netherlands: developments in the 1970s. *Sociologia Neerlandica* 21, 51–62.

DE MEO, G., editor, 1991: *L'agricoltura italiana e i initamenti dello scenario economico internazionale*. Bologna: Guaderni della Pivista di Economica Agraria 15.

DE SCHOUTHEETE, P. 1990: The European Community and its subsystems. In W. Wallace (ed.) *The dynamics of European integration*. London: Pinter, 106–124.

DE SCHOUTHEETE, P. 1992: Sovereignty and supranationality. In M.A.G. van Meerhaeghe (ed.) *Belgium and EC membership evaluated*. London: Pinter, 109–117.

DELORME, H. 1991: Agricultural policy in the internationalization of trade. In M. Tracy (ed.) *Farmers and politics in France*. Enstone: Arkleton Trust, 13–34.

DEMATTEIS, G. 1982: Repeuplement et revalorisation des espaces périphériques: le cas de l'Italie. *Revue Géographique des Pyrénées et du Sud Ouest* 53, 129–143.

DEMATTEIS, G. and PETSIMERIS, P. 1989: Italy: counterurbanization as a transitional phase in settlement reorganization. In A.G. Champion (ed.) *Counterurbanization*. London: Edward Arnold, 187–206.

DENOON, D. 1983: *Settler capitalism: the dynamics of dependent development in the southern hemisphere*. Oxford: Clarendon.

DEPARTMENT OF THE ENVIRONMENT 1991: *Digest of environmental statistics*. London: HMSO.

DEVERRE, C. 1995: Rare birds and flocks: agriculture and social legitimization of environmental protection. In A. Jansen (ed.) *Agricultural restructuring and rural change in Europe*. Wageningen: Wageningen University Press, forthcoming.

DIAS, J. 1961: *A Portuguese contribution to cultural anthropology*. Johannesburg: University of Witswatersrand Press.

DICKEN, P. 1992: *Global shift: industrial change in a turbulent world*. second edition. London: Harper and Row.

DICKEN, P. 1994: Global – local tensions: firm and state in the global space economy. *Economic Geography* 70, 101–128.

DION, R. 1934: *Essai sur la formation du paysage rural français*. 1991 edition. Paris: Durier.

DODGSHON, R.A. 1987: *The European past: social evolution and spatial order*. Basingstoke: Macmillan.

DOLFUS, O. 1979: *L'espace géographique*. Paris: Presses Universitaires de France.

DOUGLASS, C. B. 1992: 'Europe', 'Spain' and the bulls. *Journal of Mediterranean Studies* 2, 67–76.

DOUGLASS, W.A. 1976: Serving girls and sheepherders: emigration and continuity in a Spanish Basque village. In J.B. Aceves and W.A. Douglass (eds.) *The changing faces of rural Spain*. Cambridge, Massachusetts: Schenkman, 45–61.

DOUGLASS, W.A. 1984: *Emigration in a south Italian town: an anthropological history*. New Brunswick, New Jersey: Rutgers University Press.

DRAIN, M. 1985: *Portugal: la réforme agraire*. Paris: Universalia.

DRAIN, M. 1990: Agricultures et sociétés rurales. In J. Beaujeu-Garnier and A. Gamblin (eds.) *Dossiers des images économiques du monde 12: la CEE mediterranéenne*. Paris: Sedes, 141–176.

DRUDY, P.J. 1978: Depopulation in a prosperous agricultural region. *Regional Studies* 12, 49–60.

DUCHÊNE, F., SZCZEPANIK, E. and LEGG, W. 1985: *New limits on European agriculture: politics and the Common Agricultural Policy.* Totowa, New Jersey: Rowman and Allanheld.

DUFFEY, E. 1983: Nature conservation in national parks in Western Europe. In A. Warren and F. Goldsmith (eds.) *Conservation in perspective.* London: Wiley, 447–463.

DUNCAN, S.S. 1989: What is locality? In R. Peet and N.J. Thrift (eds.) *New models in geography: volume two.* London: Unwin Hyman, 221–252.

DUNCAN, S.S. 1991: The geography of gender divisions of labour in Britain. *Transactions of the Institute of British Geographers* 16, 420–439.

DUNFORD, M. 1993: Regional disparities in the European Community: evidence from the REGIO databank. *Regional Studies* 27, 727–743.

DUNFORD, M. and PERRONS, D. 1984: Integration and unequal development: towards an understanding of the impact of economic integration on Ireland and southern Italy. Brighton: University of Sussex Urban and Regional Studies Working Paper 36.

DUNFORD, M. and PERRONS, D. 1994: Regional inequality, regimes of accumulation and economic development in contemporary Europe. *Transactions of the Institute of British Geographers* 19, 163–182.

DUNLEAVY, P.J. and O'LEARY, B. 1987: *Theories of the state.* London: Macmillan.

EADINGTON, W.R. and SMITH, V.L. 1992: The emergence of alternative forms of tourism. In V.L. Smith and W.R. Eadington (eds.) *Tourism alternatives.* Philadelphia: University of Pennsylvania Press, 1–12.

ECCLES, J. and FULLER, G.W. 1970: Tomato production in Portugal promoted by the Heinz Company. In A.H. Bunting (ed.) *Change in agriculture.* London: Duckworth, 257–262.

ECONOMIC AND SOCIAL CONSULTATIVE ASSEMBLY 1990: *Upland areas: basic data and statistics.* Luxembourg: Office for Official Publications of the European Communities.

ECONOMOU, D. 1993: New forms of geographical inequalities and spatial problems in Greece. *Society and Space* 11, 583–598.

EDELSTEIN, M. 1982: *Overseas investment in the age of high imperialism: the United Kingdom 1850–1914.* London: Methuen.

EDWARDES, M. 1991: *The nabobs at home.* London: Constable.

EFSTRATOGLOU-TODOULOU, S. 1990: Pluriactivity in different socioeconomic contexts: a test of the push-pull hypothesis in Greek farming. *Journal of Rural Studies* 6, 407–413.

ELLIS, F. 1988: *Peasant economics.* Cambridge: Cambridge University Press.

ELSON, M. 1986: *Green belts.* London: Heinemann.

ERRINGTON, A. and HARRISON, L. 1990: Employment in the food system. In T.K. Marsden and J. Little (eds.) *Political, social and economic perspectives on the international food system.* Aldershot: Avebury, 115–140.

ERRINGTON, A. and TRANTER, R.B. 1991: Getting out of farming: part two – the farmers. Reading: University of Reading Farm Management Unit Study 27.

ESCALONA ORCAO, A. and ESCOLONO UTRILLA, S. 1993: Los trabajadores extranjeros en al provincia de Zaragoza. *Geographicalia* 30, 155–76.

ETXEZARRETA, M. 1992: Transformation of the labour system and work processes in a rapidly modernizing agriculture: the evolving case of Spain. In T.K. Marsden, P.D. Lowe and S.J. Whatmore (eds.) *Labour and locality*. London: David Fulton, 44–67.

ETXEZARRETA, M. and VILADOMIU, L. 1989: The restructuring of Spanish agriculture and Spain's accession to the EEC. In D. Goodman and M. Redclift (eds.) *The international farm crisis*. Basingstoke: Macmillan, 156–182.

EUROPE ENVIRONNEMENT 1993: Mesures agri-environnementales en Allemagne et Royaume Uni. *Europe Environnement* 419, 4.

EUROSTAT 1990: *Environmental statistics 1989*. Luxembourg: Office for Official Publications of the European Communities.

EUROSTAT 1992: *Europe in figures*. third edition. Luxembourg: Office for Official Publications of the European Communities.

EUROSTAT monthly: *Industrial trends: monthly statistics*. Luxembourg: Office for Official Publications of the European Communities.

EVANS, N.J. and ILBERY, B.W. 1992: Farm-based accommodation and the restructuring of agriculture: evidence from three English counties. *Journal of Rural Studies* 8, 85–96.

EVANS, P.B., RUESCHEMEYER, D. and SKOCPOL, T., editors, 1985: *Bringing the state back in*. Cambridge: Cambridge University Press.

FABIANI, G. 1986: *L'agricoltura italiana fra sviluppo e crisi 1945–1985*. Bologna: Il Mulino.

FAIRBROTHER, N. 1970: *New lives, new landscapes*. Harmondsworth: Penguin.

FANARIOTOU, I. and SKOURAS, D. 1991: Fertilizer use and environmental policy in agriculture: a socio-economic study in Greece. *Environmental Conservation* 18, 137–142.

FAO annual: *Production yearbook*. Roma: Food and Agriculture Organization.

FEATHERSTONE, K. 1989: Socialist parties in southern Europe and the enlarged European Community. In T. Gallagher and A.M. Williams (eds.) *Southern European socialism*. Manchester: Manchester University Press, 247–270.

FEATHERSTONE, M. 1990: Perspectives on consumer culture. *Sociology* 24, 5–22.

FENNELL, R. 1987: *The Common Agricultural Policy of the European Community*. second edition. Oxford: BSP Professional Books.

FERKISS, V. 1993: *Nature, technology and society*. London: Adamantine.

FIELDING, A.J. 1982: Counterurbanization in Western Europe. *Progress in Planning* 17, 1–52.

FIELDING, A.J. 1990: Counterurbanization: threat or blessing? In D.A. Pinder (ed.) *Western Europe: challenge and change*. London: Belhaven, 226–239.

FIELDING, A.J. 1992: Migration and social mobility: South East England as an escalator region. *Regional Studies* 26, 1–15.

FIRST-DILIC, R. 1978: The productive roles of farm women in Yugoslavia. *Sociologia Ruralis* 17, 125–139.

FISCHER, M.M., FRÖHLICH, J. and GASSLER, H. 1994: An exploration into the determinants of patent activities: some empirical evidence from Austria. *Regional Studies* 28, 1–12.

FLAUBERT, G. 1976: *The dictionary of received ideas*. Harmondsworth: Penguin.

FOPPEN, J.W. 1989: The Netherlands and the crisis as a policy challenge. In E. Damgaard, P. Gerlich and J.J. Richardson (eds.) *The politics of economic crisis: lessons from western Europe.* Aldershot: Avebury, 89–106.

FORREST, R. and MURIE, A. 1992: Change on a rural council estate: an analysis of dwelling histories. *Journal of Rural Studies* 8, 53–65.

FORREST, R. and MURIE, A. 1994: The dynamics of the owner-occupied housing market in southern England in the late 1980s. *Regional Studies* 28, 275–289.

FORSYTHE, D.E. 1980: Urban incomers and rural change: the impact of migrants from the city on life in an Orkney community. *Sociologia Ruralis* 20, 287–305.

FORSYTHE, D.E. 1983: Planning implications of urban-rural migration. Gloucester: Gloucestershire College of Arts and Technology Paper in Local and Regional Planning 21.

FOTHERGILL, S., MONK, S. and PERRY, M. 1987: *Property and industrial redevelopment.* London: Hutchinson.

FOURNY, M-C. 1995: L'autre dans le territoire. *Études Rurales* forthcoming.

FRANKENBERG, R. 1957: *Village on the border: a social study of religion, politics and football in a north Wales community.* London: Cohen and West.

FRANKLIN, S. 1969: *The European peasantry: the final phase.* London: Methuen.

FRANZMEYER, F., HRUBESCH, P., SEIDEL, B. and WEISE, C. 1991: *The regional impact of Community policies.* Luxembourg: European Parliament Directorate-General for Research, Regional Policy and Transport Series 17.

FRIEDL, E. 1959: The role of kinship in the transmission of national culture to rural villages in mainland Greece. *American Anthropologist* 61, 30–38.

FRIEDMANN, G., editor, 1953: *Villes et campagnes: civilisation urbaine et civilisation rurale en France.* Paris: Colin.

FROUWS, J. 1988: State and society with respect to agriculture and the rural environment in the Netherlands. In J. Frouws and W. de Groot (eds.) *Environment and agriculture in the Netherlands.* Leiden: Centrum voor Milieukunde CML-Mededeling 47, 10–25.

FROUWS, J. 1989: Agriculture and environmental pollution in the Netherlands. *Tijdschrift voor Sociaal Wetenschappelijke Onderzoek van de Landbouw* 3, 258–274.

FULLER, A. 1990: From part-time farming to pluriactivity: a decade of change in rural Europe. *Journal of Rural Studies* 6, 361–371.

FURLONG, P. 1994: *Modern Italy: representation and reform.* London: Routledge.

FÚSTER, L.F. 1985: *Introducción a la teoría y técnica del turismo.* Madrid: Alianza Editorial.

GAMBLE, A. 1988: *The free economy and the strong state: the politics of Thatcherism.* London: Macmillan.

GANE, M. 1991: *Baudrillard: critical and fatal theory.* London: Routledge.

GANNON, A. 1994: Rural tourism as a factor in rural community economic development for economies in transition. *Journal of Sustainable Tourism* 2, 51–60.

GARNER, B.J. 1975: The effect of local government reform on access to public services: a case study from Denmark. In R. Peel, M. Chisholm and P. Haggett (eds.) *Processes in physical and human geography.* London: Heinemann, 319–338.

GAROFOLI, G. 1994: New firm formation and regional development: the Italian case. *Regional Studies* 28, 381–393.

GASSON, R.M. 1966: *The influence of urbanization on farm-ownership and practice.* Ashford: Wye College Studies in Rural Land Use 7.

GASSON, R.M. 1969: Occupational immobility of small farmers. *Journal of Agricultural Economics* 20, 279–288.

GASSON, R.M. 1981: Opportunities for women in agriculture. Ashford: Wye College Department of Environmental Studies and Countryside Planning Occasional Paper 5.

GASSON, R.M. 1992: Farmers' wives – their contribution to the farm business. *Journal of Agricultural Economics* 43, 74–87.

GATT 1993: *International trade 1993: statistics.* Geneva: General Agreement on Tariffs and Trade.

GAVIGNAUD, G. 1990: *Les campagnes en France au XXᵉ siècle.* Paris: Ophrys.

GAY, F. AND WAGRET, P. 1986: *L'économie de l'Italie.* Paris: Presses Universitaires de France.

GAZE, J. 1988: *Figures in a landscape: a history of the National Trust.* London: Barrie and Jenkins.

GENERAL SECRETARIAT OF THE COUNCIL OF THE EUROPEAN COMMUNITIES annual: *Review of the Council's work (the Secretary–General's report).* Luxembourg: Office for Official Publications of the European Communities.

GEORGE, S. 1985: *Politics and policy in the European Community.* Oxford: Clarendon.

GEORGIOU, G.A. 1994: The implementation of EC regional programmes in Greece: a critical review. *European Planning Studies* 2, 321–336.

GERVAIS, M., JOLLIVET, M. and TAVERNIER, Y. 1977: *Depuis 1914: histoire de la France rurale IV.* Paris: Seuil.

GETIMIS, P. and KAFKALAS, G. 1992: Local development and forms of regulation: fragmentation and hierarchy of spatial policies in Greece. *Geoforum* 23, 73–83.

GETZ, D. 1981: Tourism and rural settlement policy. *Scottish Geographical Magazine* 97, 158–168.

GEVAERT, P. 1994: *L'avenir sera rural.* Paris: Le Courrier du Livre.

GIANNITSIS, T. 1994: Trade effects, the balance of payments and implications for the productive system. In P. Kazakos and P.C. Ioakimidis (eds.) *Greece and EC membership evaluated.* London: Pinter, 36–55.

GILG, A.W. 1992: *Countryside planning policies for the 1990s.* Wallingford: CAB International.

GILL, S. and LAW, D. 1988: *The global political economy.* Hemel Hempstead: Harvester.

GILLIE, O. 1993: Rural sites 'few can reach' receive £7.5m. *The Independent* 18 March, 6.

GILLMOR, D.A. 1986: Rural industrialization. In P. Breathnach and M.E. Cawley (eds.) *Change and development in rural Ireland.* Maynooth: Geographical Society of Ireland Special Publication 1, 25–33.

GINER, S. 1985: Political economy, legitimation and the state in southern Europe. In R. Hudson and J.R. Lewis (eds.) *Uneven development in southern Europe.* London: Methuen, 309–350.

GLENNIE, P. and THRIFT, N.J. 1992: Modernity, urbanism and modern consumption. *Society and Space* 10, 423–443.

GMELCH, G. 1986: The readjustment of return migrants in western Ireland. In R.L. King (ed.) *Return migration and regional economic problems.* London: Croom Helm, 152–170.

GOLDE, G. 1975: *Catholics and protestants: agricultural modernization in two German villages.* New York: Academic Press.

GOODMAN, D. and REDCLIFT, M. 1991: *Refashioning nature: food, ecology and culture.* London: Routledge.

GOODMAN, D. and WATTS, M.J. 1994: Reconfiguring the rural and fording the divide? capitalist restructuring and the global agro-food system. *Journal of Peasant Studies* 22, 1–49.

GOODMAN, D. and WILKINSON, J. 1990: Patterns of research and innovation in the modern agro-food system. In P.D. Lowe, T.K. Marsden and S.J. Whatmore (eds.) *Technological change and the rural environment.* London: David Fulton, 127–146.

GOODMAN, D., SORJ, B. and WILKINSON, J. 1987: *From farming to biotechnology: a theory of agro-industrial development.* Oxford: Blackwell.

GOULD, A. and KEEBLE, D.E. 1984: New firms and rural industrialization in East Anglia. *Regional Studies* 18, 189–201.

GOURDOMICHALIS, A. 1991: Women and the reproduction of family farms: change and continuity in the region of Thessaly, Greece. *Journal of Rural Studies* 7, 57–62.

GOW, D. 1991: Germany registers first trade deficit in 10 years. *The Guardian* 11 June, 11.

GRAFTON, D.J. 1984: Small scale growth centres in remote rural areas. *Applied Geography* 4, 29–46.

GRAHL, J. and TEAGUE, P. 1990: *1992 – the big market: the future of the European Community.* London: Lawrence and Wishart.

GRANADOS CABEZAS, V. 1992: Autonomous development as a strategy for regional policy: a balance-sheet of Spanish experience with special reference to Andalusia. In G. Garofoli (ed.) *Endogenous development in southern Europe.* Aldershot: Avebury, 185–193.

GRANDE, E. 1989: West Germany: from reform policy to crisis management. In E. Damgaard, P. Gerlich and J.J. Richardson (eds.) *The politics of economic crisis: lessons from western Europe.* Aldershot: Avebury, 50–69.

GRANT, G. 1965: *Lament for a nation: the defeat of Canadian nationalism.* Toronto: McClelland and Stewart.

GRANT, G. 1984: The rural-urban dichotomy in social care: rhetoric and reality. Bangor: University College of North Wales Department of Social Theory and Institutions Care Networks Project Working Paper 31.

GRANT, R. 1993: Against the grain: agricultural trade policies of the US, the European Community and Japan at the GATT. *Political Geography* 12, 247–262.

GRAZIANO, L. 1978: Centre-periphery relations and the Italian crisis. In S. Tarrow, P.J. Katzenstein and L. Graziano (eds.) *Territorial politics in industrial nations.* New York: Praeger, 290–326.

GREEN, N. 1990: *The spectacle of nature: landscape and bourgeois culture in nineteenth century France.* Manchester: Manchester University Press.

GREEN, P. 1964: Drymen: village growth and community problems. *Sociologia Ruralis* 4, 52–62.

GREENWOOD, D.J. 1976: *Unrewarding wealth: the commercialization and collapse of agriculture in a Spanish Basque town*. Cambridge: Cambridge University Press.

GRIFFITH-JONES, S. 1984: *International finance and Latin America*. London: Croom Helm.

GRIGG, D.B. 1974: *The agricultural systems of the world*. Cambridge: Cambridge University Press.

GRIGG, D.B. 1993: *The world food problem*. second edition. Oxford: Blackwell.

GRIGNON, C. and WEBER, F. 1993: Sociologie et ruralisme, ou les séquelles d'une mauvaise rencontre. *Cahiers d'Economie et Sociologie Rurales* 29, 59–74.

GROENENDIJK, J.G. 1988: The Netherlands. In P.J. Cloke (ed.) *Policies and plans for rural people*. London: Unwin Hyman, 47–68.

GROVE, A.T. 1994: EC policy strategies and their impact on desertification and land degradation. In P. Mariota and N. Geeson (eds.) EC policy strategies for land-use in the Mediterranean states. London: King's College London Department of Geography MEDALUS Working Paper 17, 13–26.

GUESNIER, B. 1994: Regional variations in new firm formation in France. *Regional Studies* 28, 347–358.

GUY, C.M. 1991: Urban and rural contrasts in food prices and availability – a case study in Wales. *Journal of Rural Studies* 7, 31–325.

GUYOMARD, G. 1981: *Associations de protection de l'environnement et systèmes politico-administratives locaux*. Rennes: Presses Universitaires.

GUYOMARD, H., MAHÉ, L.P., MONK, K.J. and RUE, T.L. 1993: Agriculture in the Uruguay Round: ambitions and realities. *Journal of Agricultural Economics* 44, 245–263.

HADJIMICHALIS, C. 1994: The fringes of Europe and EU integration: a view from the South. *European Urban and Regional Studies* 1, 19–30.

HADJIMICHALIS, C. and PAPAMICHOS, N. 1990: 'Local' development in southern Europe: toward a new mythology. *Antipode* 22, 181–210.

HALFACREE, K.H. 1993: Locality and social representation: space, discourse and alternative definitions of the rural. *Journal of Rural Studies* 9, 23–37.

HALL, P.G. and HAY, D. 1980: *Growth centres in the European urban system*. London: Heinemann.

HALL, P.G., BREHENY, M., MCQUAID, R. and HART, D. 1987: *Western sunrise: the genesis and growth of Britain's major high tech corridor*. London: Allen and Unwin.

HALL, P.G., GRACEY, H., DREWETT, R. and THOMAS, R. 1973: *The containment of urban England*. London: Allen and Unwin.

HANNIGAN, K. 1994: National policy, European structural funds and sustainable tourism: the case of Ireland. *Journal of Sustainable Tourism* 2, 179–192.

HARDING, S.F. 1984: *Remaking Iberia: rural life in Aragon under Franco*. Chapel Hill: University of North Carolina Press.

HARPER, S. 1987: The rural-urban interface in England. *Transactions of the Institute of British Geographers* 12, 284–302.

HARPER, S. 1991: People moving into the countryside: case studies of decision-making. In A.G. Champion and C. Watkins (eds.) *People in the countryside*. London: Paul Chapman, 22–37.

HARPER, S. 1993: The greening of rural discourse. In S. Harper (ed.) *The greening of rural policy*. London: Belhaven, 3–11.

HARRIS, R. 1972: *Prejudice and tolerance in Ulster: a study of neighbours and 'strangers' in a border community.* Manchester: Manchester University Press.

HARRISON, C. 1991: *Countryside recreation in a changing society.* London: TMS.

HARRISON, J. 1985: *The Spanish economy in the twentieth century.* London: Croom Helm.

HART, M. and GUDGIN, G. 1994: Spatial variations in new firm formation in the Republic of Ireland 1980–1990. *Regional Studies* 28, 367–380.

HART, P.W.E. 1978: Geographical aspects of contract farming, with special reference to the supply of crops to processing plants. *Tijdschrift voor Economische en Sociale Geografie* 69, 205–215.

HAUGEN, M.S. 1994: Rural women's status in family and property law: lessons from Norway. In S.J. Whatmore, T.K. Marsden and P.D. Lowe (eds.) *Gender and rurality.* London: David Fulton, 87–101.

HAVERKATE, R. and VAN HASELEN, H. 1992: *Demographic evolution through time in European regions (Demeter 2015).* Luxembourg: Office for Official Publications of the European Communities.

HEINZE, R.G. and VOELZKOW, H. 1993: Organizational problems for the German Farmers' Association and alternative policy options. *Sociologia Ruralis* 33, 25–41.

HENDRIKS, G. 1994: The national politics of international trade reform: the case of Germany. In P.D. Lowe, T.K. Marsden and S.J. Whatmore (eds.) *Regulating agriculture.* London: David Fulton, 149–162.

HERINGTON, J. and EVANS, D.M. 1979: The spatial pattern of movement in 'key' and 'non-key' settlements. Loughborough: Loughborough University Department of Geography Working Paper 3.

HERRIGEL, G.B. 1993: Power and the redefinition of industrial districts: the case of Baden-Württemberg. In G. Grabher (ed.) *The embedded firm.* London: Routledge, 227–252.

HERVIEU, B. 1991: Discontinuities in the French farming world. *Sociologia Ruralis* 31, 291–299.

HERVIEU, B. 1993: *Les champs du future.* Paris: Bourin.

HERVIEU, B. 1994: L'impératif territorial. *Sciences Humaines* Hors Série 4, 15.

HERVIEU, B. and LAGRAVE, R–M. 1992: *Les syndicats agricoles en Europe.* Paris: L'Harmattan.

HERVIEU, B. and VIARD, J. 1994: L'état sans territoire. *Le Monde* 31 October 1994, Supplement 'Heures Locales VI.

HILL, B.E. 1984: *The Common Agricultural Policy.* London: Methuen.

HILL, B.E. 1993: The 'myth' of the family farm: defining the family farm and assessing its importance in the European Community. *Journal of Rural Studies* 9, 359–370.

HILL, B.E. and YOUNG, N. 1991: Support policy for rural areas in England and Wales: its assessment and quantification. *Journal of Rural Studies* 7, 191–206.

HILL, C.M. 1980: Leisure behaviour in Norfolk rural communities. Norwich: unpublished PhD thesis, University of East Anglia.

HILLEBRAND, H. and BLOM, U. 1993: Young women on Dutch family farms. *Sociologia Ruralis* 33, 178–189.

HILLMAN, J.S. 1985: Evolution of American agricultural trade policy and European integration. In H. de Haan, G.L. Johnson and S. Tangermann (eds.) *Agriculture and international relations.* Basingstoke: Macmillan, 155–169.

HILLMAN, J.S. 1992: Confessions of a double agent in the EC-US agricultural policy argument. *Journal of Agricultural Economics* 43, 327–342.

HILLMAN, J.S. 1994: The US perspective. In K.A. Ingersent, A.J. Rayner and R.C. Hine (eds.) *Agriculture in the Uruguay Round*. Basingstoke: Macmillan, 26–54.

HOBDAY, M. 1994: Export-led technology development in the four dragons. *Development and Change* 25, 333–361.

HOBSBAWM, E.J. 1968: *Industry and empire*. Harmondsworth: Penguin.

HODGE, I. and MONK, S. 1991: In search of a rural economy: patterns and differentiation in nonmetropolitan England. Cambridge: University of Cambridge Department of Land Economy Monograph 20.

HOETJES, B. 1993: Les agriculteurs néerlandais à la campagne français. Paper to the annual conference of the Association of Ruralists Français, Rennes, May 1993.

HOGGART, K. 1988: Not a definition of rural. *Area* 20, 35–40.

HOGGART, K. 1990: Let's do away with rural. *Journal of Rural Studies* 6, 245–257.

HOGGART, K. 1993: House construction in nonmetropolitan districts: economy, politics and rurality. *Regional Studies* 27, 651–664.

HOGGART, K. 1994: Political parties and district council economic policies 1978–1990. *Urban Studies* 31, 59–77.

HOGGART, K. 1995: Political parties and the implementation of homeless legislation by nonmetropolitan districts in England and Wales, 1985–1990. *Political Geography* 14, 59–79.

HOGGART, K. and BULLER, H. 1987: *Rural development: a geographical perspective*. London: Croom Helm.

HOGGART, K. and BULLER, H. 1994: Property agents as gatekeepers in British house purchases in rural France. *Geoforum* 25, 173–187.

HOGGART, K. and BULLER, H. 1995a: Retired British home owners in rural France. *Ageing and Society*, 15, forthcoming.

HOGGART, K. and BULLER, H. 1995b: British home owners and housing change in rural France. *Housing Studies* 10, 179–198.

HOGGART, K. and BULLER, H. 1995c: Geographical differences in British property acquisitions in rural France. *Geographical Journal* 161, 69–78.

HOKWERDA, H. 1992: The ideological role of the countryside and the city in Dimitris Chatzis' literary work. *Journal of Mediterranean Studies* 2(2), 213–225.

HOLLEY, H.A. 1987: *Developing country debt: the role of the commercial banks*. London: Royal Institute of International Affairs Chatham House Paper 35.

HOLMES, D.R. 1989: *Cultural disenchantments: worker peasants in northeast Italy*. Princeton, New Jersey: Princeton University Press.

HOMMES, P.M. 1990: The Common Agricultural Policy. In M. Wolters and P. Coffey (eds.) *The Netherlands and EC membership evaluated*. London: Pinter, 41–47.

HOSKINS, W.G. 1955: The *making of the English landscape*. London: Hodder and Stoughton.

HOUÉE, P. 1979: *Les politiques de développement rurale*. Paris: Economica.

HOUSE OF LORDS 1990: *The future of rural society*. London: HMSO (report to the Select Committee on the European Communities).

HOWKINS, A. 1986: The discovery of rural England. In R. Colls and P. Dodd (eds.) *Englishness, politics and culture 1880–1920*. London: Croom Helm, 62–88.

ILBERY, B.W. and BOWLER, I.R. 1993: Land diversion and farm business diversi-fication in EC agriculture. In E.C.A. Bolsius, G. Clark and J.G. Groenendijk (eds.) The *retreat: rural land-use and European agriculture*. Utrecht: Nederlandse Geografische Studies 172, 15–27.

ILBERY, B.W. and BOWLER, I.R. 1994: International competition in agriculture and intervention: the case of horticulture in England. *Geography* 79, 361–366.

INGERSENT, K.A., RAYNER, A.J. and HINE, R.C. 1994a: The EC perspective. In K.A. Ingersent, A.J. Rayner and R.C. Hine (eds.) *Agriculture in the Uruguay Round*. Basingstoke: Macmillan, 55–87.

INGERSENT, K.A., RAYNER, A.J. and HINE, R.C., editors, 1994b: *Agriculture in the Uruguay Round*. Basingstoke: Macmillan.

INGHAM, G. 1984: *Capitalism divided? the city and industry in British social development*. Basingstoke: Macmillan.

INSEE 1992: *Recensement général de la population: évolutions démographiques*. Paris: Institut National de la Statistique et des Etudes Economiques.

INSTITUTO DE ESTADISTICA DE ANDALUCÍA annual: *Migraciones Andalucía*. Sevilla.

IOAKIMIDIS, P.C. 1984: Greece: from military dictatorship to socialism. In A.M. Williams (ed.) *Southern Europe transformed*. London: Harper and Row, 33–60.

ISTITUTO NAZIONALE ECONOMIA AGRARIA 1992: *Italian agriculture 1991*. Roma.

JAEGER, C. and DÜRRENBERGER, G. 1991: Services and counterurbanization: the case of central Europe. In P.W. Daniels (ed.) *Services and metropolitan develop-ment*. London: Routledge, 107–128.

JANSEN, A.J. 1991: The future of the periphery of the periphery: the case of a Sicilian agrotown. *Sociologia Ruralis* 31, 122–129.

JEANS, D.N. 1990: Planning and the myth of the English countryside, in the interwar period. *Rural History* 1, 249–264.

JENKINS, R. 1979: *The road to Alto: an account of peasants, capitalists and the soil in the mountains of southern Portugal*. London: Pluto.

JENKINS, R. 1992: *Pierre Bourdieu*. London: Routledge.

JOHANSEN, H.C. 1987: *The Danish economy in the twentieth century*. London: Croom Helm.

JOHNSON, R.W. 1972: The nationalisation of English rural politics: Norfolk South West 1945–1970. *Parliamentary Affairs* 26, 8–55.

JOHNSTON, R.J. 1991: *A question of place*. Oxford: Blackwell.

JOLLIVET, M., editor, 1988: *Pour une agriculture diversifiée*. Paris: L'Harmattan.

JOLLIVET, M., editor, 1994: *Bilan des recherches en sciences sociales sur les problèmes d'environnement en milieu rural dans les pays européens*. Paris: CNRS Groupe de Recherches sur les Mutations des Sociétés Européennes.

JOLLIVET, M. 1994: France. In M. Jollivet (ed.) *Bilan des recherches en sciences sociales sur les problèmes d'environnement en milieu rural dans les pays européens*. Paris: CNRS Groupe de Recherches sur les Mutations des Sociétés Européennes, 31–37.

JOLLIVET, M. and MENDRAS, H. 1971: *Les collectivités rurales françaises*. Paris: Colin.

JONAS, A. 1988: A new regional geography of localities?. *Area* 20, 101–110.

JONES, H., CAIRD, J.B., BERRY, W. and DEWHURST, J. 1986: Peripheral counter–urbanization: findings from an integration of census and survey data in northern Scotland. *Regional Studies* 20, 15–26.

JONES, M.A. 1960: *American immigration*. Chicago: University of Chicago Press.

JONES, W.G. 1986: *Denmark: a modern history*. London: Croom Helm.

JOUGANATOS, G.A. 1992: *The development of the Greek economy 1950–1991*. Westport, Connecticut: Greenwood.

JUSSILA, H., LOTVONEN, E. and TYKKYLÄINEN, M. 1992: Business strategies of rural shops in a peripheral region. *Journal of Rural Studies* 8, 185–192.

JUST, F. 1992: Denmark, la gestion d'une agriculture intensive. In B. Hervieu and R-M. Lagrave (eds.) *Les syndicats agricoles en Europe*. Paris: L'Harmattan, 49–71.

JUST, F. 1994: Agriculture and corporatism in Scandinavia. In P.D. Lowe, T.K. Marsden and S.J. Whatmore (eds.) *Regulating agriculture*. London: David Fulton, 31–52.

KALAORA, B. and SAVOYE, A. 1985: La protection des régions de montagne au XIX siècle. In A. Cadoret (ed.) *Protection de la nature: histoire et idéologie de la nature à l'environnement*. Paris: L'Harmattan, 6–23.

KAMPP, Aa.H. 1975: *An agricultural geography of Denmark*. Budapest: Akadémiai Kiadó.

KATZNELSON, I. 1986: Working class formation: constructing cases and comparisons. In I. Katznelson and A.R. Zolberg (eds.) *Working-class formation*. Princeton, New Jersey: Princeton University Press, 3–41.

KAYSER, B. 1990: *La renaissance rurale*. Paris: Armand Colin.

KAYSER, B. 1991: Country planning, development policies and the future of rural areas. *Sociologia Ruralis* 31, 262–268.

KAYSER, B., editor, 1993: *Naissance de nouvelles campagnes*. Paris: DATAR.

KAYSER, B. 1994: Angoisses et espérances de la France rurale. *Sciences Humaines* Hors Série 4, 16–21.

KAYSER, B., BRUN, A., CAVAILHES, J. and LACOMBE, P. 1994: *Pour une ruralité choisie*. Paris: Editions de l'Aube.

KEEBLE, D.E. 1989: Core-periphery disparities, recession and new regional dynamics in the European Community. *Geography* 74, 1–11.

KEEBLE, D.E. and WALKER, S. 1994: New firms, small firms and dead firms: spatial patterns and determinants in the United Kingdom. *Regional Studies* 28, 411–428.

KEEBLE, D.E., OFFORD, J. and WALKER, S. 1988: *Peripheral regions in a Community of twelve member states*. Luxembourg: Office for Official Publications of the European Communities.

KEEBLE, D.E., OWENS, P.L. and THOMPSON, C. 1983: The urban-rural manufacturing shift in the European Community. *Urban Studies* 20, 405–418.

KEEBLE, D.E., TYLER, P., BROOM, G. and LEWIS, J. 1992: *Business success in the countryside: the performance of rural enterprise*. London: HMSO.

KEITH, W. J. 1975: *The rural tradition*. Hassocks: Harvester.

KENDALL, D. 1963: Portrait of a disappearing English village. *Sociologia Ruralis* 3, 157–165.

KENNY, M. 1960: Patterns of patronage in Spain. *Anthropological Quarterly* 33, 14–23.

KERSTEN, J.W. 1990: *Policies for rural peripheral regions in the European Community*. Saarbrücken: Verlag Breitenbach (Nijmegen Studies in Development and Cultural Change 4).

KEUR, J.Y. and KEUR, D.L. 1955: *The deeply rooted: a study of a Drents community in the Netherlands*. Assen: Van Gorcum.

KIELSTRA, N. 1985: The rural Languedoc: periphery to 'relictual space'. In R. Hudson and J.R. Lewis (eds.) *Uneven development in southern Europe*. London: Methuen, 246–262.

KILJUNEN, K. 1992: *Finland and the new international division of labour*. Basingstoke: Macmillan.

KING, R.L. 1982: Southern Europe: dependency or development? *Geography* 67, 221–234.

KING, R.L. 1984: Population mobility: emigration, return migration and internal migration. In A.M. Williams (ed.) *Southern Europe transformed*. London: Harper and Row, 145–178.

KING, R.L. 1987: *Italy*. London: Harper and Row.

KING, R.L. 1991: Italy: multi-faceted tourism. In A.M. Williams and G. Shaw (eds.) *Tourism and economic development: western European experiences*. second edition. London: Belhaven, 61–83.

KING, R.L. 1992: Italy: from sick man to rich man of Europe. *Geography* 77, 153–169.

KING, R.L., editor, 1993: *Mass migrations in Europe*. London: Belhaven.

KING, R.L. and KILLINGBECK, J. 1989: Carlo Levi, the Mezzogiorno and emigration: fifty years of demographic change at Aliano. *Geography* 74, 128–143.

KING, R.L. and RYBACZUK, C. 1993: Southern Europe and the international division of labour: from emigration to immigration. In R.L. King (ed.) *The new geography of European migrations*. London: Belhaven, 175–206.

KING, R.L. and SHUTTLEWORTH, I. 1988: Ireland's new wave of emigration in the 1980s. *Irish Geography* 21, 104–108.

KING, R.L., STRACHAN, A. and MORTIMER, J. 1986: Gastarbeiter go home: return migration and economic change in the Italian Mezzogiorno. In R.L. King (ed.) *Return migration and regional economic problems*. London: Croom Helm, 3868.

KING, R.L., MORTIMER, J., STRACHAN, A. and TRONO, A. 1985: Return migration and rural economic change: a south Italian case study. In R. Hudson and J.R. Lewis (eds.) *Uneven development in southern Europe*. London: Methuen, 101–122.

KJELDAHL, R. and TRACY, M., editors, 1994: *Renationalization of the Common Agricultural Policy?* Genappe: Agricultural Policy Studies.

KLATZMANN, J. 1971: *Géographie agricole de la France*. Paris: Presses Universitaires de France.

KLATZMANN, J. 1994: Trois impossibilités pour la population rurale: la mesurer, suivre son évolution, faire des comparaisons internationales. In V. Rey (ed.) *Géographies et campagnes*. Paris: Ecole Normale Supérieure, 257–260.

KNOEPFEL, P. 1990: La Suisse. In M. Bodiguel (ed.) *Produire ou préserver l'environnement*. Paris: L'Harmattan, 29–65.

KOCH, R. 1980: Counterurbanisation auch in Westeuropa? *Informationen zur Reumertwichlung* 2, 65–73.

KOENIGSBERGER, H. and MOSSE, G. 1971: *Europe in the sixteenth century*. London: Longman.

KOFMAN, E. 1985: Dependent development in Corsica. In R. Hudson and J.R. Lewis (eds.) *Uneven development in southern Europe*. London: Methuen, 263–283.

KOKKINOS, D. 1992: The influx of Albanian refugees to Greece. Paper to the third International Research and Advisory Panel on Refugees and Displaced Persons, University of Oxford, January 1992.

KOLYVAS, M. and GOTSINAS, M. 1990: Greece. In CEPFAR (ed.) *L'état des lieux de la ruralité: les réalités nationales face à l'approche communitaire*. Brussels: CEPFAR, 120–142.

KONSALAS, N. 1992: EEC regional policy and the Integrated Mediterranean Programs. In M. Tykkyläinen (ed.) *Development issues and strategies in the new Europe*. Aldershot: Avebury, 41–52.

KONTULY, T. and VOGELSANG, R. 1988: Explanation for the intensification of counterurbanization in the Federal Republic of Germany. *Professional Geographer* 40, 42–53.

KOSTROWICKI, J. 1991: Trends in the transformation of European agriculture. In F.M. Brouwer, A.J. Thomas and M.J. Chadwick (eds.) *Land-use changes in Europe*. Dordrecht: Kluwer, 21–47.

KOUSIS, M. 1994: Environment and the state in the EU periphery: the case of Greece. In S. Baker, K. Milton and S. Yearly (eds.) *Protecting the periphery*. London: Frank Cass, 118–135.

KYRIAKIDOU-NESTOROS, A. 1985: L'idée de peuple dans les théories folklorique de la Grèce moderne. In M–P. Canapa and associates *Paysans et nations d'Europe centrale et Balkanique*. Paris: Maisoneuve et Larose, 63–69.

LA CALLE DOMINGUEZ, J. 1994: Espagne. In M. Jollivet (ed.) *Bilan des recherches en sciences sociales sur les problèmes d'environnement en milieu rural dans les pays européens*. Paris: CNRS Groupe de Recherches sur les Mutations des Sociétés Européennes, 25–29.

LA SPINA, A. and SCIORTINO, G. 1993: Common agenda, southern rules: European integration and environmental change in the Mediterranean states. In J.D. Liefferink, P.D. Lowe and A.P. Mol (eds.) *European integration and environmental policy*. London : Belhaven, 217–236.

LAINE, M. 1993: Le développement rural: une priorité pour l'Europe. *Initiatives Rurales* 12, 18–20.

LAMBERT, A.M. 1963: Farm consolidation in Western Europe. *Geography* 48, 31–48.

LAMBERT, C., JEFFERS, S., BURTON, P. and BRAMLEY, G. 1992: *Homelessness in rural areas*. Salisbury: Rural Development Commission.

LARRUE, C. 1988a: *La pollution des eaux par les nitrates: comportements agricoles et politique publique: deux études de cas dans la bassin Seine Normandie*. Paris: Ministère de l'Environnement.

LARRUE, C. 1988b: Pollutions d'origine agricole: législations en Europe. In M. Mormont (ed.) *Environment and agriculture*. Arlon: Fondation Universitaire Luxembourgeoise, 65–76.

LAST, F. 1988: Introduction. In M.C. Whitby and J. Ollerenshaw (eds.) *Land-use and the European environment*. London: Belhaven, 1–4.

LAUBER, V. 1989: France and the economic crisis 1974–1987. In E. Damgaard, P. Gerlich and J.J. Richardson (eds.) *The politics of economic crisis: lessons from Western Europe*. Aldershot: Avebury, 107–124.

LAURENT, C. 1994: L'agriculteur paysagiste: du discours aux réalités. *Natures Sciences, Sociétés* 2, 231–242.

LAVOUX, T. 1991: Les quatre Europes de l'environnement. *Cahiers Français* 250, 82–89.

LAVOUX, T., BALDOCK, D., CONRAD, J. and DER KLEY, L. 1988: *L'action pour la réduction de la pollution de l'eau par les nitrates en provenance des activités agricoles*. Paris: Institut pour une Politique Européenne de l'Environnement.

LAZERSON, M. 1993: Factory or putting out? Knitting networks in Modena. In G. Grabher (ed.) *The embedded firm*. London: Routledge, 203–226.

LE COZ, J., editor, 1990: Espaces méditerranéens et dynamiques agricoles: état territorial et communautés rurales. *Options Méditerranéennes* B2.

LE HERON, R. 1988: Food and fibre production under capitalism. *Progress in Human Geography* 12, 409–430.

LE HERON, R. 1994: *Globalized agriculture: political choice*. Oxford: Pergamon.

LE MEUR, P-Y. 1993: Possibilités et limites d'une politique agricole régionale: le case de la Vallée d'Aoste. *Revue de Géographie Alpine* 81, 85–100.

LE MONDE 1992: Un paysage sans paysans. *Le Monde des Débats* 1992 (2). Paris.

LEBEAU, R. 1991: *Les grands types de structures agraires dans le monde*. Paris: Masson.

LÉGER, D. and HERVIEU, B. 1979: *Le retour à la nature*. Paris: Seuil.

LEEUWIS, C. 1989: *Marginalization misunderstood: different patterns of farm development in the west of Ireland*. Wageningen: Agricultural University of Wageningen Studies in Sociology 22.

LEFEVRE, H. 1970: *Du rural à l'urbain*. Paris: Anthropos.

LENORMAND, P. 199+: L'emploi rural dans la compétition économique: contraintes, aménités, rentes différentielles. In N. Mathieu (ed.) *Propositions méthodologiques pour une prospective sur les espaces ruraux français: emploi, nouvelles ressources*. Paris: CNRS–AGRAL, 33–40.

LEVY, J. 1994: Oser le désert. *Sciences Humaines* Hors Série 4, 6–12.

LEWIS, G.J. and SHERWOOD, K.B. 1991: Unravelling the counterurbanization process in lowland Britain: a case study in East Northamptonshire. *Cambria* 16, 58–77.

LEWIS, J.R and TOWNSEND, A., editors, 1989: *The north–south divide: regional change in Britain in the 1980s*. London: Paul Chapman.

LEWIS, J.R. and WILLIAMS, A.M. 1987: Production decentralization or indigenous growth? small manufacturing enterprises and regional development in central Portugal. *Regional Studies* 21, 343–361.

LEWIS, J.R. and WILLIAMS, A.M. 1988: Factories in the field: small manufacturing firms in rural southern Europe. In G.J.R. Linge (ed.) *Peripheralisation and industrial change*. London: Croom Helm, 113–130.

LEYNAUD, E. 1985: *L'état et la nature: l'exemple des parcs nationaux français*. Florac: Parc National des Cevennes.

LIJFERING, J.H.W. 1974: Socio-structural changes in relation to rural out-migration. *Sociologia Ruralis* 14, 3–14.

LIMOUZIN, P. 1980: Les facteurs de redynamisme des communes rurales françaises. *Annales de Géographie* 495, 547–587.

LITTLE, J. 1984: Social change in rural areas: a planning perspective. unpublished PhD thesis, University of Reading.

LITTLE, J. 1987: Rural gentrification and the influence of local level planning. In P.J. Cloke (ed.) *Rural planning: policy into action?* London: Harper and Row, 185–199.

LITTLE, J. 1990: Women's employment in the food system. In T.K. Marsden and J. Little (eds.) *Political, social and economic perspectives on the international food system*. Aldershot: Avebury, 141–156.

LITTLE, J. 1994: Gender relations and the rural labour process. In S.J. Whatmore, T.K. Marsden and P.D. Lowe (eds.) *Gender and rurality*. London: David Fulton, 11–30.

LITTLEJOHN, J. 1963: *Westrigg: the sociology of a Cheviot parish*. London: Routledge and Kegan Paul.

LITTLEWOOD, P. 1981: Patrons or bigshots? Paternalism, patronage and clientist welfare in southern Italy. *Sociologia Ruralis* 21, 1–17.

LOPREATO, J. 1967: *Peasants no more: social class and social change in an underdeveloped society*. San Francisco: Chandler.

LOUKISSAS, P.J. 1977: The impact of tourism on regional development: a comparative analysis of Greek islands. Ithaca, New York: unpublished PhD thesis, Cornell University.

LOULOUDIS, L. 1985: Socio-economic and environmental side effects of technological change: the case of olive oil mills in rural Greece. In D. Hall, N. Myers and N. Margaris (eds.) *Economics of ecosystems management*. Dordrecht: Junk, 199–206.

LOVERING, J. 1989: The restructuring debate. In R. Peet and N.J. Thrift (eds.) *New models in geography: volume one*. London: Unwin Hyman, 198–223.

LOWE, P.D. 1989: The rural idyll defended: from preservation to conservation. In G.E. Mingay (ed.) *The rural idyll*. London: Routledge, 113–131.

LOWE, P.D. 1992: Industrial agriculture and environmental regulation: a new agenda for rural sociology. *Sociologia Ruralis* 32, 4–10.

LOWE, P.D. 1995: The rural economy: regulation or chaos. Public lecture series, Centre for Rural Economy, University of Newcastle-upon-Tyne, 17 January 1995.

LOWE, P.D. and BULLER, H. 1990: Historical and cultural contexts. In P.D. Lowe and M. Bodiguel (eds.) *Rural studies in Britain and France*. London: Belhaven, 3–20.

LOWE, P.D. and GOYDER, J. 1983: *Environmental groups in politics*. London: Allen and Unwin.

LOWE, P.D., COX, G., MACEWEN, M., O'RIORDAN, T. and WINTER, M. 1986: *Countryside conflicts: the politics of farming, forestry and conservation*. Aldershot: Gower.

LOWENTHAL, D. 1991: British national identity and the English landscape. *Rural History* 2, 205–230.

LOWENTHAL, D. and PRINCE, H.C. 1965: English landscape tastes. *Geographical Review* 55, 186–222.

LUCY, D.I.F. and KALDOR, D.R. 1969: *Rural industrialization: the impact of industrialization on two rural communities in western Ireland*. London: Geoffrey Chapman.

MCCORRISTON, S. 1993: Rebalancing EC cereals protection and the GATT Uruguay Round. *Journal of Agricultural Economics* 44, 14–24.

MACDONALD, J.S. 1963: Agricultural organization, migration and labour militancy in rural Italy. *Economic History Review* 16, 61–75.

MACEWEN, A. and MACEWEN, M. 1982: *National parks: conservation or cosmetics?* London: Allen and Unwin.

MCEWEN, J. 1977: *Who owns Scotland? A study in land ownership*. Edinburgh: EUSPB.

MACKAY, G.A. and LAING, G. 1982: *Consumer problems in rural areas*. Glasgow: Scottish Consumer Council.

MACKINNON, N., BRYDEN, J.M., BELL, C., FULLER, A.M., and SPEARMAN, M. 1991: Pluriactivity, structural change and farm household vulnerability in western Europe. *Sociologia Ruralis* 31, 58–71.

MCMICHAEL, P. 1994: GATT, global regulation and the construction of a new hegemonic order. In P.D. Lowe, T.K. Marsden and S.J. Whatmore (eds.) *Regulating agriculture*. London: David Fulton, 163–190.

MAFF 1994: Government announces rural white paper. News Release 371/94, 12 October 1994.

MAIR, A. 1993: New growth poles? just-in-time manufacturing and local economic development strategies. *Regional Studies* 27, 207–221.

MANGANARA, J. 1977: Some social aspects of the return movement of Greek migrant workers from West Germany to rural Greece. *Greek Review of Social Research* 29, 65–75.

MANN, M. 1986: *The sources of social power: a history of power from the beginning to AD 1760*. Cambridge: Cambridge University Press.

MANN, M. 1988: *States, war and capitalism*. Oxford: Blackwell.

MANSVELT BECK, J. 1988: *The rise of the subsidized periphery in Spain*. Utrecht: Nederlandse Geografische Studies 69.

MARAVEYAS, N. 1994: The Common Agricultural Policy and Greek agriculture. In P. Kazakos and P.C. Ioakimidis (eds.) *Greece and EC membership evaluated*. London: Pinter, 56–73.

MARCHENA GOMEZ, M. 1984: *La distribución de la población en Andalucía 1960–1981*. Sevilla: Diputacion Provincial de Sevilla.

MARGARIS, N. 1987: Desertification in the Aegean islands. *Ekistics* 323/324, 132–136.

MARGARIS, N. 1988: *La marginalization des espaces montagneux en Grèce*. Paris: Institut Nationale de Recherche Agronomique.

MARSDEN, T.K. 1992: Exploring a rural sociology for the Fordist tradition. *Sociologia Ruralis* 32, 209–230.

MARSDEN, T.K., MUNTON, R.J. and WARD, 1992a: Incorporating social trajectories into uneven agrarian development: farm businesses in upland and lowland Britain. *Sociologia Ruralis* 32, 408–430.

MARSDEN, T.K., MURDOCH, J. and WILLIAMS, S. No: 1992b: Regulating agriculture in deregulating economies: emerging trends in the uneven development of agriculture. *Geoforum* 23, 333–345.

MARSDEN, T.K., MURDOCH, J., LOWE, P.D., MUNTON, R.J.C. and FLYNN, A. 1993: *Constructing the countryside*. London: UCL Press.

MARSHALL, J.N. 1988: *Services and uneven development*. Oxford: Oxford University Press.

MARSHALL, J.N. and RAYBOULD, S. 1993: New corporate structure and the evolving geography of white collar work. *Tijdschrift voor Economische en Sociale Geografie* 84, 362–377.

MARTIN, S. 1993: The Europeanisation of local authorities: challenges for rural areas. *Journal of Rural Studies* 9, 153–161.

MASSA, I. 1988: The opening of the Finnish north: resource-based development and agrarian change. In T. Ingold (ed.) *The social implications of agrarian change in northern and eastern Finland*. Helsinki: Suomen Antropologinen Seura Toimituksia 22, 24–47.

MASSEY, D.B. 1986: The legacy lingers on: the impacts of Britain's international role on its internal geography. In R.L. Martin and B. Rowthorn (eds.) *The geography of de-industrialisation*. Basingstoke: Macmillan, 31–52.

MATA-PORRAS, M. 1994: Impact of the CAP on socio–economic and land degradation. In P. Mariota and N. Geeson (eds.) EC policy strategies for land-use in the Mediterranean states. London: King's College London Department of Geography MEDALUS Working Paper 17, 47–60.

MATHIEU, N. and JOLLIVET, M., editors, 1989: Du *rural à l'environnement*. Paris: L'Harmattan.

MAYNARD, G. 1988: *The economy under Mrs Thatcher*. Oxford: Blackwell.

MELLOR, R.E.H. 1978: *The two Germanies: a modern geography*. London: Harper and Row.

MENDONSA, E.L. 1982: Benefits of migration as a personal strategy in Nazaré, Portugal. *International Migration Review* 16, 635–645.

MENDOZA-GOMEZ, J. 1994: Dossier Espagne. Paper to INRA/CNRS seminar Agriculture et Environnement en Europe, Vaucluse, September 1994.

MENDRAS, H. 1967: *La fin des paysans*. Arles: Actes Sud.

MENDRAS, H. 1976: *Les sociétés paysannes*. Paris: Colin.

MESSENGER, J.C. 1969: *Inis Beag: isle of Ireland*. New York: Holt, Rinehart and Winston.

MEUS, J.H.A., WIJERMANS, M.P. and VROOM, M.J. 1990: Agricultural landscapes in Europe and their transformation. *Landscape and Urban Planning* 18, 289–352.

MICHEL, R. 1986: *Sociologie rurale*. Paris: Presses Universitaires de France.

MILLAR, J. 1992: *The socio-economic situation of solo women in Europe*. Luxembourg: Office for Official Publications of the European Communities (Women of Europe Supplement 41).

MILLER, K.E. 1991: *Denmark: a troubled welfare state*. Boulder, Colorado: Westview.

MILLWARD, H. 1993: Public access in the West European countryside: a comparative survey. *Journal of Rural Studies* 9, 39–51.

MILWARD, A.S. 1992: *The European rescue of the nation-state*. London: Routledge.

MINGAY, G.E. 1976: *The gentry: the rise and fall of a ruling class*. London: Longman.

MINGAY, G.E., editor, 1977: *The agricultural revolution: changes in agriculture 1650–1850*. London: A & C Black.

MINGAY, G.E. 1990: *A social history of the English countryside*. London: Routledge.

MINISTÈRE DE L'ENVIRONNEMENT 1994: *L'environnement en France*. Paris: Dunod.

MINISTERO DELL'AGRICULTURA E DELLE FORESTE 1991: *Italian agriculture in figures 1991*. Roma: Istituto Nazionale di Economia Agraria.

MOLLE, W.T.M. 1988: Industry and services. In W.T.M. Molle and R. Cappellin (eds.) *Regional impact of Community policies in Europe*. Aldershot: Avebury, 45–64.

MOLLE, W.T.M. 1990: Regional policy. In M. Wolters and P. Coffey (eds.) *The Netherlands and EC membership evaluated*. London: Pinter, 87–92.

MONTENARI, A. 1991: The southern region of the EEC on the threshold of 1992: environment, agriculture and economic development. In A. Montenari (ed.) *Growth and perspectives of the agrarian sector in Portugal, Italy, Greece and Turkey*. Napoli: Edizioni Scientifiche Italiane, 9–32.

MONTEUX, D. 1993: Géographes, crise de l'agriculture et dégradation des paysages agraires. *La Pensée* 292, 23–40.

MOORE, K. 1976: Modernization in a Canary island village. In J.B. Aceves and W.A. Douglass (eds.) *The changing faces of rural Spain*. Cambridge, Massachusetts: Schenkman, 17–27.

MOPU 1987: *El paisaje*. Madrid: Ministerio de Obras Publicas y Urbanismo.

MOPU 1989: *Contaminación agraria difusa*. Madrid: Ministerio de Obras Publicas y Urbanismo.

MORAN, W. 1993: Rural space as intellectual property. *Political Geography* 12, 263–277.

MØRKENBERG, H. 1978: Working conditions of women married to self-employed farmers. *Sociologia Ruralis* 18, 95–105.

MORMONT, M. 1987: Rural nature and urban natures. *Sociologia Ruralis* 17, 3–20.

MORMONT, M. 1990: Who is rural? Or, how to be rural: towards a sociology of the rural. In T.K. Marsden, P.D. Lowe and S.J. Whatmore (eds.) *Rural restructuring*. London: David Fulton, 21–44.

MORMONT, M. 1994a: Belgique. In M. Jollivet (ed.) *Bilan des recherches en sciences sociales sur les problèmes d'environnement en milieu rural dans les pays européens*. Paris: CNRS Groupe de Recherches sur les Mutations des Sociétés Européennes, 12–23.

MORMONT, M. 1994b: Les dispositifs de prise en charge de l'environnement par les agriculteurs. Paper to INRA/CNRS seminar Agriculture et Environnement en Europe, Vaucluse, September 1994.

MOSELEY, M.J. 1973: The impact of growth centres in rural regions: an analysis of spatial flows in East Anglia. *Regional Studies* 7, 77–94.

MOSELEY, M.J. and DARBY, J. 1978: The determinants of female activity rates in rural areas. *Regional Studies* 12, 297–309.

MOSELEY, M.J. and TOWNROE, P.M. 1973: Linkage adjustment following industrial movement. *Tijdschrift voor Economische en Sociale Geografie* 64, 137–144.

MOYER, H.W. and JOSLING, T.E. 1990: *Agricultural policy reform: politics and process in the EC and the USA*. Hemel Hempstead: Harvester Wheatsheaf.

MUCCI, A. 1992: L'agriculture: un impératif – la modernisation. *Notes et Etudes Documentaires* 4946/4947, 35–52.

MULLER, P. 1984: *Le technocrate et le paysan*. Paris: Ouvières, Economie et Humanisme.

MULLER, P. 1991: Vers une agriculture de services. *Economie Rurale* 202/203, 67–70.

MUNTON, R.J.C. 1992a: The uneven development of capitalist agriculture: the repositioning of agriculture within the food system. In K. Hoggart (ed.) *Agricultural change, environment and economy*. London: Mansell, 25–48.

MUNTON, R.J.C. 1992b: Factors of production in modern agriculture. In I.R. Bowler (ed.) *The geography of agriculture in developed market economies*. Harlow: Longman, 56–84.

MURDOCH, J. and MARSDEN, T.K. 1994: *Reconstituting rurality*. London: UCL Press.

MURDOCH, J. and PRATT, A.C. 1993: Rural studies: modernism, postmodernism and the 'post–rural'. *Journal of Rural Studies* 9, 411–427.

MYKOLENKO, L. and CALMES, E. 1985: *L'Europe agricole: une nouvelle géographie des productions*. Paris: Ellipses.

MYKOLENKO, L., DE RAYMOND, Th. and HENRY, P. 1987: *The regional impact of the Common Agricultural Policy in Spain and Portugal.* Luxembourg: Office for Official Publications of the European Communities.

NAGLE, J.C. 1976: *Agricultural trade policies.* Farnborough: Saxon House.

NALSON, J.S. 1968: *Mobility of farm families: a study of occupational and residential mobility in an upland area of England.* Manchester: Manchester University Press.

NAM, C.W. and REUTER, J. 1991: *The impact of 1992 and associated legislation on the less favoured regions of the European Community.* Brussels: European Parliament Regional Policy and Transport Series 18.

NANNESTAD, P. 1991: *Danish design or British disease? Danish economic crisis policy 1974–1979 in comparative perspective.* Aarhus: Aarhus University Press.

NASH, D. 1978: Tourism as a form of imperialism. In V.L. Smith (ed.) *Hosts and guests: the anthropology of tourism.* Oxford: Blackwell, 33–47.

NASH, E.F. and ATTWOOD, E. 1961: *The agricultural policies of Britain and Denmark.* London: Land Books.

NAYLOR, E.L. 1994: Unionism, peasant protest and the reform of French agriculture. *Journal of Rural Studies* 10, 263–273.

NEATE, S. 1987: The role of tourism in sustaining farm structures and communities on the Isles of Scilly. In M. Bouquet and M. Winter (eds.) *Who from their labours rest?.* Aldershot: Avebury, 9–21.

NEGRI, G. 1993: Les montagnes de Lombardie. *Revue de Géographie Alpine* 81, 65–83.

NERVEU, A. 1993: *Les nouveaux territoires de l'agriculture française.* Paris: Uni-Editions.

NETHERLANDS ECONOMIC INSTITUTE 1993: *New location factors for mobile investment in Europe.* Luxembourg: Office for Official Publications of the European Communities.

NEWBY, H.E. 1975: The deferential dialectic. *Comparative Studies in Society and History* 17, 139–164.

NEWBY, H.E. 1977: *The deferential worker: a study of farm workers in East Anglia.* London: Allen Lane.

NEWBY, H.E. 1979: *Green and pleasant land.* London: Hutchinson.

NEWBY, H.E. 1980: Rural sociology. *Current Sociology* 28, 3–141.

NEWBY, H.E. 1986: Locality and rurality: the restructuring of rural social relations. *Regional Studies* 20, 209–215.

NEWBY, H.E. 1987: *Country life.* London: Weidenfeld and Nicolson.

NEWBY, H.E., BELL, C., ROSE, D. and SAUNDERS, P. 1978: *Property, paternalism and power.* London: Hutchinson.

NIEDERBACHER, A. 1988: *Wine in the European Community.* second edition. Luxembourg: Office for Official Publications of the European Communities.

NIESSLER, R. 1987: Income distribution in Austrian agriculture. In Y. Léon and L. Mahé (eds.) *Income disparities among farm households and agricultural policy.* Kiel: Wissenschaftsverlag Vauk Kiel, 325–338.

NIKOLINAKOS, M. 1985: Transnationalization of production, location of industry and the deformation of regional development in peripheral countries: the case of Greece. In R. Hudson and J.R. Lewis (eds.) *Uneven development in southern Europe.* London: Methuen, 192–210.

NOIRIEL, G. 1995: Les étrangers dans la France rurale entre 1920 et 1939. *Etudes Rurales* 135/136, forthcoming.

NOOIJ, A.T.J. 1969: Political radicalism among Dutch farmers. *Sociologia Ruralis* 9, 43–61.

NOQUE I FONT, J. 1988: El fénomeno neorural. *Agricultura y Sociedad* 47, 145–176.

NORTON-TAYLOR, R. 1982: *Whose land is it anyway? agriculture, planning and land-use in the British countryside.* Wellington: Turnstone Press.

NUGENT, N. 1991: *The government and politics of the European Community.* Basingstoke: Macmillan.

NUTLEY, S.D. 1988: 'Unconventional modes' of transport in rural Britain. *Journal of Rural Studies* 4, 73–87.

O T UATHAIL, G. 1993: The new East-West conflict? Japan and the Bush Administration's 'New World Order'. *Area* 25, 127–135.

O'BRIEN, P. and KEYDER, C. 1978: *Economic growth in Britain and France: two paths to the twentieth century.* London: Allen and Unwin.

O'CINNÉIDE, M.S. and KEANE, M.J. 1987: *Community self-help, economic initiatives and development agency responses in the midwest region of Ireland.* Galway: University College Galway Social Sciences Research Centre.

O'CONNOR, J. 1973: *The fiscal crisis of the state.* New York: St. Martin's Press.

O'DONNELL, R. 1991: Regional policy. In P. Keatinge (ed.) *Ireland and EC membership evaluated.* London: Pinter, 60–75.

O'FARRELL, P.N. and CROUTCHLEY, R. 1984: An industrial spatial analysis of new firm formation in Ireland. *Regional Studies* 18, 221–236.

O'FLANAGAN, T.P. 1978: Social and political organization in Galicia: a spatial unconformity. *Finisterra* 25, 77–101.

O'FLANAGAN, T.P. 1980: Agrarian structures in northwest Iberia: responses and their implications for development. *Geoforum* 11, 157–170.

OECD 1987: *National policies and agricultural trade: country study – Austria.* Paris: Organization for Economic Cooperation and Development.

OECD 1988: *National policies and agricultural trade: country study – Sweden.* Paris: Organization for Economic Cooperation and Development.

OECD 1989a: *Agricultural and environmental policies: opportunities for integration.* Paris: Organization for Economic Cooperation and Development.

OECD 1989b: *National policies and agricultural trade: country study – Finland.* Paris: Organization for Economic Cooperation and Development.

OECD 1993a: *What future for our countryside?.* Paris: Organization for Economic Cooperation and Development.

OECD 1993b: *Environmental data.* Paris: Organization for Economic Cooperation and Development.

OECD 1994a: *Agricultural policies, markets and trade: monitoring and outlook 1994.* Paris: Organization for Economic Cooperation and Development.

OECD 1994b: *Farm employment and economic adjustment in OECD countries.* Paris: Organization for Economic Cooperation and Development.

OECD 1994c: *Trends in international migration: annual report 1993.* Paris: Organization for Economic Cooperation and Development.

OECD 1994d: *Creating rural indicators for shaping territorial policy.* Paris: Organization for Economic Cooperation and Development.

OFFE, C. 1976: Political authority and class structures. In P. Connerton (ed.) *Critical sociology*. Harmondsworth: Penguin, 388–421.

OGDEN, P.E. 1980: Migration, marriage and the collapse of traditional peasant society in France. In P.E. White and R.I. Woods (eds.) *The geographical impact of migration*. London: Longman, 152–179.

OJEDA RIVERA, J. 1989: Protection ou développement: le faux dilemme du parc national de Donāna et de sa région. In N. Mathieu and M. Jollivet (eds.) *Du rural à l'environnement*. Paris: L'Harmattan, 275–278.

ØRUM, T. 1985: *The promised land: peasant struggles, agrarian reforms and regional development in a southern Italian community*. København: Akademisk Forlag Studies in Cultural Sociology 22.

ÖSTERREICHISCHEN STATISTISCHEN ZENTRALAMTES 1993: *Statistisches jahrbuch für Republik Österreich*. Wien.

OSTI, G. 1992: Co-operative regulation: contrasting organizational models for the control of pesticides: the case of north-east Italian fruit growing. *Sociologia Ruralis* 32, 163–177.

PA CAMBRIDGE ECONOMIC CONSULTANTS LTD 1987: *A study of rural tourism*. Salisbury: Rural Development Commission.

PACIONE, M. 1980: Differential quality of life in a metropolitan village. *Transactions of the Institute of British Geographers* 5, 185–206.

PAHL, R.E. 1965: *Urbs in rure: the metropolitan fringe in Hertfordshire*. London: London School of Economics Geographical Paper 2.

PAHL, R.E. 1966: The rural-urban continuum. *Sociologia Ruralis* 6, 299–329.

PAINE, R. 1963: Entrepreneurial activity without its profits. In F. Barth (ed.) *The role of the entrepreneur in social change in northern Norway*. Bergen: Universitetsforlaget, 33–55.

PALOSCIA, R. 1991: Agriculture and diffused manufacturing in the 'Terza Italia': a Tuscan case study. In S.J. Whatmore, P.D. Lowe and T.K. Marsden (eds.) *Rural enterprise*. London: David Fulton, 34–57.

PAPADIMITROU, F. 1994: Land-use change and underdevelopment in southern Europe. In P. Mariota and N. Geeson (eds.) EC policy strategies for land-use in the Mediterranean states. London: King's College London Department of Geography MEDALUS Working Paper 17, 63–70.

PAPADOPOULUS, S. 1985: La decouverte muséologique du paysan en Grèce. In M-P. Canapa and associates *Paysans et nations d'Europe centrale et Balkanique*. Paris: Maisoneuve et Larose, 87–93.

PARK, C.C. 1987: *Acid rain: rhetoric and reality*, London: Routledge.

PARKER, G. 1977: *The Dutch revolt*. London: Allen Lane.

PARKER, G. 1979: *The countries of Community Europe: a geographical survey of contemporary issues*. Basingstoke: Macmillan.

PARR, J.B. 1966: Outmigration and the depressed rural area problem. *Land Economics* 42, 149–159.

PEARCE, D.G. 1987: Spatial patterns of package tourism in Europe. *Annals of Tourism Research* 14, 183–201.

PEARCE, D.G. 1990: Tourism in Ireland: questions of scale and organization. *Tourism Management* 11, 133–151.

PEARCE, D.G. 1995: *Tourism today*. second edition. Harlow: Longman.

PÉCHOUX, P-Y. 1975: La réforme agraire en Grèce. *Révue de Géographie de Lyon* 4, 317–332.

PECK, F. and STONE, I. 1994: Sunrise in the Costa del Sol? Global restructuring, state policies and local economic change in the electronics sector in southern Spain. Newcastle-upon-Tyne: University of Northumbria Division of Geography and Environmental Management Occasional Paper 6.

PECQUEUR, B. and SILVA, M.R. 1992: Territory and economic development: the example of diffuse industrialization. In G. Garofoli (ed.) *Endogenous development in southern Europe*. Aldershot: Avebury, 17–29.

PEET, R. 1969: The spatial expansion of commercial agriculture in the nineteenth century: a von Thünen interpretation. *Economic Geography* 45, 283–301.

PENROSE, J. 1993: Reification in the name of change: the impact of nationalism on social constructions of nation, people and place in Scotland and the United Kingdom. In P.A. Jackson and J. Penrose (eds.) *Constructions of race, place and nation*. London: UCL Press, 27–49.

PERRY, P.J. 1969: The structural revolution in French agriculture: the role of Sociétés d'Améngement Foncier et d'Establissement Rural. *Revue de Géographie de Montreal* 23, 137–151.

PERRY, R., DEAN, K. and BROWN, B. 1986: *Counterurbanization: international case studies of socio-economic change in rural areas*. Norwich: Geo Books.

PERSSON, L.O. 1992: Rural labour markets – meeting urbanization and the arena society: new challenges for policy and planning in rural Scandinavia. In T.K. Marsden, P.D. Lowe and S.J. Whatmore (eds.) *Labour and locality*. London: Fulton, 68–94.

PERSSON, L.O. and WESTHOLM, E. 1993: Turmoil in welfare system reshapes rural Sweden. *Journal of Rural Studies* 9, 397–404.

PERSSON, L.O. and WESTHOLM, E. 1994: Vers une nouvelle mosaïque de la Suède rurale. *Economie Rurale* 223, 20–26.

PESCATELLO, A. 1976: *Power and the pawn: the female in Iberian societies and cultures*. Westport, Connecticut: Greenwood.

PETERSON, M. 1990: Paradigmatic shift in agriculture: global effects and the Swedish response. In T.K. Marsden, P.D. Lowe and S.J. Whatmore (eds.) *Rural restructuring*. London: David Fulton, 77–100.

PETTERSSON, O. 1993: Scandinavian agriculture in a changing environment. In S. Harper (ed.) *The greening of rural policy*. London: Belhaven, 82–98.

PFAU-EFFINGER, B. 1994: The gender contract and part-time paid work by women – Finland and Germany compared. *Environment and Planning* A26, 1355–1376.

PFEFFER, M.J. 1989: The feminization of production on part-time farms in the Federal Republic of Germany. *Rural Sociology* 54, 60–73.

PHILIP, A.B. 1990: The European Community: balancing international and domestic demands. In G. Skogstad and A.F. Cooper (eds.) *Agricultural trade: domestic pressures and international tensions*. Halifax, Nova Scotia: Institute for Research on Public Policy, 69–85.

PHILLIPS, M. 1993: Rural gentrification and the processes of class colonisation. *Journal of Rural Studies* 9, 123–140.

PHILLIPS, P.W.B. 1990: *Wheat, Europe and the GATT*. London: Pinter.

PIGRAM, J.J. 1993: Planning for tourism in rural areas: bridging the policy implementation gap. In D.G. Pearce and R.W. Butler (eds.) *Tourism research: critiques and challenges*. London: Routledge, 156–174.

PIJNENBURG, B. 1989: Belgium in crisis: political and policy responses. In E. Damgaard, P. Gerlich and J.J. Richardson (eds.) *The politics of economic crisis: lessons from western Europe.* Aldershot: Avebury, 29–49.

PINA-CABRAL, J. 1986: *Sons of Adam, daughters of Eve: the peasant worldview of the Alto Minho.* Oxford: Clarendon.

PINTO, A.S., AVILLEZ, F., ALBUQUERQUE, L. and GOMES, L.F. 1984: *A agricultura portuguesa no período 1950–1980.* Lisboa: Imprensa Nacional/Casa da Moeda.

PITMAN, J.I. 1992: Changes in crop productivity and water quality in the United Kingdom. In K. Hoggart (ed.) *Agricultural change, environment and economy.* London: Mansell, 89–121.

PITT-RIVERS, J.A. 1960: Social class in a French village. *Anthropological Quarterly* 33, 1–13.

PITT-RIVERS, J.A. 1971: *The people of the sierra.* second edition. Chicago: University of Chicago Press.

PIVOT, C. 1993: Costs and benefits of the Common Agricultural Policy for France. In F-G. Dreyfus, J. Morizet and M. Peyrard (eds.) *France and EC membership evaluated.* London: Pinter, 59–67.

PLASKOVITIS, I. 1994: EC regional policy in Greece: ten years of structural fund intervention. In P. Kazakos and P.C. Ioakimidis (eds.) *Greece and EC membership evaluated.* London: Pinter, 116–127.

POMPILI, T. 1994: Structure and performance of less developed regions in the EC. *Regional Studies* 28, 679–693.

PONGRATZ, H. 1990: Cultural tradition and social change in agriculture. *Sociologia Ruralis* 30, 5–12.

POTTER, C. and GASSON, R.M. 1988: Farmer participation in voluntary diversion schemes. *Journal of Rural Studies* 4, 365–375.

POTTER, C. and LOBLEY, M. 1992: Elderly farmers as countryside managers. In A.W. Gilg (ed.) *Restructuring the countryside: environmental policy in practice.* Aldershot: Avebury, 54–68.

POUJADE, R. 1975: *Le ministère de l'impossible.* Paris: Calmann-Lévy.

PRECEDO LEDO, A. and GRIMES, S. 1991: Urban transition and local developments in a peripheral region: the case of Galicia. In M.J. Bannon, L.S. Bourne and R. Sinclair (eds.) *Urbanization and urban development.* Dublin: University College Dublin Department of Urban and Regional Planning, 97–107.

PREEBLE, J. 1963: *The highland clearances.* Harmondsworth: Penguin.

PRIDHAM, G. 1994: National environmental policy-making in the European framework: Spain, Greece and Italy in comparison. In S. Baker, K. Milton and S. Yearley (eds.) *Protecting the periphery.* London: Frank Cass, 80–101.

PRONDZYNSKI, I. 1989: *Women in statistics.* Luxembourg: Office for Official Publications of the European Communities (Women in Europe Supplement 30).

PROST, B. and VANDENBROUKE, M. 1993: Evolution des campagnes et monde paysan: la diffusion des modèles agricoles en Italie du nord. *Revue Géographique des Pyrénées et du Sud Ouest* 63, 279–288.

PUGLIESE, E. 1985: Farm workers in Italy: agricultural working class, landless peasants or clients of the welfare state? In R. Hudson and J.R. Lewis (eds.) *Uneven development in southern Europe.* London: Methuen, 123–139.

PUMAIN, D. and SANT-JULIEN, T. 1992: *The statistical concept of the town in Europe.* Luxembourg: Office for Official Publications of the European Communities.

PUYOL ANTOLÍN, R. 1989: *La población española*. Madrid: Editorial Sintesis.

RAMBAUD, P. 1969: *Sociologie rurale et urbanisation*. Paris: Seuil.

REDCLIFT, M.R. 1973: The effects of socio-economic changes in a Spanish pueblo on community cohesion. *Sociologia Ruralis* 13, 1–14.

REECE, J.E. 1979: Internal colonialism: the case of Brittany. *Ethnic and Racial Studies* 2, 275–309.

REES, G. 1986: 'Coalfield culture' and the 1984–1985 miners' strike. *Society and Space* 4, 469–476.

REES, P.H. 1992: Elderly migration and population redistribution in the United Kingdom. In A. Rogers (ed.) *Elderly migration and population redistribution*. London: Belhaven, 203–225.

REID, M. 1992: L'Irlande. Les *espaces du développement local en milieu rural: exemples européens*. Paris: Peuple et Culture, 92(2), 3–9.

REIF, K. 1993: Cultural convergence and cultural diversity as factors in European identity. In S. García (ed.) *European identity and the search for legitimacy*. London: Pinter, 131–153.

REIS, M. and NAVE, J.G. 1986: Emigrating peasants and returning emigrants. *Sociologia Ruralis* 26, 20–35.

RENARD, J. 1991: Les changements sociaux dans les campagnes de la France de l'Ouest. In *France et Grande Bretagne rurales/rural France and Great Britain*. Caen: Université de Caen Centre de Publications, 409–419.

REPASSY, H. 1991: Changing gender roles in Hungarian agriculture. *Journal of Rural Studies* 7, 23–29.

REYNOLDS, P., STOREY, D. and WESTHEAD, P. 1994: Cross-national comparisons of the variation in new firm formation rates. *Regional Studies* 28, 443–456.

RIBEIRO, O. 1986: *Portugal o Mediterrâneo e o Atlântico : escboço de relaçoes geográficos*. Lisboa: Livreria Sá da Costa Editora.

RICHEZ, G. 1991: *Parcs nationaux et tourisme en Europe*. Paris: L'Harmattan.

RILEY, R.C. and ASHWORTH, G.J. 1975: *Benelux: an economic geography of Belgium, the Netherlands and Luxembourg*. London: Chatto and Windus.

RITAINE, E. 1994: Territoire et politique en Europe du Sud. *Revue Français de Science Politique* 44, 75–98.

RIVIERE, D. 1990: L'Italie entre l'Europe du nord et l'Europe méditerranéenne. In J. Beaujeu–Garnier and A. Gamblin (eds.) *La CEE méditerranéenne*. Paris: Sedes, 261–273.

ROBIC, M-C., editor, 1992: Du *milieu à l'environnement*. Paris: Economica.

ROBINSON, G. 1990: *Conflict and change in the countryside*. London: Belhaven.

ROBINSON, G. 1994: Beyond MacSharry: the road to CAP reform. In E.C.A. Bolsius, G. Clark and J.G. Groenendijk (eds.) *The retreat: rural land-use and European agriculture*. Utrecht: Nederlandse Geografische Studies 172, 28–41.

RODRÍGUEZ-POSE, A. 1994: Socio–economic restructuring and regional change: rethinking growth in the European Community. *Economic Geography* 70, 325–343.

RODWIN, L. and SAZANAMI, H., editors, 1991: *Industrial change and regional economic transformation: the experience of western Europe*. London: Harper Collins.

ROSE, C. 1990: *The dirty man of Europe: the great British pollution scandal*. London: Simon and Schuster.

ROSE, D., SAUNDERS, P., NEWBY, H.E. and BELL, C. 1976: Ideologies of property. *Sociological Review* 24, 699–730.

ROSENBLATT, J., MAYER, T., BARTHOLDY, K., DEMEKAS, D., GUPTA, S. and LIPSCHITZ, L. 1988: *The Common Agricultural Policy of the European Community*. Washington D.C.: International Monetary Fund Occasional Paper 62.

ROVAN, J. 1994: Identité et territoire dans l'histoire de la France et de l'Allemagne. *Herodote* 72/73, 22–29.

ROWTHORN, R.E. and WELLS, J.R. 1987: *Deindustrialization and foreign trade*. Cambridge: Cambridge University Press.

RUDE, S. and FREDERIKSEN, B.S. 1994: *National and EC nitrate policies: agricultural aspects for seven EC countries*. København: Statens Jordbrusøkonomiske Institut Rapport 77.

RUFFORD, N., LEPPARD, D. and BURRELL, I. 1991: Wealthy landowners get handouts for doing nothing. *Sunday Times* 1 December 1991, 1.

RYDIN, Y. 1986: *Housing land policy*. Aldershot: Gower.

RYING, B. 1988: *Denmark: a history*. København: Royal Danish Ministry of Foreign Affairs.

SACK, R.D. 1992: *Place, modernity and the consumer's world: a relational framework for geographical analysis*. Baltimore: Johns Hopkins University Press.

SACKMANN, R. and HÄUSSERMANN, H. 1994: Do regions matter? Regional differences in female labour-market participation in Germany. *Environment and Planning* A26, 1377–1396.

SACKS, P.M. 1976: *The Donegal Mafia: an Irish political machine*. New Haven, Connecticut: Yale University Press.

SÁENZ LORITE, M. 1988: *Geografía agraria: introducción a los paisajes rurales*. Madrid: Editorial Síntesis.

SAHLINS, M. 1981: *Historical metaphors and mythical realities*. Ann Arbor: University of Michigan Press.

SALMON, K.G. 1992: *Andalucía: an emerging regional economy in Europe*. Sevilla: Junta de Andalucía Consejería de Economia y Hacienda.

SALT, J. and CLOUT, H.D. 1976: International labour migration: the source of supply. In J. Salt and H.D. Clout (eds.) *Migration in post-war Europe: geographical essays*. Oxford: Oxford University Press, 126–167.

SAN JUAN MESONADA, C. 1993: Agricultural policy. In A.A. Barbado (ed.) *Spain and EC membership evaluated*. London: Pinter, 49–59.

SANDERSON, S.E. 1986: The emergence of the 'world steer': internationalization and foreign domination in Latin American cattle production. In F.L. Tullis and W.L. Hollist (eds.) *Food, the state and the international political economy*. Lincoln: University of Nebraska Press, 123–148.

SARACENO, E. 1994: The modern function of the small farm system: an Italian experience. *Sociologia Ruralis* 34, 308–328.

SAUNDERS, P., NEWBY, H.E., BELL, C. and ROSE, D. 1978: Rural community and rural community power. In H.E. Newby (ed.) *International perspectives in rural sociology*. Chichester: Wiley, 55–85.

SAVAGE, M., BARLOW, J., DICKENS, P. and FIELDING, A.J. 1992: *Property, bureaucracy and culture: middle-class formation in contemporary Britain*. London: Routledge.

SAVAGE, M., BARLOW, J., DUNCAN, S.S. and SAUNDERS, P. 1987: Locality research. *Quarterly Journal of Social Affairs* 3, 27–51.

SAVILLE, J. 1957: *Rural depopulation in England and Wales 1851–1951*. London: Routledge and Kegan Paul.

SCHEDVIN, C.B. 1990: Staples and regions of Pax Britannica. *Economic History Review* 43, 533–559.

SCHMITT, M. 1994: Women farmers and the influence of ecofeminism on the greening of German agriculture. In S.J. Whatmore, T.K. Marsden and P.D. Lowe (eds.) *Gender and rurality*. London: David Fulton, 102–116.

SCHMITTER–HEISLER, B. 1986: Immigrant settlement and the structure of emergent immigrant communities in western Europe. *Annals of the American Association of Political and Social Science* 485, 76–86.

SCHWEIZER, P. 1988: *Shepherds, workers, intellectuals: culture and centre-periphery relationships in a Sardinian village*. Stockholm: Stockholm Studies in Social Anthropology 18.

SCOTT, A.J. 1988: *New industrial spaces: flexible production organization and regional development in North America and Western Europe*. London: Pion.

SCOTT, J. 1985: *Corporations, classes and capitalism*. second edition. London: Hutchinson.

SCULLY, M.A. 1990: *A contemporary history of Austria*. London: Routledge.

SECRETARY OF STATE FOR THE ENVIRONMENT and others 1990: *This common inheritance: Britain's environmental strategy*. London: HMSO.

SEGALEN, M. 1991: *Fifteen generations of Bretons: kinship and society in lower Brittany 1720–1980*. Cambridge: Cambridge University Press.

SELF, P. 1977: *Administrative theories and politics*. second edition. London: Allen and Unwin.

SEROW, W.J. 1991: Recent trends and future prospects for urban-rural migration in Europe. *Sociologia Ruralis* 31, 269–280.

SHAPOSHNIKOV, A.N. 1990: The problem of developing independent peasant initiative in Russia. *International Social Science Journal* 42(124), 193–207.

SHAW, G. and WILLIAMS, A.M. 1990: Tourism and development. In D.A. Pinder (ed.) *Western Europe: challenge and change*. London: Belhaven, 240–257.

SHAW, G. and WILLIAMS, A.M. 1994: *Critical issues in tourism: a geographical perspective*. Oxford: Blackwell.

SHIEL, M.J. 1988: *Rural electrification in Ireland*. London: Panos Publications.

SHOARD, M. 1987: *This land is our land: the struggle for Britain's countryside*. London: Grafton.

SHORT, J.R. 1991: *Imagined country*. London: Routledge.

SHORTFALL, S. 1992: Power analysis and farm wives: an empirical study of power relationships affecting women on Irish farms. *Sociologia Ruralis* 32, 431–451.

SHUCKSMITH, D.M. 1990a: *Housebuilding in Britain's countryside*. London: Routledge.

SHUCKSMITH, D.M. 1990b: A theoretical perspective on rural housing: housing classes in rural Britain. *Sociologia Ruralis* 30, 210–229.

SHUCKSMITH, D.M. and LLOYD, M. 1983: Rural planning in Scotland. In A.W. Gilg (ed.) *Countryside planning yearbook: volume four*. Norwich: Geo Books, 103–128.

SIEPER, M. 1993: L'aménagement rural en Autriche. *Revue de Géographie Alpine* 81, 11–29.

SIISKONEN, P. 1988: The role of farmers and farm wives in agrarian change. In T. Ingold (ed.) *The social implications of agrarian change in northern and eastern Finland*. Helsinki: Suomen Antropologinen Seura Toimituksia 22, 91–98.

SILVERMAN, S.F. 1965: Patronage and community-nation relationships in central Italy. *Ethnology* 4, 172–189.

SINGLETON, F. 1986: *The economy of Finland in the twentieth century*. Bradford: University of Bradford Press.

SIVIGNON, M. 1989: Essai de classement des espace ruraux en Grèce. *Espace Rural* 20, 133–141.

SIVIGNON, M. 1990: Géographie économique de la Grèce. In J. Beaujeu-Garnier and A. Gamblin (eds.) *La CEE méditerranéenne*. Paris: Sedes, 289–298.

SIVIGNON, M. 1993: La diffusion des modèles agricoles essai d'interprétation des agriculteurs de l'est et du sud de l'Europe. *Revue Géographique des Pyrénées et du Sud Ouest* 63, 133–154.

SJØHOLT, P. 1988: Sweden. In P.J. Cloke (ed.) *Policies and plans for rural people*. London: Unwin Hyman, 69–97.

SMITH, A.D. 1991: *National identity*. Harmondsworth: Penguin.

SNOWDEN, F.M. 1979: From sharecropper to proletarian: the background to facism in rural Tuscany 1880–1920. In J.A. Davis (ed.) *Gramsci and Italy's passive revolution*. London: Croom Helm, 136–171.

SPINDLER, G.D. 1973: *Burgbach: urbanization and identity in a German village*. New York: Holt, Rinehart and Winston.

STATISTISCHES BUNDESAMT (HERAUSGEBER) 1994: *Statistiches Jahrbuch 1994 für die Bundesrepublik Deutschland*. Wiesbaden.

STATISTISKA CENTRALBYRÅ 1994: *Statistik årsbok för Sverge '94*. Stockholm.

STEWART, J.C. 1976: Linkages and foreign direct investment. *Regional Studies* 10, 245–258.

STOCKDALE, A. 1993: Rural service provisions and the impact of a population revival. *Area* 25, 365–378.

STOECKEL, A.B. and BRECKLING, J. 1989: Some economic effects of agricultural policies in the European Community. In A.B. Stoeckel, D. Vincent and S. Cuthbertson (eds.) *Macroeconomic consequences of farm support policies*. Durham, North Carolina: Duke University Press, 94–124.

STÖHR, W.B. 1992: Local initiative networks as an instrument for the development of peripheral areas. In M. Tykkyläinen (ed.) *Development issues and strategies in the new Europe*. Aldershot: Avebury, 203–210.

STONE, M.K. 1990: *Rural childcare*. Salisbury: Rural Development Commission.

STORPER, M. and WALKER, R.A. 1989: *The capitalist imperative: territory, technology and industrial growth*. Oxford: Blackwell.

STRIJKER, D. 1993: Dutch agriculture in a European context. *Sociologia Ruralis* 33, 137–146.

STRIJKER, D. and DE VEER, J. 1988: Agriculture. In W. Molle and R. Cappellin (eds.) *Regional impact of Community policies in Europe*. Aldershot: Avebury, 23–44.

STRIJKER, D. and DEINUM, T. 1992: Relative regional welfare differences in the EC. In M. Tykkyläinen (ed.) *Development issues and strategies in the new Europe*. Aldershot: Avebury, 53–64.

SWANN, D. 1984: *Competition and industrial policy in the European Community*. London: Methuen.

SYRETT, S. 1994: Local power and economic policy: local authority economic initiatives in Portugal. *Regional Studies* 28, 53–67.

SZUCS, J. 1985: *Les trois Europes*. Paris: L'Harmattan.

TACET, D. 1992: *Un monde sans paysans*. Paris: Hachette.

TANGERMANN, S. 1992: European integration and the Common Agricultural Policy. In C.E. Barfield and M. Perlman (eds.) *Industry, services and agriculture: the United States faces a united Europe*. Washington DC: AEI Press, 407–450.

TARRANT, J.R. 1982: EEC food aid. *Applied Geography* 2, 127–141.

TARRANT, J.R. 1992: Agriculture and the state. In I.R. Bowler (ed.) *The geography of agriculture in developing market economies*. Harlow: Longman, 239–274.

TARROW, S. 1977: *Between center and periphery: grassroots politicians in Italy and France*. New Haven, Connecticut: Yale University Press.

TAYLOR, P.J. 1989: Britain's century of decline: a world-systems interpretation. In J. Anderson and A. Cochrane (eds.) *A state of crisis: the changing faces of British politics*. London: Hodder and Stoughton, 8–26.

TAYLOR, P.J. 1993: *Political geography*. third edition. Harlow: Longman.

TERLUIN, I.J. 1991: *Production, prices and incomes in European Community agriculture: an analysis of the economic accounts for agriculture 1973–1988*. Luxembourg: Office for Official Publications of the European Communities.

TERRASSON, F. 1988: *La peur de la nature*. Paris: Sang de la Terre.

THERKILDSEN, K., VESTERLUND, J., JAKOBSEN, H. and SLIPSAGER, S. 1990: Denmark. In CEPFAR (ed.) *L'état des lieux de la ruralité: les réalités nationales face à l'approche communautaire*. Brussels: CEPFAR, 31–60.

THIEME, G. 1983: Agricultural change and its impact in rural areas. In M. T. Wild (ed.) *Urban and rural change in West Germany*. London: Croom Helm, 220–247.

THISSEN, F. 1992: Restructuring of rural areas in the Netherlands. In P. Huigen, L. Paul and K. Volkers (eds.) The *changing function and position of rural areas in Europe*. Utrecht: Nederlandse Geografische Studies 153, 75–86.

THOMAS, K. 1983: Ma*n and the natural world*. London: Allen and Lane.

THOMPSON, I.B. 1970: *Modern France*. London: Butterworth.

THOMPSON, P.B. 1994: *The spirit of the soil: agriculture and environmental ethics*. London: Routledge.

THRIFT, N.J. 1987: Manufacturing rural geography? *Journal of Rural Studies* 3, 77–81.

THRIFT, N.J. and LEYSHON, A. 1992: In the wake of money: the City of London and the accumulation of value. In L. Budd and S. Whimster (eds.) *Global finance and urban living*. London: Routledge, 282–311.

TICKELL, A. and PECK, J.A. 1992: Accumulation, regulation and the geographies of post–Fordism. *Progress in Human Geography* 16, 190–218.

TILASTKESKUS STATISTISKCENTRALEN 1994: *Suomen tilastollinen vuosikirja 1994*. Helsinki.

TODOIR, P. 1994: Transactions of professional business services and spatial systems. *Tijdschrift voor Economische en Sociale Geografie* 85, 322–332.

TOOK, L. 1986: Land tenure, return migration and rural change in the Italian province of Chieti. In R.L. King (ed.) *Return migration and regional economic problems*. London: Croom Helm, 79–99.

TOUJAS-PINÈDE, C. 1990: *L'immigration étrangère en Quercy aux XIXe et XXe siècles*. Paris: Privat.

TOUT, D. 1990: The horticulture industry of Almeria Province, Spain. *Geographical Journal* 156, 304–312.

TOVEY, H. 1994: Agriculture and environment in Ireland – a brief discussion. Paper to INRA/CNRS seminar Agriculture et Environnement en Europe, Vaucluse, September 1994.

TOWNSEND, A.R. 1986: The location of employment growth after 1978: the surprising significance of dispersed centres. *Environment and Planning* A18, 529–545.

TOWNSEND, A.R. 1993: The urban-rural cycle in the Thatcher growth years. *Transactions of the Institute of British Geographers* 18, 207–221.

TRACY, M. 1989: *Government and agriculture in Western Europe 1880–1988.* third edition. Hemel Hempstead: Harvester Wheatsheaf.

TRIANTAFILLOU, J. and MESTHENEOS, E. 1994: Pathways to care for the elderly in Greece. *Social Science and Medicine* 38, 575–582.

TRICART, J. 1993: La Galice: problèmes d'intégration à la CEE d'une région attardée et marginale. *Annales de Géographie* 574, 614–623.

TROUFLEAU, P. 1992: L'espace du logement en France. *Espace Géographique* 21, 36–46.

TSOUKALIS, L. 1993: *The new European economy: the politics and economics of integration.* Oxford: Oxford University Press.

TSOULVOUVIS, L. 1987: Aspects of statism and planning in Greece. *International Journal of Urban and Regional Research* 11, 500–521.

TUSSIE, D. 1991: Trading in fear? US hegemony and the open world economy in perspective. In C.N. Murphy and R. Tooze (eds.) *The new international political economy.* Boulder, Colorado: Lynne Rienner, 79–95.

UNGER, K. 1986: Return migration and regional characteristics: the case of Greece. In R.L. King (ed.) *Return migration and regional economic problems.* London: Croom Helm, 129–151.

URRY, J. 1981: Localities, regions and social class. *International Journal of Urban and Regional Research* 5, 455–473.

URRY, J. 1984: Capitalist restructuring, recomposition and the regions. In T. Bradley and P.D. Lowe (eds.) *Locality and rurality.* Norwich: Geo Books, 45–64.

URRY, J. 1986: Class, space and disorganized capitalism. In K. Hoggart and E. Kofman (eds.) *Politics, geography and social stratification.* London: Croom Helm, 16–32.

URRY, J. 1987: Society, space and locality. *Society and Space* 5, 435–444.

URWIN, D.W. 1980: *From ploughshare to ballotbox: the politics of agrarian defence in Europe.* Oslo: Universitetsforlaget.

VALENZUELA, M. 1991: Spain: the phenomenon of mass tourism. In A.M. Williams and G. Shaw (eds.) *Tourism and economic development: Western European experiences.* second edition. London: Belhaven, 40–60.

VALERO ESCANDELL, J.R. 1992: *La inmigración extranjera en Alicante.* Alicante: Instituto de Cultura Juan Gil-Albert.

VAN DE PLOEG, J.D. and LONG, A., editors, 1994: *Born from within: practice and perspectives of endogenous rural development.* Assen: Van Gorcum.

VAN DEN BERG, L. 1989: Rural land-use planning in the Netherlands: integration or segregation of functions. In P.J. Cloke (ed.) *Rural land-use planning in developed nations.* London: Unwin and Hyman, 47–75.

VAN DEN BERG, L. 1993: New uses for disused farm buildings in the Netherlands. In E.C.A. Bolsius, G. Clark and J.G. Groenendijk (eds.) *The retreat: rural land-use and European agriculture*. Utrecht: Nederlandse Geografische Studies 172, 83–95.

VAN DINTEREN, J., BONAMY, J., SAVIAT, C. and WOOD, P. 1994: Business services and networks. *Tijdschrift voor Economische en Sociale Geografie* 85, 291–293.

VAN DONINCK, B. 1992: Agricultural policy. In M.A.G. van Meerhaeghe (ed.) *Belgium and EC membership evaluated*. London: Pinter, 27–39.

VARTIAINEN, P. 1989: The end of drastic depopulation in rural Finland: evidence of counterurbanization? *Journal of Rural Studies* 5, 123–136.

VÀZQUEZ-BARQUERO, A. 1992a: Local development and the regional state in Spain. In G. Garofoli (ed.) *Endogenous development in southern Europe*. Aldershot: Avebury, 103–116.

VÀZQUEZ-BARQUERO, A. 1992b: Local development initiatives under incipient regional autonomy: the Spanish experience in the eighties. In G. Garofoli (ed.) *Endogenous development in southern Europe*. Aldershot: Avebury, 165–183.

VERA REBELLO, J.F., PONCE HERRERO, G.J., DÁVILA LINARES, J.M. and RAMÓN MORTE, A. 1990: Evaluación del grado de especialización turística de los municipios litorales Valencianos. *Investigaciones Geográficas* 8, 83–112.

VERMEULEN, H. 1981: Conflict and peasant protest in the history of a Macedonian village 1900–1936. In S. Damaniakos (ed.) *Aspects du changement social dans la campagne grecque*. Athens: Greek Review of Social Research special edition, 93–103.

VERGOPOULOS, K. 1978: Capitalism and peasant productivity. *Journal of Peasant Studies* 5, 446–465.

VEYRET, Y. and PECH, P. 1993: *L'homme et l'environnement*. Paris: Presses Universitaires de France.

VIARD, D. 1990: *Le tiers espace: essai sur la nature*. Paris: Méridiens Klincksieck.

VINCENT, J.A. 1980: The political economy of alpine development: tourism or agriculture in St Maurice. *Sociologia Ruralis* 20, 250–271.

VOLKER, K. 1988: The social impact of land consolidation with particular reference to the Achterhoek area of the Netherlands. *Tijdschrift voor Sociaal Wetenschappelijke Onderzoek van de Landbouw* 1, 241–257.

VOLKER, K. 1989: Private and public rights and responsibilities in the management of the Dutch environment. Wageningen: Instituut voor Onderzoek van het Landelijk Gebied Staring Centrum Working Paper.

VOLKER, K. 1992: Adapted farming systems for a rural landscape: a social typology of Dutch farmers. *Sociologia Ruralis* 32, 146–162.

VON CRAMON-TAUBADEL, S. 1993: The reform of the CAP from a German perspective. *Journal of Agricultural Economics* 44, 394–409.

WADE, R. 1971: Political behaviour and world view in a central Italian village. In F.G. Bailey (ed.) *Gifts and poison: the politics of reputation*. Oxford: Blackwell, 252–280.

WADE, R. 1979: Fast growth and slow development in south Italy. In D. Seers, B. Schaffer and M-L. Kiljunen (eds.) *Underdeveloped Europe*. Hassocks: Harvester, 197–221.

WALL, K. 1984: Mulheres que partem e mulheres que ficam: uma primeira análise da função social e económica das mulheres no processo migratório. *Ler História* 3, 53–63.

WALLMAN, S. 1977: The shifting sense of 'us': boundaries against development in the western Alps. In S. Wallman (ed.) *Perceptions of development*. Cambridge: Cambridge University Press, 17–29.

WALSH, J.A. 1986: Agricultural change and development. In P. Breathnach and M.E. Cawley (eds.) *Change and development in rural Ireland.* Maynooth: Geographical Society of Ireland Special Publication. 1, 11–24.

WALSH, J.A. 1991: The turnaround of the turnaround in the population of the Republic of Ireland. *Irish Geography* 24, 117–125.

WARD, N. and MUNTON, R.J. c.1992: Conceptualizing agriculture-environment relations: combining political economy and sociocultural approaches to pesticide pollution. *Sociologia Ruralis* 32, 127–145.

WARD, N., BULLER, H. and LOWE, P.D. 1995: Implementing European environmental policy at the local level: a sociopolitical analysis of the UK experience with water quality directives. Newcastle-upon-Tyne: University of Newcastle–upon–Tyne Centre for Rural Economy Research Report, forthcoming.

WARD, N., LOWE, P.D. and BULLER, H. 1994: Implementing environmental policy in rural Europe. Paper to European Consortium of Political Research Green Politics Standing Group Conference, University of Crete, October 1994.

WARDE, A. 1985: Comparable localities: some problems of method. In Lancaster Regionalism Group *Localities, class and gender.* London: Pion, 54–76.

WARDE, A. 1986: Space, class and voting in Britain. In K. Hoggart and E. Kofman (eds.) *Politics, geography and social stratification.* London: Croom Helm, 33–61.

WARNES, A.M. 1991: London's population trends: metropolitan area or megalopolis?. In K. Hoggart and D.R. Green (eds.) *London: a new metropolitan geography.* London: Edward Arnold, 156–175.

WATKINS, C. and WINTER, M. 1988: *Superb conversions? farm diversification – the farm building experience.* London: Council for the Protection of Rural England.

WEATHERLEY, R.D. 1982: Domestic tourist and second homes as motors of rural development in the Sierra Morena, Spain. *Iberian Studies* 11, 40–46.

WEBER, M. 1930: *The protestant ethic and the spirit of capitalism.* London: Allen and Unwin.

WEBSTER, F.H. 1973: Agricultural cooperation in Denmark. Oxford: Plunkett Foundation for Cooperative Studies Occasional Paper 39.

WEISSGLAS, G. 1975: *Studies on service problems in the sparsely populated areas in northern Sweden.* Umeå: University of Umeå Geographical Report 5.

WENGER, G.C. 1980: *Mid-Wales: deprivation or development.* Cardiff: University of Wales Press.

WENGER, G.C. 1981: The elderly in the community: mobility and access to services. Bangor: University College of North Wales Department of Social Theory and Institutions Social Services in Rural Areas Research Project Working Paper 16.

WENTURIS, N. 1994: Political culture. In P. Kazakos and P.C. Ioakimidis (eds.) *Greece and EC membership evaluated.* London: Pinter, 225–237.

WESTMACOTT, R. and WORTHINGTON, T. 1974: *New agricultural landscapes.* London: Countryside Commission.

WHATMORE, S.J. 1991: *Farming women: gender, work and family enterprise.* Basingstoke: Macmillan.

WHATMORE, S.J. 1994: Global agro-food complexes and the refashioning of rural Europe. In A. Amin and N.J. Thrift (eds.) *Globalization, institutions and regional development in Europe.* Oxford: Oxford University Press, 46–67.

WHITBY, M.C. 1994: What future for ESAs?. In M.C. Whitby (ed.) *Incentives for countryside management.* Wallingford: CAB International, 253–272.

WHITE, P.E. 1981: Migration at the micro-scale: intra-parochial movements in rural Normandy 1946–1954. *Transactions of the Institute of British Geographers* 6, 451–470.

WHITE, P.E. 1985: Modelling population change in the Cilento region of southern Italy. *Environment and Planning* A17, 1401–1413.

WHITE, P.E. 1990: Labour migration and counterurbanization in France. In J.H. Johnson and J. Salt (eds.) *Labour migration*. London: David Fulton, 99–111.

WHITE, R.L. and WATTS, H.D. 1977: The spatial evolution of an industry: the example of broiler production. *Transactions of the Institute of British Geographers* 2, 175–191.

WICKHAM, C. 1988: Historical materialism, historical sociology. *New Left Review* 171, 63–78.

WIENER, M.J. 1981: *English culture and the decline of the industrial spirit 1850–1980*. Cambridge: Cambridge University Press.

WIJKMAN, P.M. 1990: Patterns of production and trade. In W. Wallace (ed.) *The dynamics of European integration*. London: Pinter, 89–105.

WILD, T. 1983: The residential dimension to rural change. In T. Wild (ed.) *Urban and rural change in West Germany*. London: Croom Helm, 161–199.

WILD, T. and JONES, P.N. 1994: Spatial impacts of German unification. *Geographical Journal* 160, 1–16.

WILLIAMS, A.M. 1987: *The western European economy: a geography of post-war development*. London: Hutchinson.

WILLIAMS, A.M. 1991: *The European Community: the contradictions of integration*. Oxford: Blackwell.

WILLIAMS, A.M., SHAW, G. and GREENWOOD, J. 1989: From tourist to tourism entrepreneur, from consumption to production: evidence from Cornwall, England. *Environment and Planning* A21, 1639–1653.

WILLIAMS, A.M., SHAW, G. and HUBER, M. 1995: The arts and economic development: regional and urban-rural contrasts in UK local authority policies for the arts. *Regional Studies* 29, 73–80.

WILLIAMS, C.H. 1980: Ethnic separatism in Western Europe. *Tijdschrift voor Economische en Sociale Geografie* 71, 142–158.

WILLIAMS, R. 1973: *The country and the city*. London: Chatto and Windus.

WILLIAMS, W.M. 1956: *The sociology of an English village: Gosforth*. London: Routledge and Kegan Paul.

WILLIAMS, W.M. 1963: *A west country village, Ashworthy: family, kinship and land*. London: Routledge and Kegan Paul.

WILSON, G. 1978: Farmers organizations in advanced societies. In H.E. Newby (ed.) *International perspectives on rural sociology*. Chichester: Wiley, 31–53.

WILSON, G.A. 1994: German agri-environmental schemes. *Journal of Rural Studies* 10, 27–45.

WILSON, O. 1992: Landownership and rural development in the north Pennines. *Journal of Rural Studies* 8, 145–158.

WILTSHIRE, R. 1992: Rural development and processes of agricultural change in Japan. In K. Hoggart (ed.) *Agricultural change, environment and economy*. London: Mansell, 49–67.

WINCHESTER, H.P.M. 1993: *Contemporary France*. Harlow: Longman.

WINTOUR, P. 1994: Portillo says no to EU's training cash. *The Guardian* 19 August, 4.

WISHLADE, F.G. 1994: Achieving coherence in European Community approaches to area designation. *Regional Studies* 28, 79–87.

WOLF, E.R. 1966: *Peasants*. Englewood Cliffs, New Jersey: Prentice-Hall.

WOOLLETT, S. 1982: *Alternative rural services*. London: Bedford Square Press.

WRENCH, J. and SOLOMOS, J., editors, 1993: *Racism and migration in western Europe*. Oxford: Berg.

WRIGHT, G. 1955: Peasant politics in the Third French Republic. *Political Science Quarterly* 70, 75–86.

WRIGHT, G. 1964: *Rural revolution in France: the peasantry in the twentieth century*. Palo Alto, California: Stanford University Press.

WTO 1993: *Tourism trends worldwide and in Europe 1980–1992*. Madrid: World Tourism Organization.

WYLIE, L. 1966: *Chanzeaux: a village in Anjou*. Cambridge, Massachusetts: Harvard University Press.

WYLIE, L. 1974: *Village in the Vaucluse*. third edition. Cambridge, Massachusetts: Harvard University Press.

YOUNG, K. 1988: Rural prospects. In R. Jowell, S. Witherspoon and L. Brook (eds.) *British social attitudes 1988*. Aldershot: Gower, 155–174.

Place index

Subject index